Sandra Tisdell-Clifford

DEVELOPMENTAL MATHEMATICS BOOK 1

Founding authors
Allan Thompson · Effie Wrightson

Series editor
Robert Yen

Australia • Brazil • Japan • Korea • Mexico • Singapore • Spain • United Kingdom • United States

Developmental Mathematics Book 1
5th Edition
Sandra Tisdell-Clifford

Publishing editor: Robert Yen
Project editor: Alan Stewart
Editors: Elaine Cochrane and Anna Pang
Art direction: Luana Keays
Text design: Sarah Hazell
Cover design: Sarah Hazell
Cover image: iStockphoto/shuoshu; shutterstock/javarman
Permissions researcher: Miriam Allen
Production controller: Erin Dowling
Typeset by: Cenveo Publisher Services

Any URLs contained in this publication were checked for currency during the production process. Note, however, that the publisher cannot vouch for the ongoing currency of URLs.

For product information and technology assistance,
in Australia call **1300 790 853**;
in New Zealand call **0800 449 725**

For permission to use material from this text or product, please email
aust.permissions@cengage.com

National Library of Australia Cataloguing-in-Publication Data
Tisdell-Clifford, Sandra, author.
Developmental maths. Book 1 / Sandra Tisdell-Clifford.

5th edition.
9780170350969 (paperback)
For secondary school age.

Mathematics--Australia--Textbooks.

510.76

Cengage Learning Australia
Level 7, 80 Dorcas Street
South Melbourne, Victoria Australia 3205

Cengage Learning New Zealand
Unit 4B Rosedale Office Park
331 Rosedale Road, Albany, North Shore 0632, NZ

For learning solutions, visit **cengage.com.au**

Printed in China by 1010 Printing International Limited.
5 6 7 8 25 24 23 22 21

CONTENTS

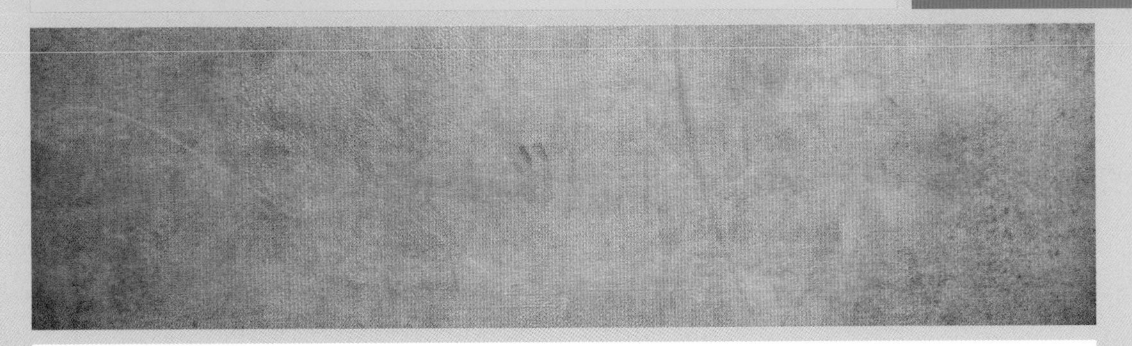

Preface/About the author vi
Features of this book vii
Curriculum grids viii
Series overview x

CHAPTER 1
INTEGERS AND THE NUMBER PLANE

1-01 Numbers above and below zero 2
1-02 Ordering integers 4
1-03 Adding integers 6
1-04 Subtracting integers 8
1-05 Coordinates on maps 10
1-06 The number plane 12
1-07 The number plane with negative numbers 14
Practice Test 1 17

CHAPTER 2
ANGLES

2-01 Naming angles 20
2-02 Types of angles 22
2-03 Measuring and drawing angles 24
2-04 Complementary and supplementary angles 26
2-05 Angles at a point and vertically opposite angles 28
2-06 Corresponding angles on parallel lines 30
2-07 Alternate angles on parallel lines 32
2-08 Co-interior angles on parallel lines 34
Practice Test 2 37

CHAPTER 3
WORKING WITH NUMBERS

3-01 Rounding and estimating 40
3-02 Adding numbers 42
3-03 Subtracting numbers 44
3-04 Multiplication facts 46
3-05 Lowest common multiple (LCM) 48
3-06 Multiplying numbers 50
Practice Test 3 53

CHAPTER 4
FACTORS AND PRIMES

4-01 Division facts 56
4-02 Dividing numbers 58
4-03 Division with remainders 60
4-04 Divisibility tests 62
4-05 Highest common factor (HCF) 64
4-06 Prime and composite numbers 66
4-07 Factor trees 68
Practice Test 4 71

CHAPTER 5
POWERS AND DECIMALS

5-01 Powers 74
5-02 Square and square root 76
5-03 Cube and cube root 78
5-04 Order of operations 80
5-05 Decimals 82
5-06 Ordering decimals 84
5-07 Adding and subtracting decimals 86
Practice Test 5 89

CHAPTER 6
MULTIPLYING AND DIVIDING DECIMALS

6-01 Multiplying and dividing decimals by powers of 10 92
6-02 Multiplying decimals by whole numbers 94
6-03 Multiplying decimals 96
6-04 Dividing decimals by whole numbers 98
6-05 Dividing decimals 100
6-06 Rounding decimals 102
6-07 Best buys 104
6-08 Decimal problems 106
Practice Test 6 109

CHAPTER 7
FRACTIONS

7-01 Fractions 112
7-02 Fractions and decimals 114
7-03 Equivalent fractions 116
7-04 Simplifying fractions 118
7-05 Improper fractions and mixed numerals 120
7-06 Ordering fractions 124
7-07 Adding and subtracting fractions 126
7-08 Adding and subtracting mixed numerals 128
Practice Test 7 131

CHAPTER 8
MULTIPLYING AND DIVIDING FRACTIONS

8-01 Fraction of a quantity 134
8-02 Multiplying fractions 136
8-03 Reciprocals 138
8-04 Dividing fractions 140
8-05 Multiplying mixed numerals 142
8-06 Dividing mixed numerals 144
Practice Test 8 147

CHAPTER 9
ALGEBRA AND EQUATIONS

9-01 The laws of arithmetic 150
9-02 Variables 152
9-03 From words to algebraic expressions 154
9-04 Substitution 156
9-05 Equations 158
9-06 One-step equations 160
9-07 Two-step equations 162
Practice Test 9 165

CHAPTER 10
SHAPES AND SYMMETRY

10-01 Polygons 168
10-02 Translation and reflection 171
10-03 Rotation 174
10-04 Composite transformations 176
10-05 Line symmetry 179
10-06 Rotational symmetry 181
10-07 Prisms and pyramids 183
Practice Test 10 186

ISBN 9780170350969

CHAPTER 11
GEOMETRY

11-01 Types of triangles 190
11-02 Angle sum of a triangle 192
11-03 Exterior angle of a triangle 194
11-04 Types of quadrilaterals 196
11-05 Angle sum of a quadrilateral 198
11-06 Properties of quadrilaterals 200
Practice Test 11 205

CHAPTER 12
LENGTH AND TIME

12-01 The metric system 208
12-02 Measuring length 210
12-03 Perimeter 212
12-04 Perimeter of composite shapes 214
12-05 Time 216
12-06 24-hour time 218
12-07 Time calculations 220
12-08 Timetables 222
Practice Test 12 225

CHAPTER 13
AREA AND VOLUME

13-01 Area 228
13-02 Area of a rectangle 230
13-03 Area of a triangle 232
13-04 Area of a parallelogram 234
13-05 Area of composite shapes 236
13-06 Volume 238
13-07 Volume of a rectangular prism 240
13-08 Volume and capacity 242
Practice Test 13 245

CHAPTER 14
STATISTICAL GRAPHS

14-01 Picture graphs 248
14-02 Column graphs 250
14-03 Line graphs 252
14-04 Divided bar graphs 256
14-05 Sector graphs 258
14-06 Misleading graphs 260
Practice Test 14 263

CHAPTER 15
ANALYSING DATA

15-01 Frequency tables 266
15-02 Frequency histograms and polygons 268
15-03 Dot plots 270
15-04 Stem-and-leaf plots 272
15-05 The mean and mode 274
15-06 The median and range 276
15-07 Analysing dot plots and stem-and-leaf plots 278
Practice Test 15 281

CHAPTER 16
PROBABILITY

16-01 The language of chance 284
16-02 Sample spaces 286
16-03 Probability 288
16-04 The range of probability 290
16-05 Experimental probability 292
Practice Test 16 295

CHAPTER 17
PERCENTAGES AND RATIOS

17-01 Percentages 298
17-02 Percentages and fractions 300
17-03 Percentages and decimals 302
17-04 Percentage of a quantity 304
17-05 Ratios 306
17-06 Equivalent ratios 308
17-07 Simplifying ratios 310
Practice Test 17 313

Answers 315
Index 347

PREFACE

In schools for over four decades, *Developmental Mathematics* has been a unique, well-known and trusted Years 7–10 mathematics series aimed at developing key numeracy and literacy skills. This 5th edition of the series has been revised for the new Australian curriculum as well as the NSW syllabus Stages 4 and 5.1. The four books of the series contain short chapters with worked examples, definitions of key words, graded exercises, a language activity and a practice test. Each chapter covers a topic that should require about two weeks of teaching time.

Developmental Mathematics supports students with mathematics learning, encouraging them to experience more confidence and success in the subject. This series presents examples and exercises in clear and concise language to help students master the basics and improve their understanding. We have endeavoured to equip students with the essential knowledge required for success in junior high school mathematics, with a focus on basic skills and numeracy.

Developmental Mathematics Book 1 is written for students in Years 7–8, covering the Australian curriculum (mostly Year 7 content) and NSW syllabus (see the curriculum grids on the following pages and the teaching program on the NelsonNet teacher website). This book presents concise and highly structured examples and exercises, with each new concept or skill on a double-page spread for convenient reading and referencing.

Students learning mathematics need to be taught by dynamic teachers who use a variety of resources. Our intention is that teachers and students use this book as their primary source or handbook, and supplement it with additional worksheets and resources, including those found on the *NelsonNet* teacher website (access conditions apply). We hope that teachers can use this book effectively to help students achieve success in secondary mathematics. Good luck!

ABOUT THE AUTHOR

Sandra Tisdell-Clifford teaches at Newcastle Grammar School and was the Mathematics coordinator at Our Lady of Mercy College (OLMC) in Parramatta for 10 years. Sandra is best known for updating *Developmental Mathematics* for the 21st century (4th edition, 2003) and writing its blackline masters books. She also co-wrote *Nelson Senior Maths 11 General* for the Australian curriculum, teaching resources for the NSW senior series *Maths in Focus* and the Years 7–8 homework sheets for *New Century Maths/NelsonNet*.

Sandra expresses her thanks and appreciation to the Headmaster and staff of Newcastle Grammar School and dedicates this book to her husband, Ray Clifford, for his support and encouragement. She also thanks series editor **Robert Yen** and editors **Elaine Cochrane**, **Anna Pang** and **Alan Stewart** at Cengage Learning for their leadership on this project.

Original authors **Allan Thompson** and **Effie Wrightson** wrote the first three editions of *Developmental Mathematics* (published 1974, 1981 and 1988) and taught at Smith's Hill High School in Wollongong. Sandra thanks them for their innovative pioneering work, which has paved the way for this new edition for the Australian curriculum.

ISBN 9780170350969

FEATURES OF THIS BOOK

- Each chapter begins with a table of contents and list of chapter outcomes
- Each teaching section of a chapter is presented clearly on a double-page spread

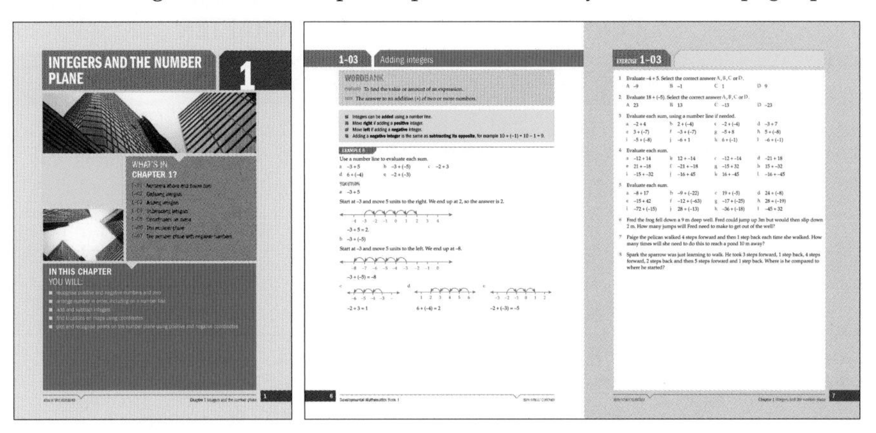

- The left page contains explanations, worked examples, and, if appropriate, a Wordbank of mathematical terminology and a fact box
- The right page contains an exercise set, including multiple-choice questions, scaffolded solutions and realistic applications of mathematics
- Each chapter concludes with a **Language activity** (puzzle) that reinforces mathematical terminology in a fun way, and a **Practice test** containing non-calculator questions on general topics and topic questions grouped by chapter subheading

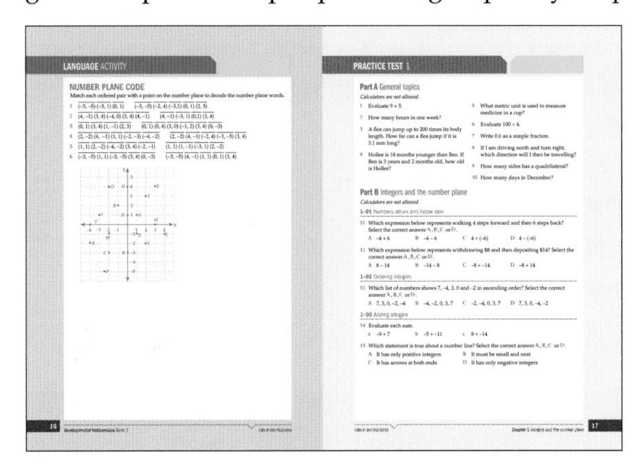

- **Answers** and **index** are at the back of the book
- Additional teaching resources can be downloaded from the NelsonNet teacher website at **www.nelsonnet.com.au**: worksheets, puzzle sheets, skillsheets, video tutorials, technology worksheets, teaching program, curriculum grids, chapter PDFs of this book
- Note: NelsonNet access is available to teachers who use Developmental Mathematics as a core educational resource in their classroom. Contact your sales representative for information about access codes and conditions.

CURRICULUM GRID
AUSTRALIAN CURRICULUM

STRAND AND SUBSTRAND	DEVELOPMENTAL MATHEMATICS BOOK 1 CHAPTER	DEVELOPMENTAL MATHEMATICS BOOK 2 CHAPTER
NUMBER AND ALGEBRA		
Number and place value	1 Integers and the number plane 3 Working with numbers 4 Factors and primes 5 Powers and decimals 6 Multiplying and dividing decimals 9 Algebra and equations	1 Working with numbers 2 Primes and powers 4 Integers
Real numbers	5 Powers and decimals 6 Multiplying and dividing decimals 7 Fractions 8 Multiplying and dividing fractions 17 Percentages and ratios	5 Decimals 11 Fractions 12 Percentages 16 Ratios and rates
Money and financial mathematics	6 Multiplying and dividing decimals	12 Percentages
Patterns and algebra	9 Algebra and equations	6 Algebra 15 Further algebra
Linear and non-linear relationships	1 Integers and the number plane 9 Algebra and equations	15 Further algebra 17 Graphing lines
MEASUREMENT AND GEOMETRY		
Using units of measurement	12 Length and time 13 Area and volume	9 Length and time 10 Area and volume
Shape	10 Shapes and symmetry	
Location and transformation	10 Shapes and symmetry	7 Angles and symmetry
Geometric reasoning	2 Angles 11 Geometry	7 Angles and symmetry 8 Triangles and quadrilaterals
Pythagoras and trigonometry		3 Pythagoras' theorem
STATISTICS AND PROBABILITY		
Chance	16 Probability	14 Probability
Data representation and interpretation	14 Statistical graphs 15 Analysing data	13 Investigating data AC

ISBN 9780170350969

CURRICULUM GRID
AUSTRALIAN CURRICULUM

STRAND AND SUBSTRAND	DEVELOPMENTAL MATHEMATICS BOOK 3 CHAPTER	DEVELOPMENTAL MATHEMATICS BOOK 4 CHAPTER
NUMBER AND ALGEBRA		
Real numbers	2 Whole numbers and decimals 3 Integers and fractions 6 Percentages 7 Indices 16 Ratios and rates	1 Working with numbers 2 Percentages 7 Ratios and rates
Money and financial mathematics	6 Percentages	3 Earning and saving money
Patterns and algebra	4 Algebra 7 Indices	4 Algebra 10 Indices
Linear and non-linear relationships	9 Equations 14 Graphing lines	13 Equations and inequalities 15 Coordinate geometry 16 Graphing lines and curves
MEASUREMENT AND GEOMETRY		
Using units of measurement	12 Length and time 13 Area and volume	9 Length and time 11 Area and volume
Geometric reasoning	8 Geometry	8 Congruent and similar figures
Pythagoras and trigonometry	1 Pythagoras' theorem 5 Trigonometry	5 Pythagoras' theorem 6 Trigonometry
STATISTICS AND PROBABILITY		
Chance	15 Probability	14 Probability
Data representation and interpretation	11 Investigating data	12 Investigating data AC

ISBN 9780170350969

SERIES OVERVIEW

BOOK 1

1 Integers and the number plane
2 Angles
3 Working with numbers
4 Factors and primes
5 Powers and decimals
6 Multiplying and dividing decimals
7 Fractions
8 Multiplying and dividing fractions
9 Algebra and equations
10 Shapes and symmetry
11 Geometry
12 Length and time
13 Area and volume
14 Statistical graphs
15 Analysing data
16 Probability
17 Percentages and ratios

BOOK 2

1 Working with numbers
2 Primes and powers
3 Pythagoras' theorem
4 Integers
5 Decimals
6 Algebra
7 Angles and symmetry
8 Triangles and quadrilaterals
9 Length and time
10 Area and volume
11 Fractions
12 Percentages
13 Investigating data
14 Probability
15 Further algebra
16 Ratios and rates
17 Graphing lines

BOOK 3

1 Pythagoras' theorem
2 Whole numbers and decimals
3 Integers and fractions
4 Algebra
5 Trigonometry
6 Percentages
7 Indices
8 Geometry
9 Equations
10 Earning money
11 Investigating data
12 Length and time
13 Area and volume
14 Graphing lines
15 Probability
16 Ratios and rates

BOOK 4

1 Working with numbers
2 Percentages
3 Earning and saving money
4 Algebra
5 Pythagoras' theorem
6 Trigonometry
7 Ratios and rates
8 Congruent and similar figures
9 Length and time
10 Indices
11 Area and volume
12 Investigating data
13 Equations and inequalities
14 Probability
15 Coordinate geometry
16 Graphing lines and curves

ISBN 9780170350969

INTEGERS AND THE NUMBER PLANE

1

WHAT'S IN CHAPTER 1?

1-01 Numbers above and below zero
1-02 Ordering integers
1-03 Adding integers
1-04 Subtracting integers
1-05 Coordinates on maps
1-06 The number plane
1-07 The number plane with negative numbers

IN THIS CHAPTER YOU WILL:

- recognise positive and negative numbers and zero
- arrange integers in order, including on a number line
- add and subtract integers
- find locations on maps using coordinates
- plot and recognise points on the number plane using positive and negative coordinates

Shutterstock.com/Peshkova

1-01 Numbers above and below zero

WORDBANK

negative number A number that is less than 0, for example –5.

integer A positive or negative whole number or zero.

infinite Continuing on forever or endlessly, such as our number system.

Numbers above and below zero can be used to describe everyday situations such as:

- 10 m above sea level +10
- 20 m below sea level –20
- Climbing up 12 steps +12
- Going down 5 steps –5

Shutterstock.com/Niyazz

Whole numbers that have direction and size are called **integers**.
The integers are an infinite set of numbers made up of positive integers, negative integers, and zero:
..., –4, –3, –2, –1, 0, +1, +2, +3, +4, ...

Positive integers can be written without the '+' sign. For example, '+1' is the same number as '1'.

EXAMPLE 1

Write an integer to represent each situation.

a Depositing $50 in a bank

b Losing $15

c Temperature dropping 5°C

SOLUTION

a As depositing money in the bank involves adding, it is a positive action: +50

b Losing money means you have less, so it is negative: –15

c A drop means decreasing, so it is negative: –5

EXAMPLE 2

Which integer is larger in each pair?

a –7 or +2 b –8 or –9 c +12 or +5

SOLUTION

a +2 b –8 c +12

ISBN 9780170350969

EXERCISE 1-01

1 What is an integer? Select the correct answer A, B, C or D.

A a negative number
B a positive number
C zero
D any positive or negative whole number or zero

2 Which pair of integers represents depositing $50 and then withdrawing $28? Select A, B, C or D.

A −50, +28
B +50, +28
C +50, −28
D −50, −28

3 Write an integer to represent each situation.

a Going up 4 stairs
b Falling 60 metres
c Depositing $20
d Rising 18 m above sea level
e Losing $28
f Going down 10 stairs
g Gaining $100
h Going 24 m below sea level

4 For this list of integers 0, −1, +2, −3, +4, −5, +6, −7, +8:

a write the largest integer
b write the smallest integer
c rewrite the integers in order from smallest to largest.

5 Jack was searching for his sister Emma. He started on the third floor and climbed up 12 steps to the fourth floor. He then went down 24 steps to the second floor and still couldn't find her. Jack then ascended to the fifth floor before deciding to check the ground floor for her.

a Draw a diagram of Jack's journey.
b How many steps would he need to have climbed to get to the fifth floor?
c How many steps would he then have to descend to get to the ground floor?

Alamy/beryan dale

6 List the following integers from largest to smallest: −9, +4, 0, −3, +2, −11, +5, +6

1-02 Ordering integers

WORDBANK

number line A line that shows the position and order of numbers.

ascending Going up, increasing from smallest to largest (1-2-3).

descending Going down, decreasing from largest to smallest (3-2-1).

Integers can be shown on a **number line**.

- The **positive integers** are on the right because they are greater than 0.
- The **negative integers** are on the left because they are less than 0.
- The number line extends in both directions forever so we place arrows on both ends.
- We can delete the + sign for the positive integers, so 3 is the same as +3.

EXAMPLE 3

Plot these integers on a number line: –3, 4, 0, –2, 1, –4.

SOLUTION

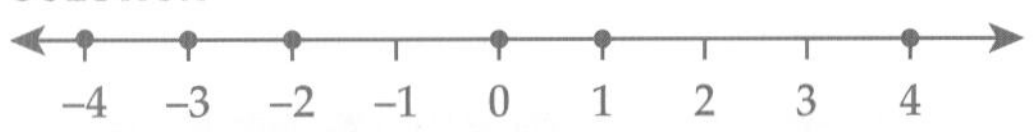

< means **is less than**, for example 7 < 10.
> means **is greater than**, for example 4 > 1.
Think of the symbol pointing to the smaller number with the other side as opening up to the larger number.

EXAMPLE 4

Use < or > signs to complete each statement.

a 3 __ –2 b –5 __ 4 c –8 __ –4

SOLUTION

a 3 > –2 b –5 < 4 c –8 < –4

EXAMPLE 5

Write the integers –6, 8, 0, 4, –3, 3, 5, –1:

a in ascending order b in descending order

SOLUTION

a –6, –3, –1, 0, 3, 4, 5, 8. ← Ascending means from smallest to largest

b 8, 5, 4, 3, 0, –1, –3, –6. ← Descending means from largest to smallest

ISBN 9780170350969

EXERCISE 1-02

1 Which of these integers 18, –4, 9, –2, –16, 21 is the smallest? Select the correct answer A, B, C or D.

A –2   B –16   C –4   D 21

2 When listing the integers in Question 1 in descending order, what is the 3rd integer? Select A, B, C or D.

A 9   B 18   C –2   D –4

3 Plot each set of integers on a number line.

a –3, 0, 4, –2, 1
b –7, 4, 2, –3, –1
c 10, –3, 8, 0, –2
d –12, 3, –7, 1, –4
e –24, –18, –13, –9, –2

4 For the list of integers 26, –5, 13, 0, 8, –16, 11:

a write the largest integer
b write the smallest integer
c rewrite the integers from smallest to largest.

5 Plot each set of integers on a separate number line from –8 to 8.

a the odd integers between 2 and 8
b the integers less than 5
c the even negative and positive integers greater than –6

6 Use the number line below to decode each message. Each number on the number line has a letter above it to use in the code.

a –1 –2 0 –3 2 –3 1 4
b –4 1 –3
c –1 –2 0 –3 1 –3 4 0 –1 –2 2

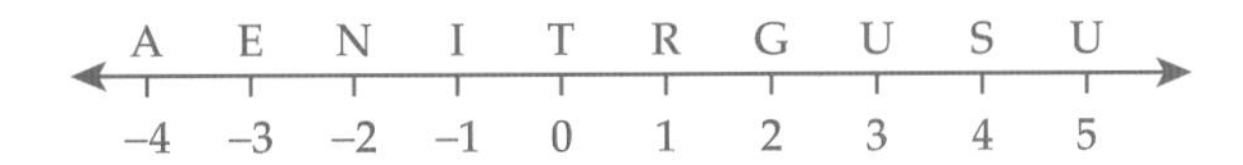

7 Copy and complete each statement using a > or < symbol.

a 3 __ 2
b 0 __ 5
c 6 __ –6
d 0 __ –2
e 8 __ –3
f –2 __ –4
g –8 __ 3
h –9 __ –3
i –12 __ 3
j –10 __ –4
k –17 __ 7
l 26 __ –6

8 Rewrite each set of integers in ascending order.

a –2, 2, 0, –1, 4
b 6, –5, 3, –4, –3
c –3, –7, –1, –12, 0
d 8, –4, 3, –1, –8
e –20, 35, –25, 5, 0
f 29, –12, –18, –23, 4

9 Rewrite each set of integers in descending order.

a 5, 3, –2, 7, –3
b –8, 4, –6, –1, 0
c 8, –1, 6, –4, –11
d 12, –9 , –15, 8, –6
e –9, –14, –28, –4, 7
f 32, 10, –90, –48, –52

1-03 Adding integers

WORDBANK

evaluate To find the value or amount of an expression.

sum The answer to an addition (+) of two or more numbers.

Integers can be **added** using a number line.

- Move **right** if adding a **positive** integer.
- Move **left** if adding a **negative** integer.
- Adding a **negative integer** is the same as **subtracting its opposite**, for example $10 + (-1) = 10 - 1 = 9$.

EXAMPLE 6

Use a number line to evaluate each sum.

a $-3 + 5$ b $-3 + (-5)$ c $-2 + 3$

d $6 + (-4)$ e $-2 + (-3)$

SOLUTION

a $-3 + 5$

Start at –3 and move 5 units to the right. We end up at 2, so the answer is 2.

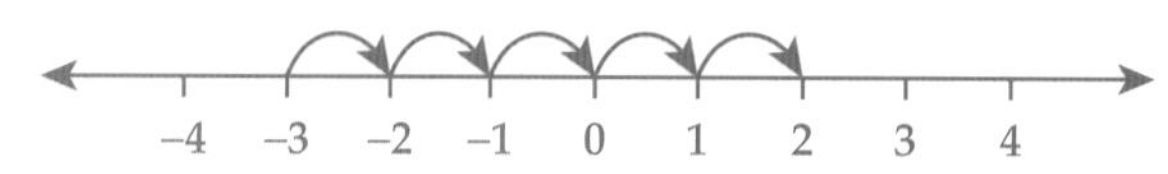

$-3 + 5 = 2.$

b $-3 + (-5)$

Start at –3 and move 5 units to the left. We end up at –8.

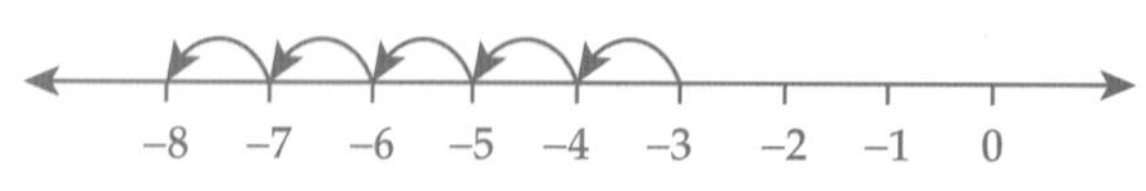

$-3 + (-5) = -8$

c

$-2 + 3 = 1$

d

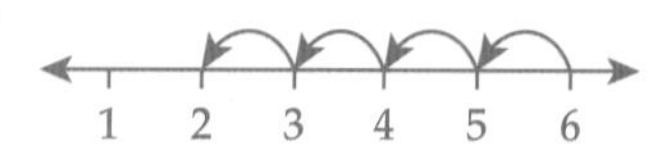

$6 + (-4) = 2$

e

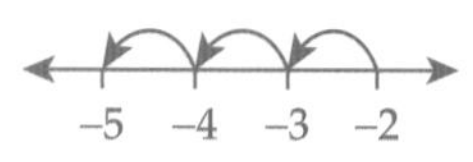

$-2 + (-3) = -5$

ISBN 9780170350969

EXERCISE 1-03

1 Evaluate $-4 + 5$. Select the correct answer A, B, C or D.

A -9 B -1 C 1 D 9

2 Evaluate $18 + (-5)$. Select A, B, C or D.

A 23 B 13 C -13 D -23

3 Evaluate each sum, using a number line if needed.

a $-2 + 4$ b $2 + (-4)$ c $-2 + (-4)$ d $-3 + 7$
e $3 + (-7)$ f $-3 + (-7)$ g $-5 + 8$ h $5 + (-8)$
i $-5 + (-8)$ j $-6 + 1$ k $6 + (-1)$ l $-6 + (-1)$

4 Evaluate each sum.

a $-12 + 14$ b $12 + -14$ c $-12 + -14$ d $-21 + 18$
e $21 + -18$ f $-21 + -18$ g $-15 + 32$ h $15 + -32$
i $-15 + -32$ j $-16 + 45$ k $16 + -45$ l $-16 + -45$

5 Evaluate each sum.

a $-8 + 17$ b $-9 + (-22)$ c $19 + (-5)$ d $24 + (-8)$
e $-15 + 42$ f $-12 + (-63)$ g $-17 + (-25)$ h $28 + (-19)$
i $-72 + (-15)$ j $28 + (-13)$ k $-36 + (-18)$ l $-45 + 32$

6 Fred the frog fell down a 9 m deep well. Fred could jump up 3 m but would then slip down 2 m. How many jumps will Fred need to make to get out of the well?

7 Paige the pelican walked 4 steps forward and then 1 step back each time she walked. How many times will she need to do this to reach a pond 10 m away?

Alamy/imageBROKER/Erhard Nerger

Shutterstock.com/Charles Brutlag

8 Spark the sparrow was just learning to walk. He took 3 steps forward, 1 step back, 4 steps forward, 2 steps back and then 5 steps forward and 1 step back. Where is he compared to where he started?

1-04 Subtracting integers

WORDBANK

minus To subtract or 'take-away' a number.

difference The result of subtracting two numbers.

Integers can be **subtracted** using a number line.

- Move **left** if subtracting a **positive** integer.
- Move **right** if subtracting a **negative** integer.
- Subtracting a **negative integer** is the same as **adding its opposite**, for example $3 - (-4) = 3 + 4 = 7$.

Think of how we speak:
A single negative means negative: 'I am *not* going to the movies'.
A double negative reverses the meaning to positive: 'I am *not not* going to the movies' means 'I am going to the movies'.

iStockphoto/Rich Legg

EXAMPLE 7

Evaluate each difference.

a $-3 - 2$　　b $5 - (-3)$　　c $1 - 4$　　d $-6 - (-6)$

SOLUTION

a −6 −5 −4 −3 −2 −1

$-3 - 2 = -5$

b 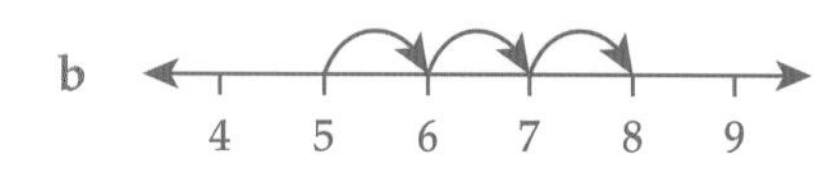

$5 - (-3) = 5 + 3 = 8$

c 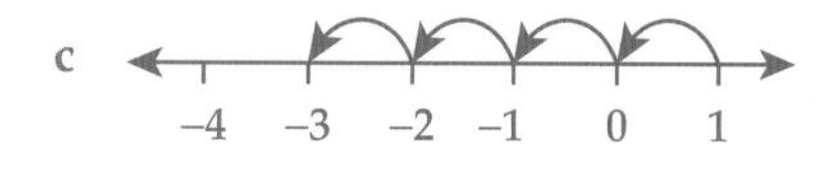

$1 - 4 = -3$

d −6 −5 −4 −3 −2 −1 0

$-6 - (-6) = -6 + 6 = 0$

ISBN 9780170350969

EXERCISE 1-04

1 Evaluate –8 – 3. Select the correct answer A, B, C or D.

A –11 B –5 C 11 D 5

2 Evaluate –11 – (–5). Select A, B, C or D.

A 6 B –16 C 16 D –6

3 Evaluate each difference, using a number line if needed.

a 4 – (–3) b –4 – 3 c –4 – (–3) d 6 – (–2)
e –6 – 2 f –6 – (–2) g 5 – (–1) h –5 – 1
i –5 – (–1) j 7 – (–4) k –7 – 4 l –7 – (–4)

4 Evaluate each difference.

a 12 – (–3) b –12 – 3 c –12 – (–3) d 21 – (–7)
e –21 – 7 f –21 – (–3) g 18 – (–8) h –18 – 8
i –18 – (–8) j 9 – (–24) k –9 – 24 l –9 – (–24)

5 Ahmed opened a new bank account and deposited $58 in it. A week later he had to withdraw $32 but then deposited another $85 the following day. How much money does he have in his account now?

6 Kate set up an automatic deposit of $64 per week into her new bank account. She then had to arrange for $120 to be withdrawn once every 4 weeks. How much money would she have in her account after 8 weeks?

7 Tyler was lost in a maze and decided to try and find his way out. He walked 25 steps forward and then realised he had missed a turn. He had to go back 12 steps, then continued forward for 15 more steps. He walked back 3 steps, forward 8 more and then saw the exit 14 steps further forward. How many steps is the exit from his starting point?

Alamy/OJO Images Ltd

1-05 Coordinates on maps

WORDBANK

coordinates Two numbers and/or letters that give the location of a point or region on a map or number plane.

EXAMPLE 8

This grid shows the location of students' desks in a classroom.

6	Alicia				Naomi
5		Carly	Elizabeth		
4	Thao			Liam	
3			Ben		
2	Sarah		Helena		
1		Joe			Orsolya
	A	**B**	**C**	**D**	**E**

Write the coordinates of each student, using a letter followed by a number.

a Helena b Liam c Orsolya d Alicia

SOLUTION

a C2 b D4 c E1 d A6

EXAMPLE 9

For the classroom grid above, who is seated at each pair of coordinates?

a B5 b C3 c E6

SOLUTION

a Carly b Ben c Naomi

EXAMPLE 10

This map shows a section of Newcastle in NSW.

Write the coordinates of each place.

a NBN Television

b Event Cinemas

c Nesca Park

SOLUTION

a D2 b D4 c C1

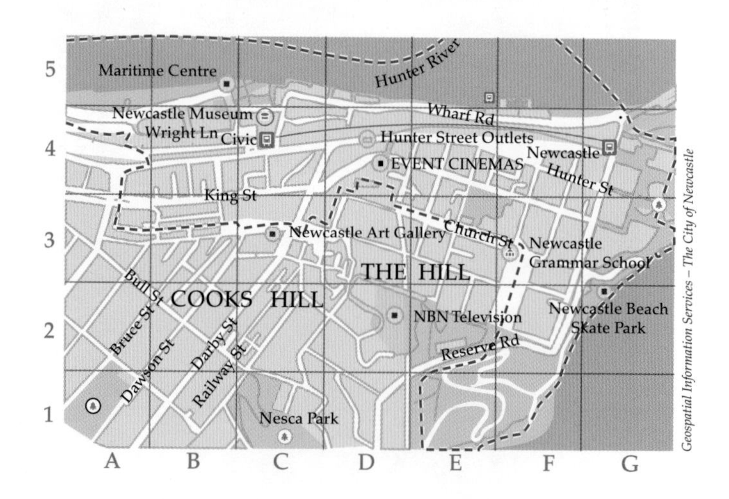

ISBN 9780170350969

EXERCISE 1–05

1 What are the coordinates of Sarah's desk in Example 8? Select the correct answer A, B, C or D.

A A3 B B2 C A2 D E6

2 Which coordinates give the location of Reserve Rd on the map of Newcastle? Select A, B, C or D.

A F3 B B3 C D4 D E2

3 a What are coordinates?

b Which direction, the vertical or horizontal, do we use first when naming coordinates?

4 The following grid shows the layout of desks and students in class 7S:

4	Sam				Katie
3		Brianna	Stuart	Tahine	
2	Jade	Alexis		Ray	Eve
1			Carol		David
	A	**B**	**C**	**D**	**E**

Write down the coordinates of each student.

a Brianna b Ray c David d Carol e Katie

5 Which student is located at each pair of coordinates?

a C3 b D3 c E2 d A2 e A4

6 Name another example of a grid that uses coordinates like the one above.

7 Write down the coordinates of the following places on the map of Newcastle.

a Maritime Centre b Newcastle Grammar School c Newcastle Art Gallery

8 The following grid has been placed on an island to help locate positions:

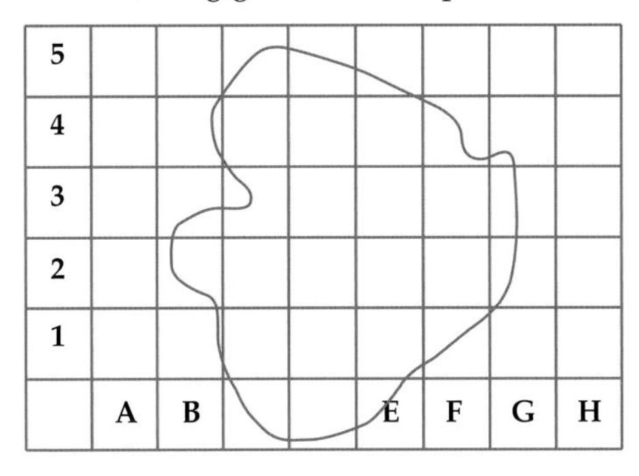

Somewhere there is a buried treasure. Follow the coordinates below to find it. Start at F3, go left 3 squares, go up to C5, go right 4 squares and down 2, left 2. This is the location of the treasure. What are the coordinates of the treasure?

1-06 The number plane

WORDBANK

number plane A grid used to plot points in mathematics, made up of two intersecting number lines.

ordered pair A pair of numbers such as (1, 6) used to locate a point on the number plane.

x-coordinate The first number in an ordered pair: the 1 in (1, 6).

y-coordinate The second number in an ordered pair: the 6 in (1, 6).

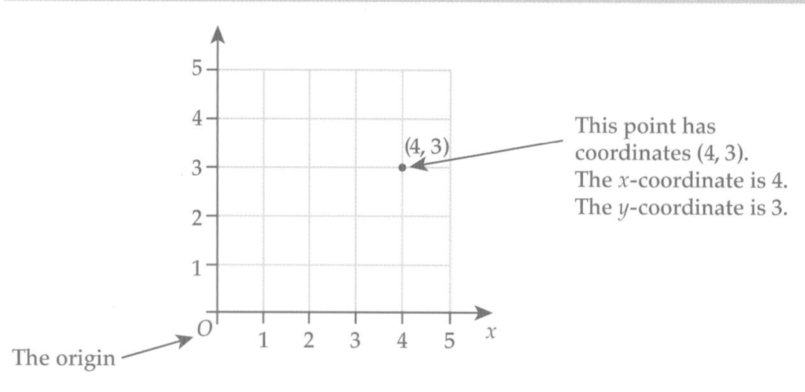

To plot the point (4, 3), start at the origin O and move 4 units right and 3 units up.

- The **horizontal** number line is called the ***x*-axis**.

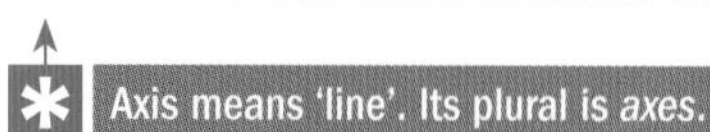

- The **vertical** number line is called the ***y*-axis**.
- The point at the corner where the x- and y-axes meet is called the **origin** and has coordinates (0, 0).
- It is easy to remember to put the x-coordinate first before the y-coordinate as x comes before y in the alphabet.

EXAMPLE 11

Draw a number plane and plot the points (2, 3) and (4, 1).

SOLUTION

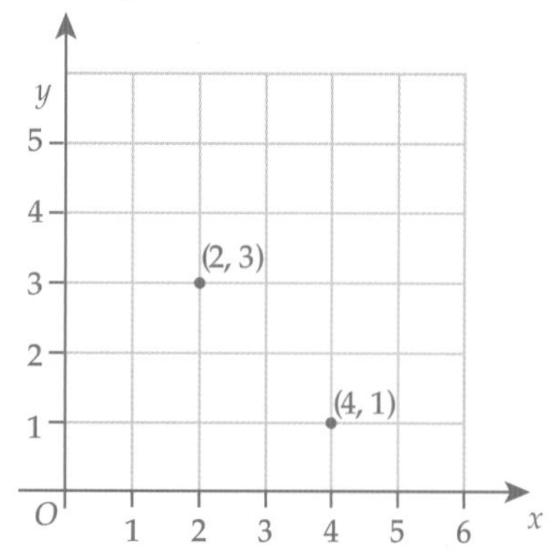

ISBN 9780170350969

EXERCISE 1-06

1 What is the x-coordinate of the point (3, 5)? Select the correct answer **A**, **B**, **C** or **D**.

A 3   B 4   C 5   D 3, 5

2 On the number plane, what is the vertical axis is called? Select **A**, **B**, **C** or **D**.

A the origin   B the y-axis   C the x-axis   D axis of symmetry

3 a Describe in words what a number plane is.

b How do we write coordinates?

c Where is the origin on a number plane?

d What are its coordinates?

4 Copy and complete:

To plot a point on a number plane, graph the __-coordinate first by going ______ and then graph the __-coordinate by going ___.

5 Find the coordinates each point on this number plane.

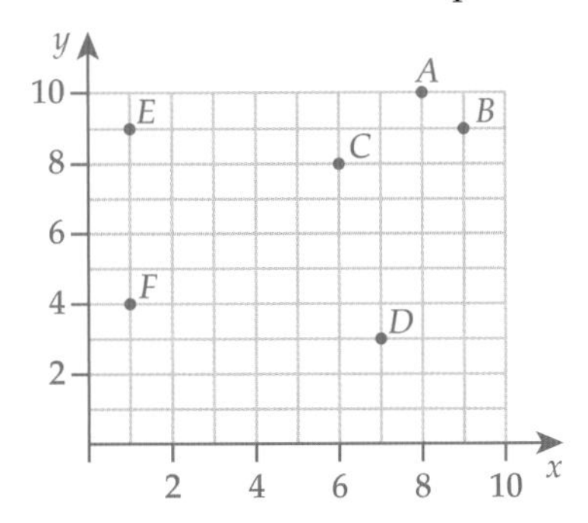

6 Draw a number plane and plot each point below.

$A(2, 3)$ $B(4, 2)$ $C(3, 5)$ $D(2, 6)$ $E(1, 8)$

$F(5, 9)$ $G(8, 4)$ $H(2, 7)$ $I(6, 8)$ $J(5, 6)$

7 Plot each set of points on a number plane and join them in the order listed. What geometrical shape is formed for each one?

a (1, 4), (6, 9), (11, 4), (1, 4)

b (2, 5), (8, 5), (8, 2), (2, 2), (2, 5)

c (4, 1), (1, 6), (8, 6), (6, 1), (4, 1)

d (3, 2), (1, 4), (3, 6), (5, 6), (7, 4), (5, 2), (3, 2)

e (2, 3), (2, 7), (5, 3), (2, 3)

f (1, 1), (3, 4), (5, 4), (3, 1), (1, 1)

8 a List the coordinates of the points A to U on the number plane.

b Plot the above points on graph paper and join them in alphabetical order. What image is formed?

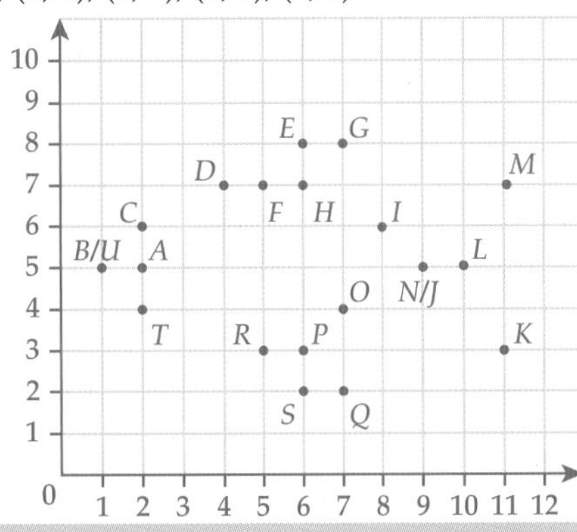

ISBN 9780170350969

1-07 The number plane with negative numbers

WORDBANK

quadrant $\frac{1}{4}$ of the number plane.

The x- and y-axes on a number plane can be extended to include **negative numbers** as well.

EXAMPLE 12

Write the coordinates of each point shown on the number plane.

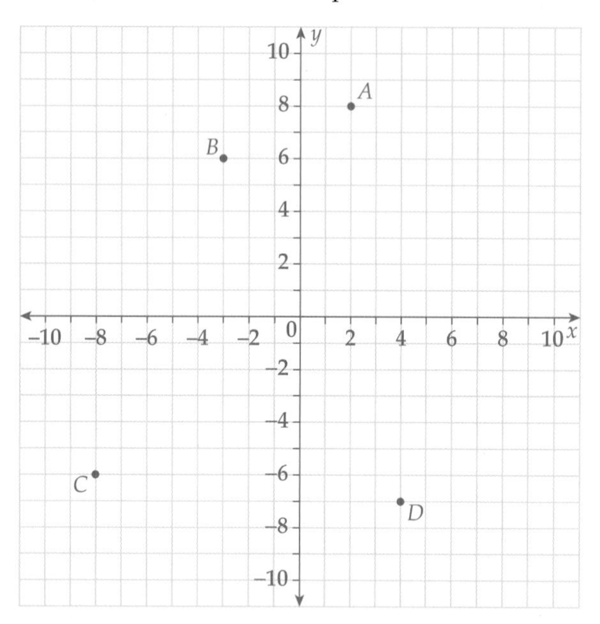

SOLUTION

$A(2, 8)$	$B(-3, 6)$	$C(-8, -6)$	$D(4, -7)$
2 right, 8 up	3 left, 6 up	8 left, 6 down	4 right, 7 down

The number plane is divided into 4 **quadrants**.

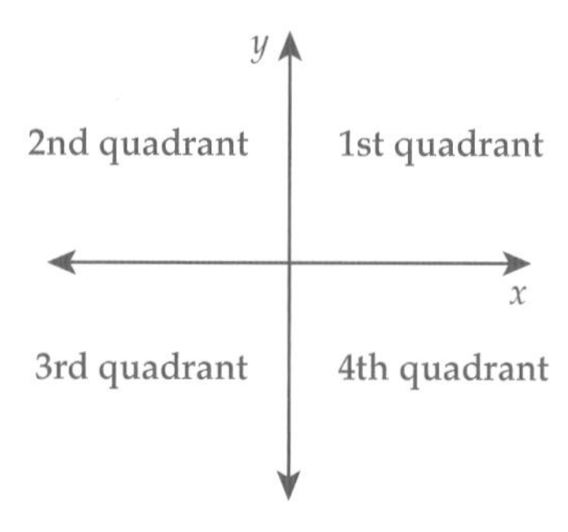

In Example 12, A is in the 1st quadrant, B in the 2nd quadrant, C in the 3rd quadrant and D is in the 4th quadrant.

ISBN 9780170350969

EXERCISE 1–07

1 Starting at the origin, how do you find the point (–3, 6)? Select the correct answer A, B, C or D.

A 3 right, 6 up B 3 left, 6 up C 3 right, 6 down D 3 left, 6 down

2 In which quadrant is the point (5, –2) located? Select A, B, C or D.

A 1st B 2nd C 3rd D 4th

3 In which direction are the quadrants numbered, clockwise or anticlockwise?

4 Write down the coordinates of each point shown on the number plane.

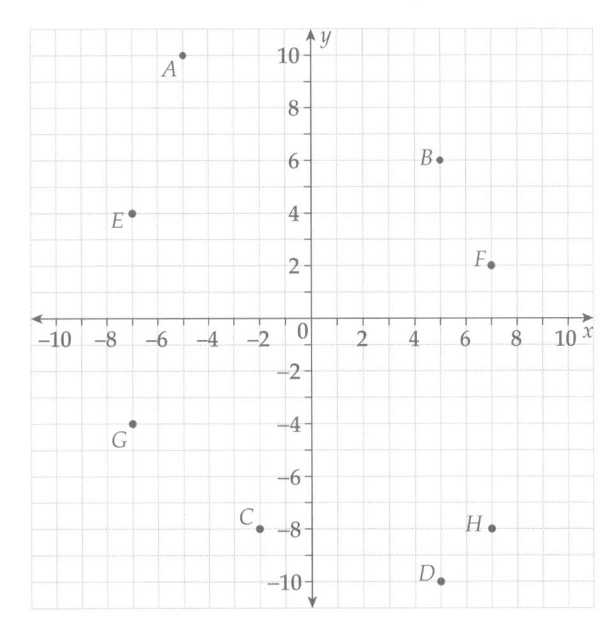

5 In the number plane above, name the quadrant that has each point.

a C b H c F d E

6 Plot each point on a number plane.

$A(-1, 4)$ $B(3, -5)$ $C(-2, -4)$ $D(3, -8)$

$E(2, 6)$ $F(-3, 0)$ $G(-5, 7)$ $H(0, -2)$

7 Which of the points in Question 6 are in the 2nd quadrant?

8 a Plot the points below in order and join them as you go. What shape is made?

(5, 3), (3, 3), (4, 2), (6, 2), (7, 3), (5, 3), (5, 8), (8, 4), (5, 4)

b In which quadrant is the shape drawn?

NUMBER PLANE CODE

Match each ordered pair with a point on the number plane to decode the number plane words.

1 (–3, –5) (–3, 1) (0, 1) (–3, –5) (–2, 4) (–3,1) (0, 1) (2, 3)

2 (4, –1) (3, 4) (–4, 0) (3, 4) (4, –1) (4, –1) (–3, 1) (0, 1) (3, 4)

3 (0, 1) (3, 4) (1, –1) (2, 3) (0, 1) (0, 4) (3, 0) (–1, 2) (3, 4) (0, –3)

4 (2, –2) (4, –1) (1, 1) (–2, –3) (–4, –2) (2, –2) (4, –1) (–2, 4) (–3, –5) (3, 4)

5 (1, 1) (2, –2) (–4, –2) (3, 4) (–2, –1) (1, 1) (1, –1) (–3, 1) (2, –2)

6 (–3, –5) (1, 1) (–3, –5) (3, 4) (0, –3) (–3, –5) (4, –1) (1, 1) (0, 1) (3, 4)

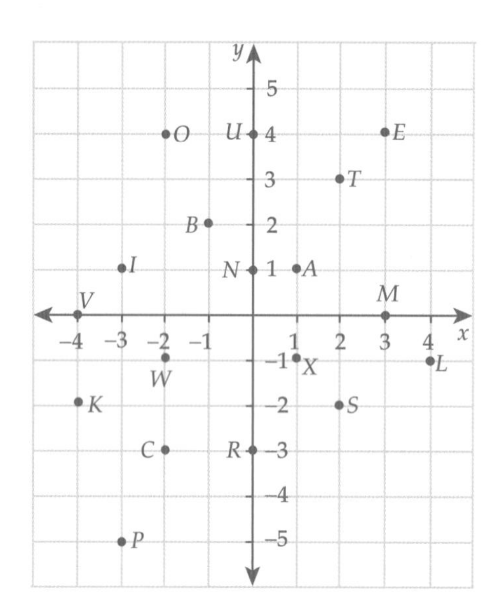

 ISBN 9780170350969

PRACTICE TEST 1

Part A General topics

Calculators are not allowed.

1 Evaluate 9×5.

2 How many hours in one week?

3 A flea can jump up to 200 times its body length. How far can a flea jump if it is 3.1 mm long?

4 Hollee is 18 months younger than Ben. If Ben is 3 years and 2 months old, how old is Hollee?

5 What metric unit is used to measure medicine in a cup?

6 Evaluate $100 \div 4$.

7 Write 0.6 as a simple fraction.

8 If I am driving north and turn right, which direction will I then be travelling?

9 How many sides does a quadrilateral have?

10 How many days in December?

Part B Integers and the number plane

Calculators are not allowed.

1-01 Numbers above and below zero

11 Which expression below represents walking 4 steps forward and then 6 steps back? Select the correct answer A, B, C or D.

A $-4 + 6$ B $-4 - 6$ C $4 + (-6)$ D $4 - (-6)$

12 Which expression below represents withdrawing \$8 and then depositing \$14? Select A, B, C or D.

A $8 - 14$ B $-14 - 8$ C $-8 + -14$ D $-8 + 14$

1-02 Ordering integers

13 Which list of numbers shows 7, –4, 3, 0 and –2 in ascending order? Select A, B, C or D.

A 7, 3, 0, –2, –4 B –4, –2, 0, 3, 7 C –2, –4, 0, 3, 7 D 7, 3, 0, –4, –2

1-03 Adding integers

14 Evaluate each sum.

a $-9 + 7$ b $-5 + (-11)$ c $8 + (-14)$

15 Which statement is true about a number line? Select A, B, C or D.

A It has only positive integers.
B It must be small and neat.
C It has arrows at both ends.
D It has only negative integers.

1-04 Subtracting integers

16 Evaluate each difference.

a $12 - 6$ b $-7 - 8$ c $-12 - (-15)$

17 a A crab walks 5 steps left, 8 steps right and then 11 steps left again. Represent this situation as a sum of integers.

b Where is the crab now compared to its starting point?

1-05 Coordinates on maps

18 This grid shows part of a aeroplane's seating.

2		Anil	Goran
1	Li	Sam	Mel
	A	**B**	**C**

a What is the position of the empty seat?

b What is Sam's position?

1-06 The number plane

19 Copy and complete:

On the number plane, the y-axis is horizontal/vertical and the x-axis is horizontal/vertical.

When plotting points we must plot the __-coordinate first and then the __-coordinate.

1-07 The number plane with negative numbers

20 Write down the coordinates of A, B and C.

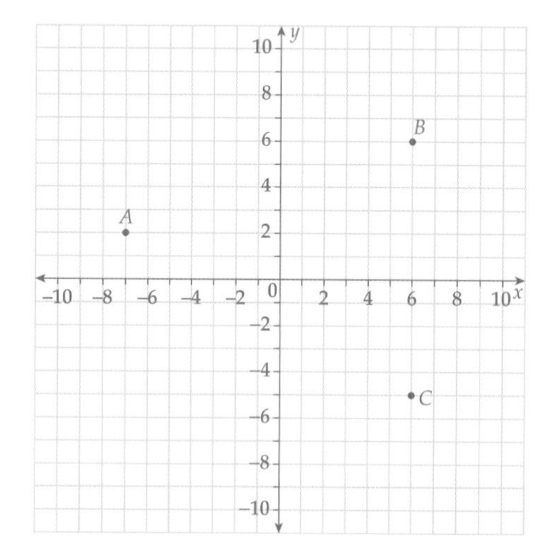

21 For the number plane in Question 20, in which quadrants do the points A, B and C lie?

22 a Plot the points $P(2, 5)$, $Q(3, -4)$, $R(-2, -4)$ and $S(-1, 5)$ on a number plane.

b What shape does $PQRS$ form?

ISBN 9780170350969

ANGLES

2

WHAT'S IN CHAPTER 2?

2-01 Naming angles
2-02 Types of angles
2-03 Measuring and drawing angles
2-04 Complementary and supplementary angles
2-05 Angles at a point and vertically opposite angles
2-06 Corresponding angles on parallel lines
2-07 Alternate angles on parallel lines
2-08 Co-interior angles on parallel lines

IN THIS CHAPTER YOU WILL:

- name angles using three letters, for example $\angle ABC$
- use a protractor to measure and draw angles
- classify angles as acute, right, obtuse, straight, reflex and revolution
- identify pairs of angles that are complementary, supplementary, adjacent and vertically opposite
- solve geometry problems involving right angles, angles on a straight line, angles at a point and vertically opposite angles
- identify pairs of angles formed by parallel lines crossed by a transversal: corresponding angles, alternate angles and co-interior angles
- solve geometry problems involving corresponding angles, alternate angles and co-interior angles

Shutterstock.com/BassKwong

ISBN 9780170350969

2-01 Naming angles

WORDBANK

arm One side or line of an angle.

vertex The corner of an angle.

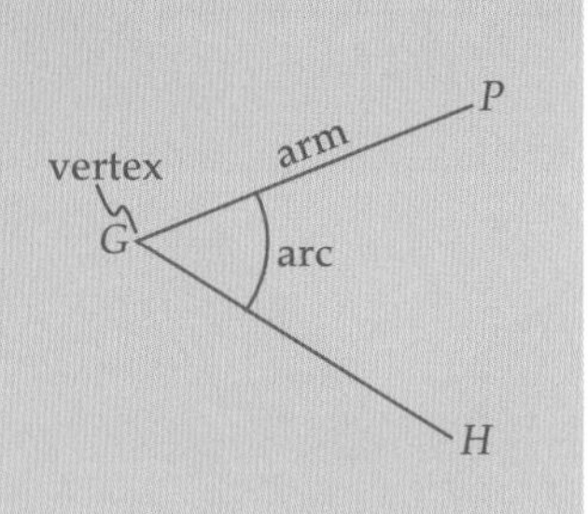

An **angle** measures how much an object turns or spins, and is measured in **degrees** (°).
An angle is usually named using three letters, with its vertex being the middle letter. The angle drawn above is named $\angle PGH$ or $\angle HGP$. It could also be named using one letter $\angle G$ but this is confusing when a diagram has more than one angle drawn at G.

Just think of the order of letters when you draw the angle: *P*–*G*–*H*.

EXAMPLE 1

Name the marked angle.

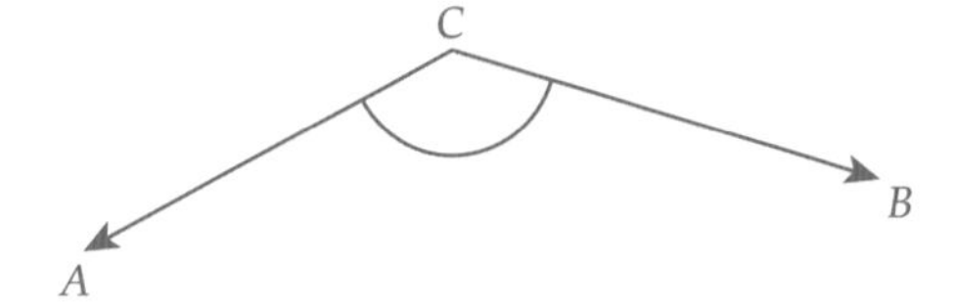

SOLUTION

$\angle ACB$ or $\angle BCA$.

Shutterstock.com/lisafx

ISBN 9780170350969

EXERCISE 2-01

1 What is an angle? Select the correct answer A, B, C or D.

A Part of a line
B The point where two lines meet
C The turn between two lines
D A corner

2 Name this angle. Select A, B, C or D.

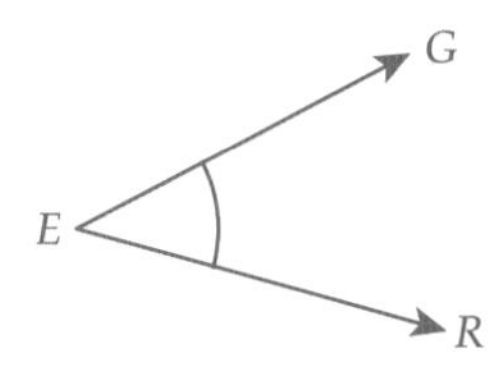

A $\angle EGR$ B $\angle GRE$ C $\angle ERG$ D $\angle GER$

3 a How many letters do we use to name an angle?

b Do we use capital letters or small letters?

c Does it matter what order we use for the letters?

4 Name each angle in two different ways using 3 letters.

a
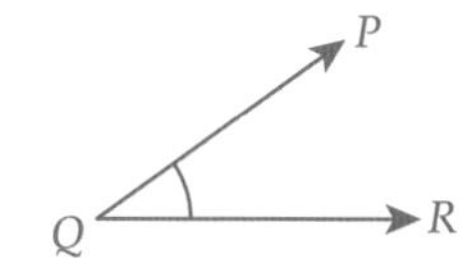

b
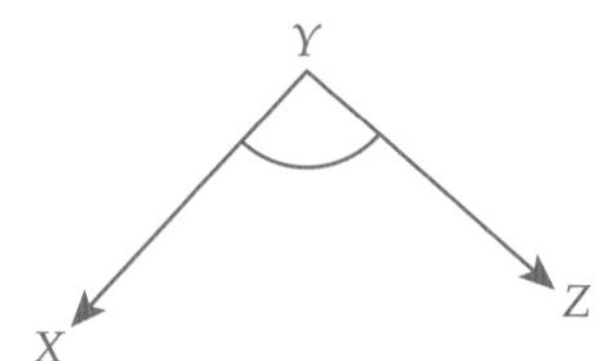

c
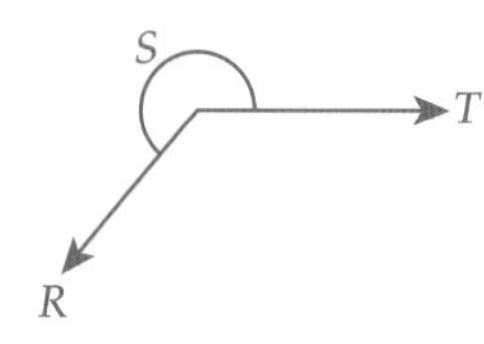

d
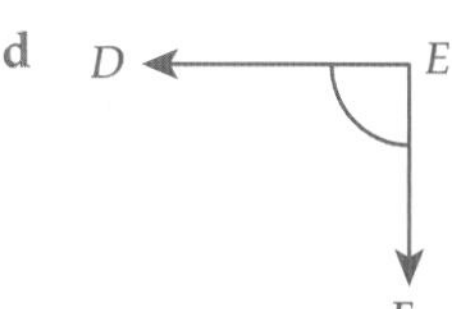

e
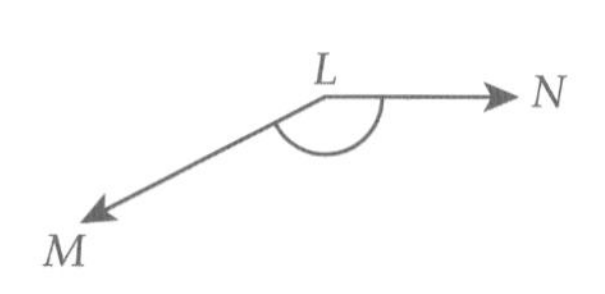

f
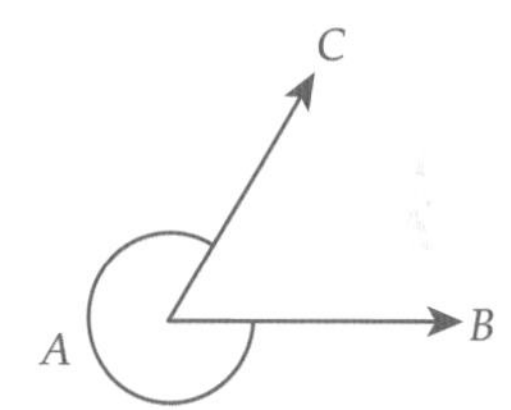

5 Draw each angle named, labelling it correctly.

a $\angle POT$ b $\angle TAF$ c $\angle RHK$

6 How can the angle marked • below be named? Select A, B, C or D.

A $\angle ABD$ B $\angle CBD$ C $\angle ABC$ D $\angle BCA$

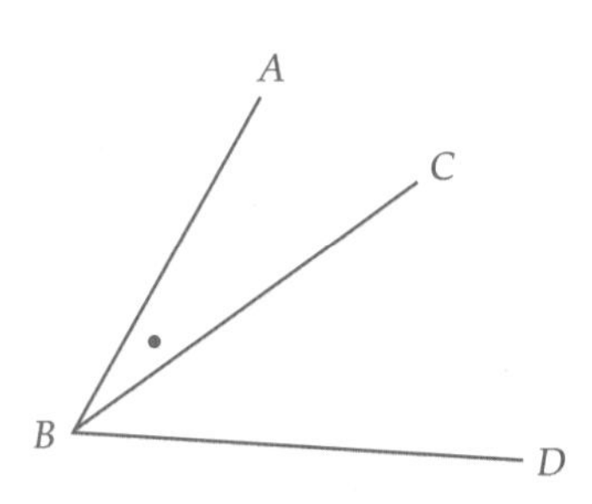

2-02 Types of angles

There are many different types of angles and they have special names.

Acute angle: less than 90° 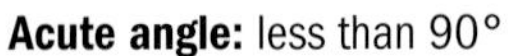 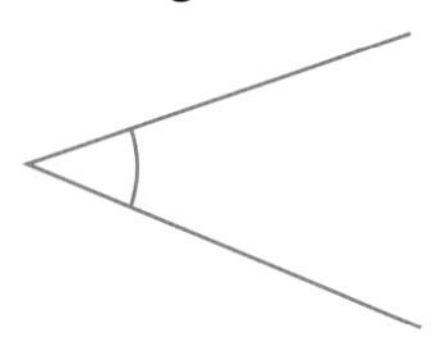	**Right angle:** 90°, a quarter-turn A right angle is marked by a small square. There are many right angles in the room where you are now.
Obtuse angle: between 90° and 180° 	**Straight angle:** 180°, a half-turn Looks like a straight line.
Reflex angle: between 180° and 360° 	**Revolution:** 360°, a complete turn 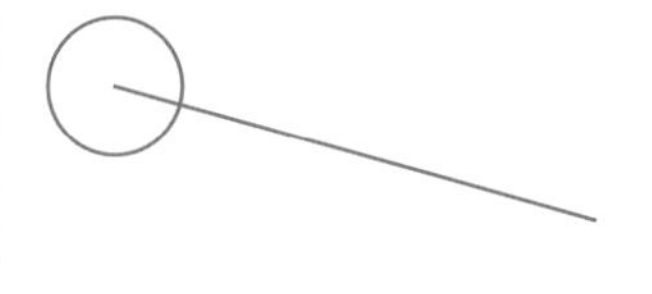

What types of angles are marked in red below?

Shutterstock.com/Jodie Johnson

istockphoto/katatonia82

Shutterstock.com/zstock

istockphoto/RobHowarth

ISBN 9780170350969

EXERCISE 2–02

1 What type of angle is 101°? Select the correct answer A, B, C or D.

A a reflex angle B an obtuse angle C a right angle D an acute angle

2 What type of angle is 234°? Select A, B, C or D.

A a reflex angle B an obtuse angle C a right angle D an acute angle

3 Classify each angle below.

a
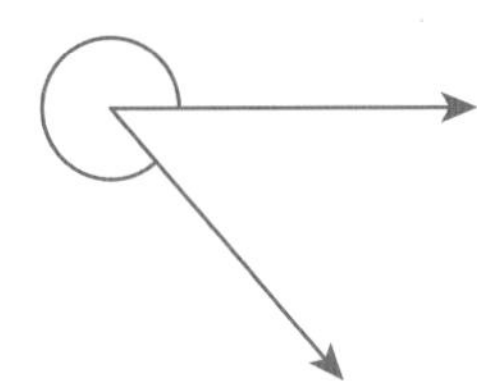

b
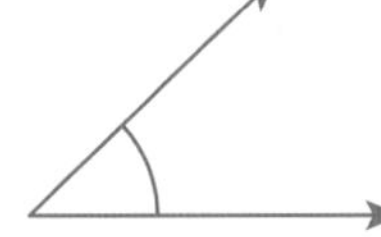

c
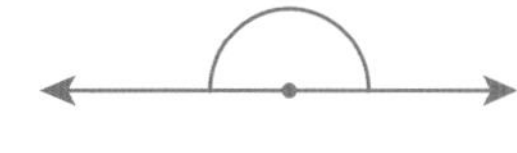

d
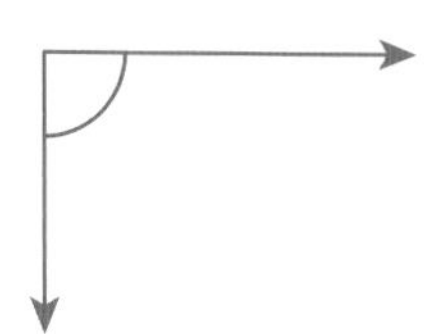

e
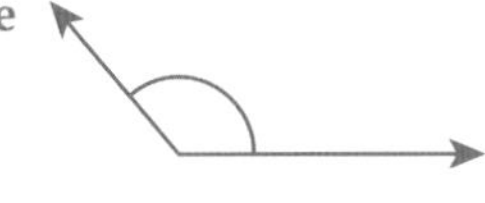

f
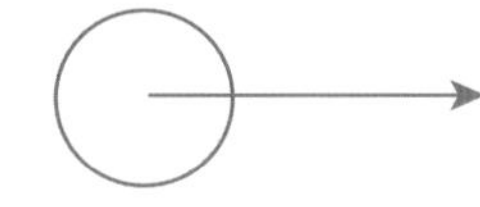

g
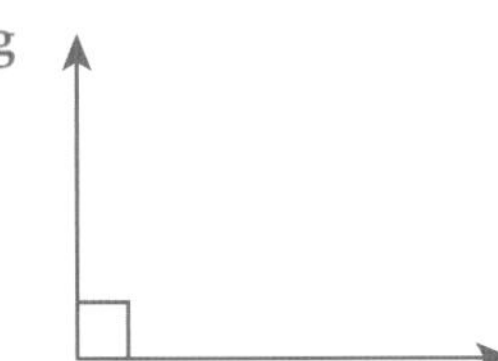

h

4 Draw:

a an acute angle b an obtuse angle c a right angle

d a reflex angle e a straight angle f a revolution

5 Classify each angle size.

a 37° b 107° c 252° d 195° e 79° f 180°

g 163° h 179° i 360° j 5° k 345° l 91°

6 Write down each type of angle that can be seen in this photo of the Royal Palace in Phnom Penh, Cambodia.

Getty Image/Marc Dozier/hemis.fr

2-03 Measuring and drawing angles

Angles are measured in degrees (°) using a **protractor**.
Each unit on a protractor has size one degree.

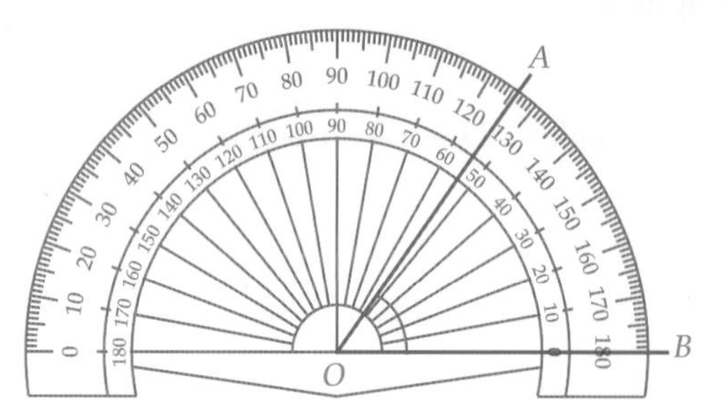

$\angle AOB$ is measured to be 54°.

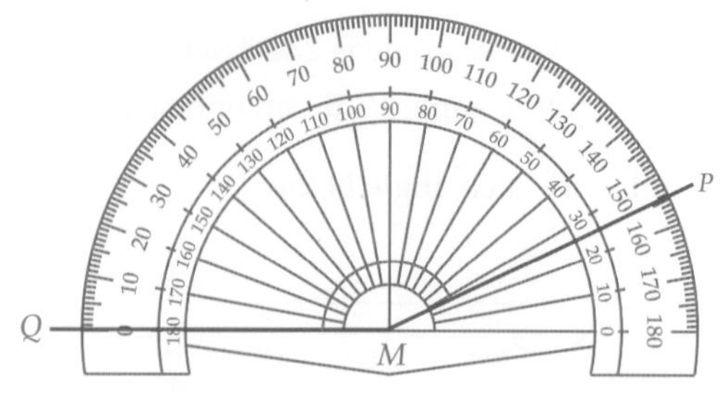

$\angle PMQ$ is measured to be 155°.

To measure an angle with a protractor:
- line up the base line of the protractor with one arm of the angle
- position the centre of the protractor on the vertex of the angle
- use the scale that begins with 0° to read off the angle size from the other arm.

EXAMPLE 2

Construct an angle $\angle KPM$ of size 76°.

SOLUTION

Draw a base line PM. Position the centre of your protractor at P, use the scale that begins with 0° and make a mark at 76°. Join this mark to point P and label it K.

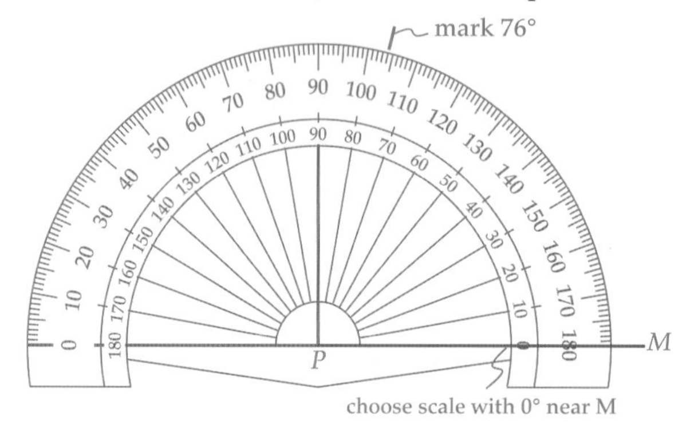

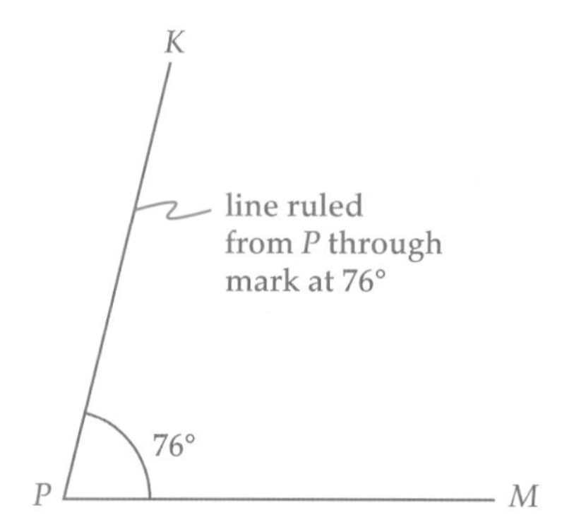

ISBN 9780170350969

EXERCISE 2-03

1 What instrument is used to measure angles? Select the correct answer A, B, C or D.

A compasses B protractor C ruler D set square

2 Estimate the size of this angle. Select A, B, C or D.

A 20° B 40° C 35° D 45°

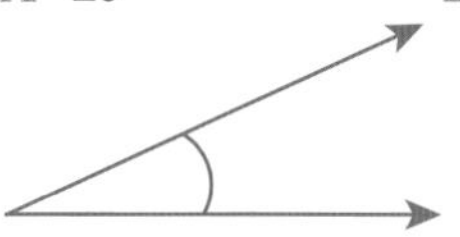

3 Use a protractor to measure each angle.

a

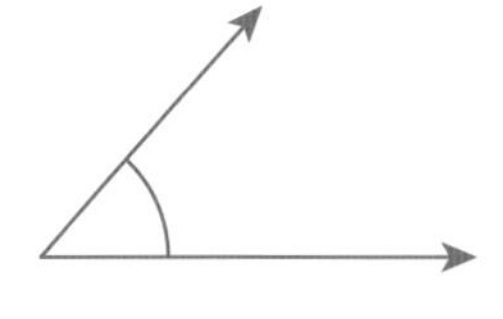

b

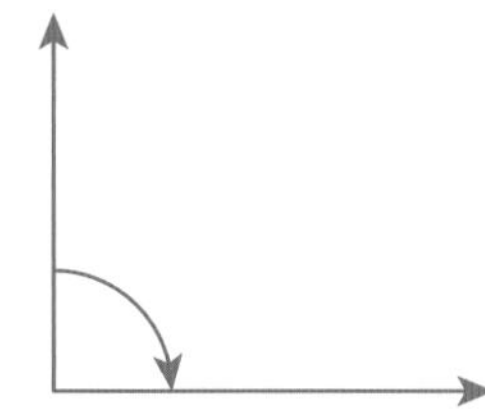

c

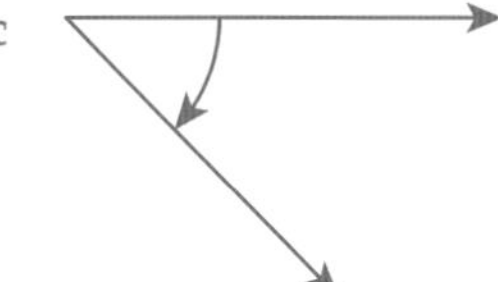

d

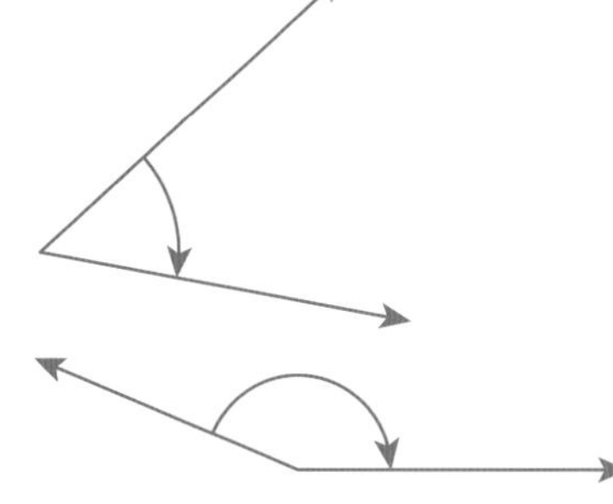

e

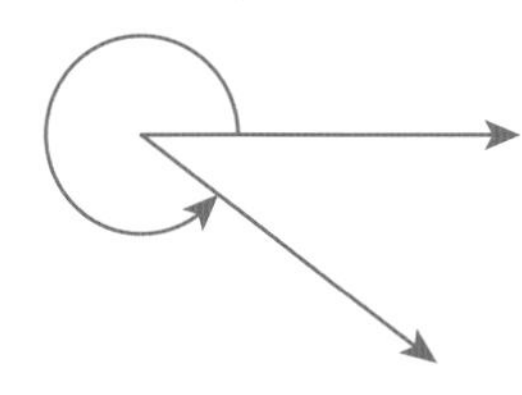

f

4 Construct an angle for each angle size.

a 50° b 75° c 110° d 160°

e 92° f 155° h 18° i 127°

5 To construct angles greater than 180°, it is easier to subtract the number of degrees from 360° (a revolution). For example, to construct an angle of 200°, construct 360° – 200° = 160° and mark the other side of the angle as 200°.

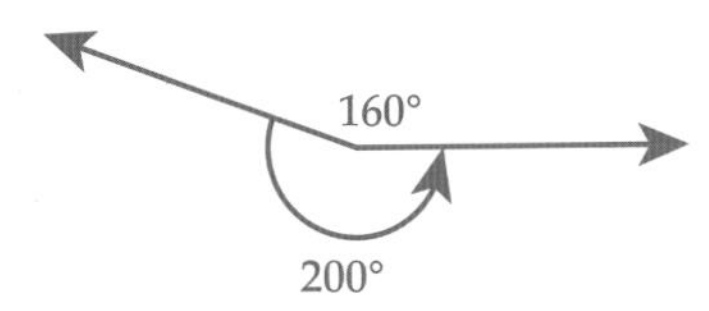

Use this method to construct an angle of size:

a 220° b 295° c 320°

2-04 Complementary and supplementary angles

WORDBANK

complementary angles Two angles that add to 90°, for example, 35° and 55°.

supplementary angles Two angles that add to 180°, for example, 40° and 140°.

adjacent angles Two angles that are joined together, sharing the same arm and vertex.

pronumeral A letter of the alphabet that stands for a number.

EXAMPLE 3

Find:

a the complement of 38°

b the supplement of 38°

SOLUTION

a $90° - 38° = 52°$

b $180° - 38° = 142°$

Angles in a right angle are complementary (add up to 90°).
Angles on a straight line are supplementary (add up to 180°).

Two angles joined together are called **adjacent angles**.

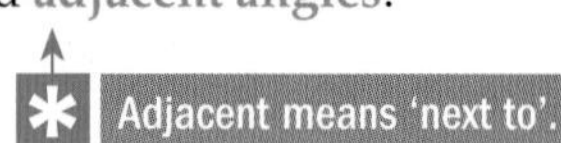

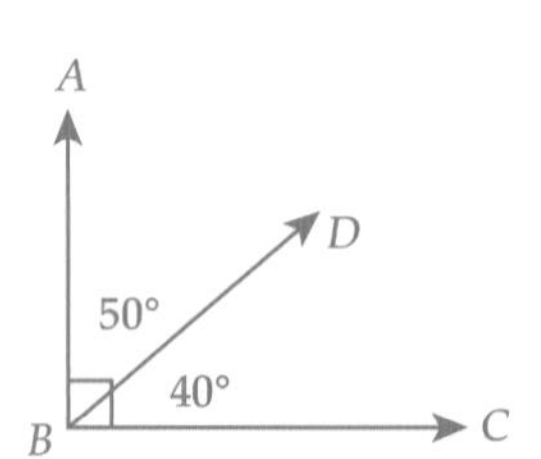

These two adjacent angles are complementary.

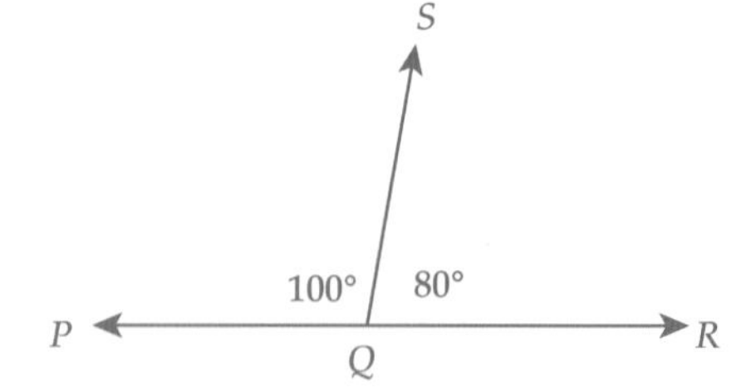

These two adjacent angles are supplementary

EXAMPLE 4

Each diagram has an angle size labelled by a letter called a **pronumeral**, which stands for a number. Find the size of each pronumeral in the diagrams below, giving reasons.

a

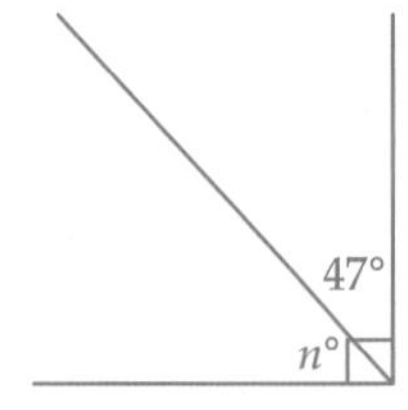

b

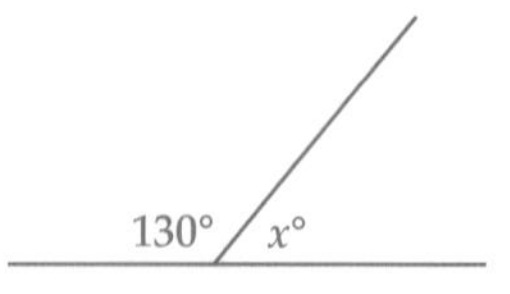

SOLUTION

a $n = 90 - 47$ (angles in a right angle)
$= 43$

b $x = 180 - 130$ (angles on a straight line)
$= 50$

ISBN 9780170350969

EXERCISE 2-04

1 What is the supplement of 27°? Select the correct answer A, B, C or D.

A 63° B 153° C 73° D 163°

2 What do adjacent angles share? Select A, B, C or D.

A two same arms

B the same vertex

C the same letters

D one same arm and the same vertex

3 Is each statement true or false?

a 50° and 130° are complementary angles.

b 25° and 155° are supplementary angles.

c 49° is the complement of 41°.

d 37° is the supplement of 143°.

e Two complementary angles must both be acute.

f Adjacent supplementary angles make up a straight angle.

4 Find the complement of each angle.

a 80° b 75° c 12° d 47°

5 Find the supplement of each angle.

a 50° b 125° c 78° d 163°

6 Draw a pair of:

a adjacent complementary angles b adjacent supplementary angles

7 Find the value of each pronumeral, giving reasons.

a

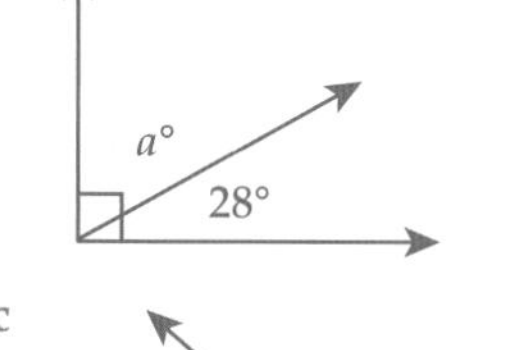

b

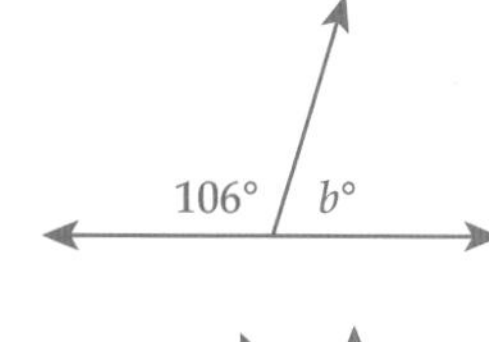

c

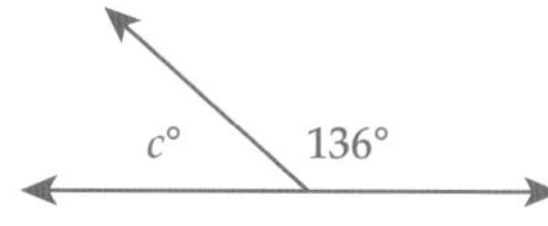

d

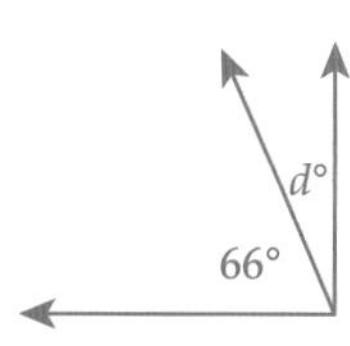

8 a Name a pair of adjacent complementary angles in this diagram.

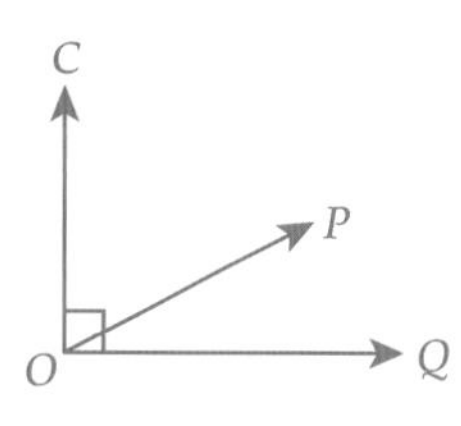

b Name a pair of adjacent supplementary angles in this diagram.

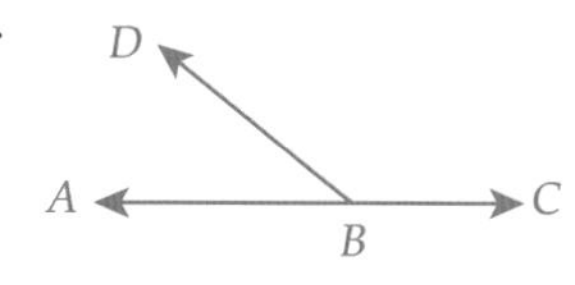

ISBN 9780170350969

2-05 Angles at a point and vertically opposite angles

WORDBANK

angles at a point Angles formed around a point to make a revolution (360°).

vertically opposite angles Pairs of equal opposite angles formed when two straight lines intersect.

Angles at a point form a revolution, so they will add to 360°. For example, in the diagram below, $a + b + c + d = 360$.

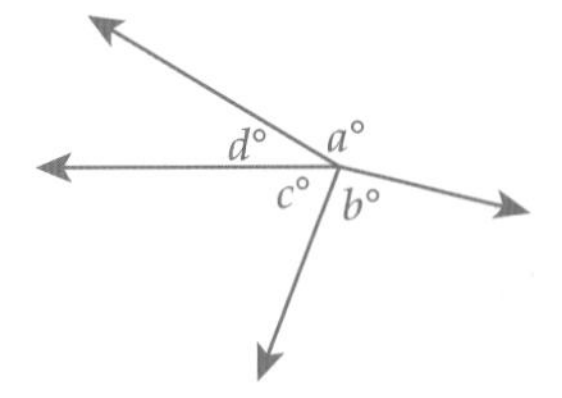

Vertically opposite angles are a pair of equal angles opposite each other when two straight lines cross. For example, in the diagram below, the angles marked * are vertically opposite and equal.

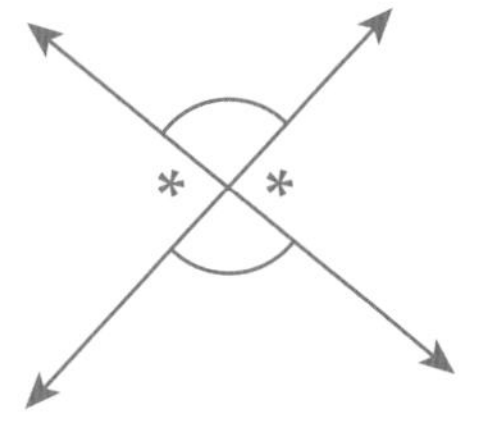

There are two pairs of vertically opposite angles. The other pair is marked with arcs.

EXAMPLE 5

Find the value of each pronumeral, giving reasons.

a

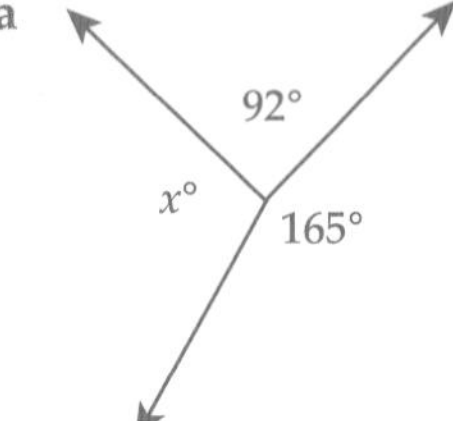

b

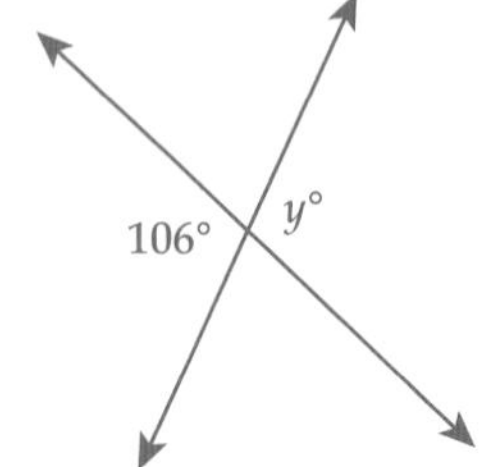

SOLUTION

a $x = 360 - 92 - 165$ (angles at a point)
$= 103$

b $y = 106$ (vertically opposite angles)

 ISBN 9780170350969

EXERCISE 2-05

1 What do angles at a point add to? Select the correct answer A, B, C or D.

A 360° B 90° C 180° D 270°

2 Which of the following is a true statement about vertically opposite angles? Select A, B, C or D.

A They are complementary. B They are equal.

C They are supplementary. D They are adjacent.

3 a Describe the rule about angles at a point and draw an example.

b Describe the rule about vertically opposite angles and draw an example.

4 Is each statement true or false?

a Vertically opposite angles are supplementary.

b Angles at a point add up to 180°.

c One example of angles at a point is 4 adjacent right angles.

d For vertically opposite angles to be formed, the lines must be straight.

5 Find the value of each pronumeral.

a

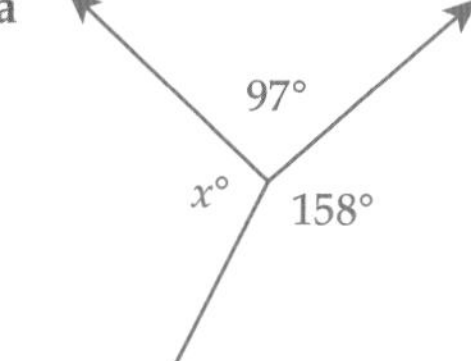

b

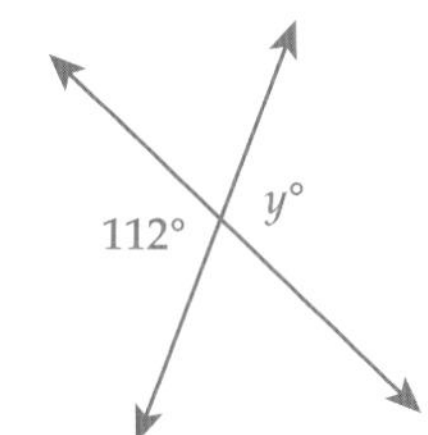

c

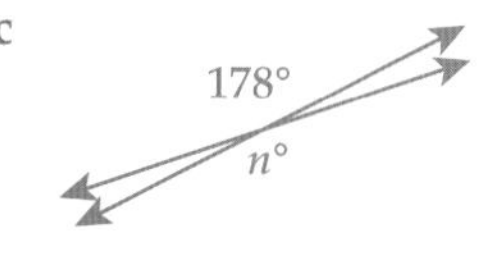

d

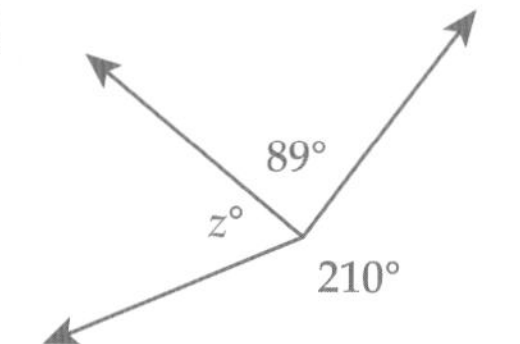

e

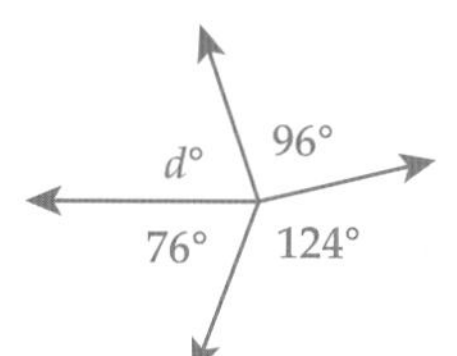

f

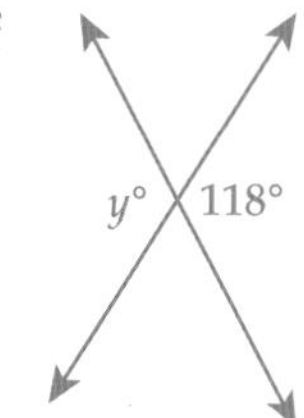

4 In this diagram name the angle that is vertically opposite:

a $\angle WKZ$ b $\angle ZKY$

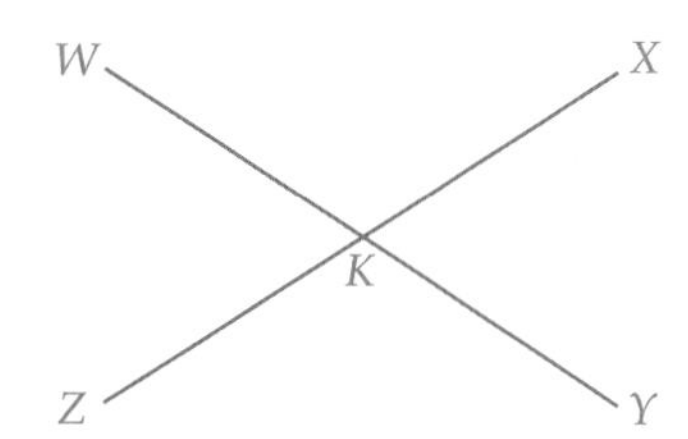

2-06 Corresponding angles on parallel lines

WORDBANK

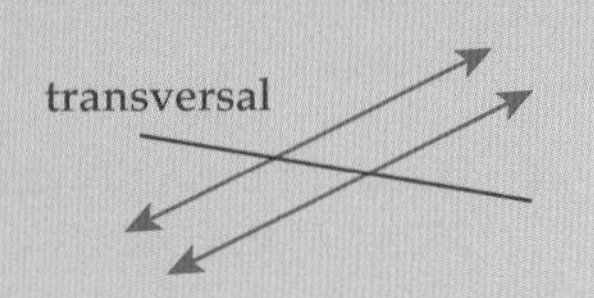

parallel lines Lines that point in the same direction and which never meet (the blue lines in this diagram).

transversal A line that cuts across two or more lines.

corresponding angles Pairs of 'matching' angles formed when a transversal crosses two or more other lines.

If a transversal crosses two *parallel* lines, the corresponding angles are equal.

Corresponding angles on parallel lines

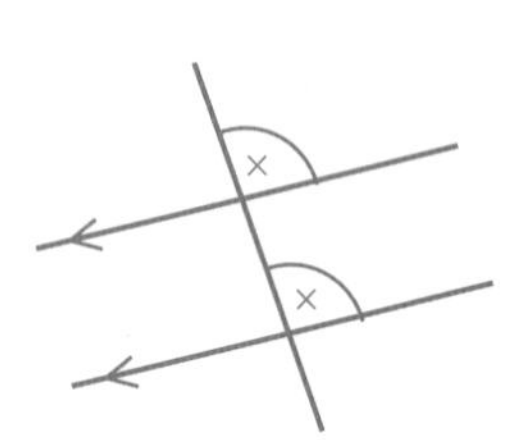

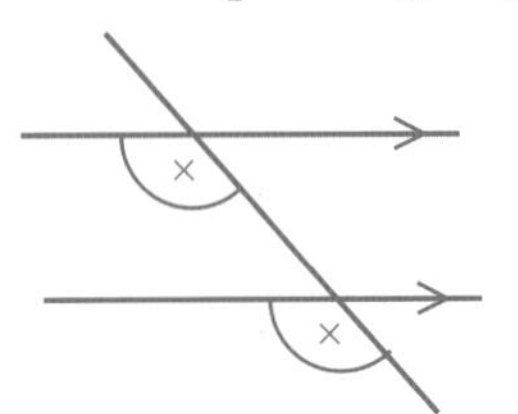

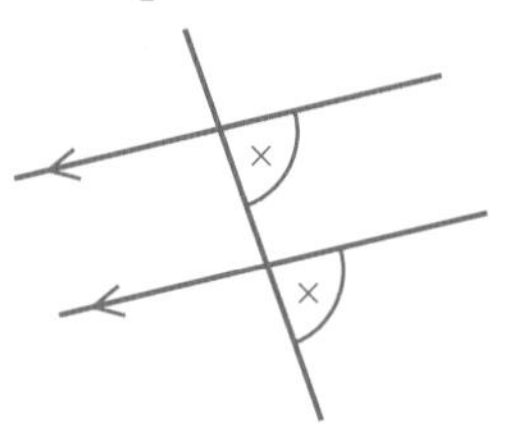

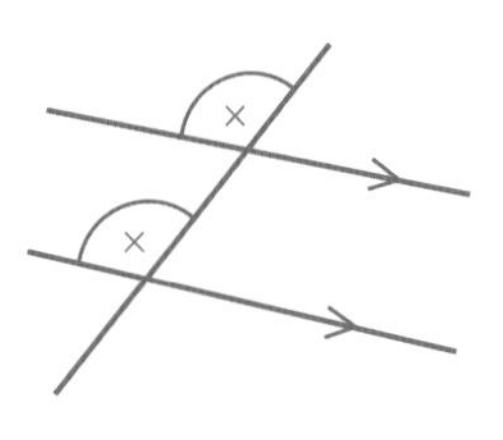

In the first diagram, the corresponding angles are both above the parallel lines and right of the transversal.
In the second diagram, they are both below the parallel lines and left of the transversal.

> Corresponding angles on parallel lines are equal.

EXAMPLE 6

Find the value of z, giving a reason.

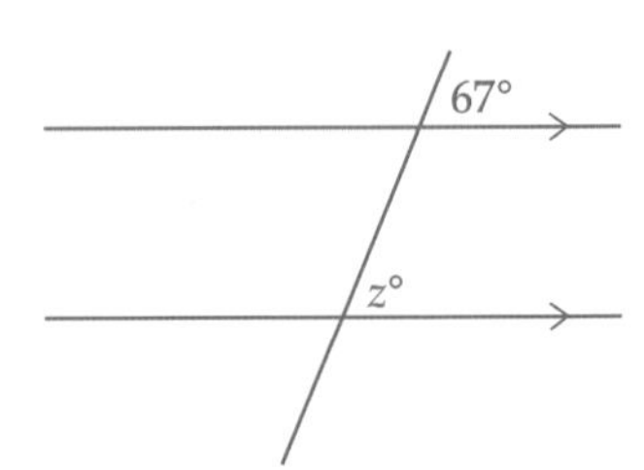

SOLUTION

$z = 67$ (corresponding angles on parallel lines)

ISBN 9780170350969

EXERCISE 2–06

1 Complete: Corresponding angles on parallel lines are __________. Select the correct answer A, B, C or D.

A adjacent B complementary C supplementary D equal

2 For each diagram, name the angle that is corresponding to the angle marked.

a

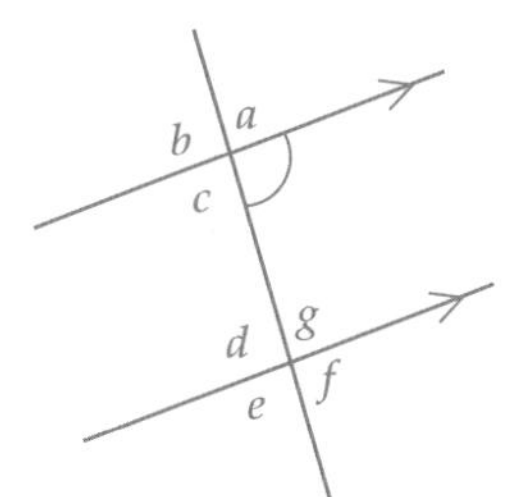

b

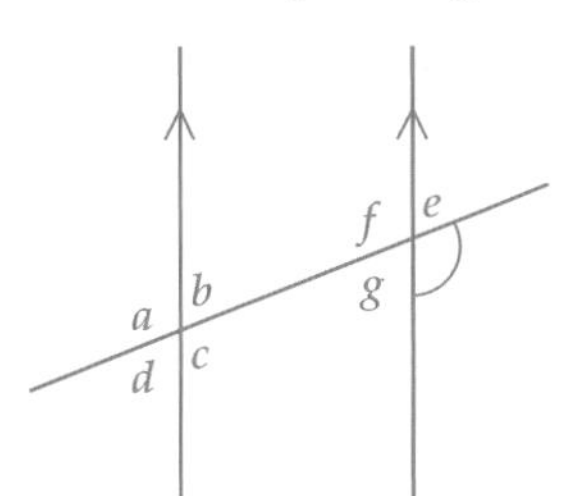

c

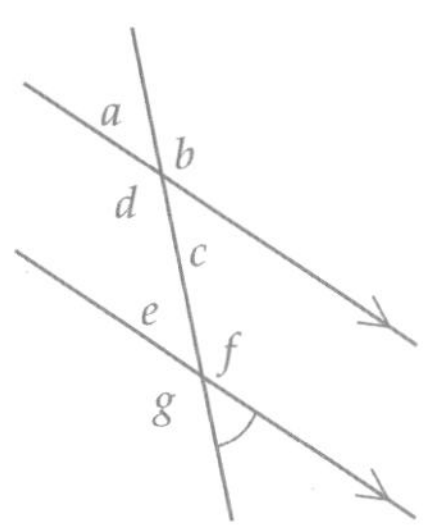

3 Is each statement true or false?

a A transversal is any line that crosses another line.

b Parallel lines point in the same direction.

c Corresponding angles are opposite each other.

d Corresponding angles are equal if the lines are parallel.

4 Copy each diagram and mark a pair of corresponding angles on each one.

a

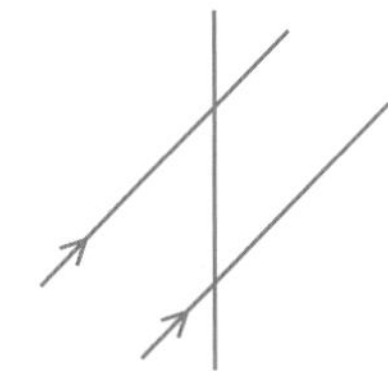

b

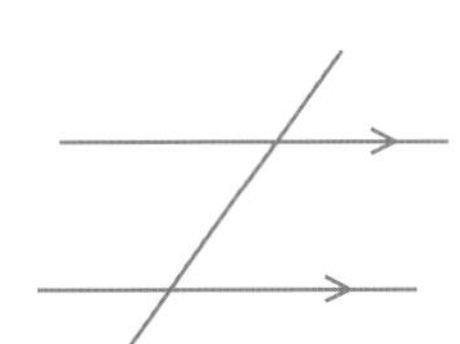

c

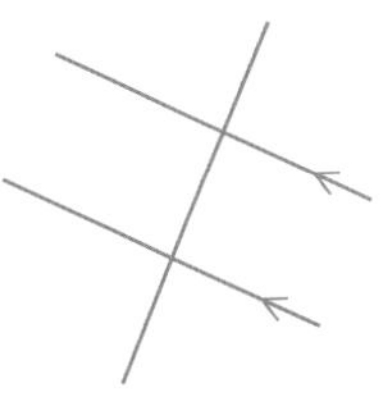

5 Find the value of each pronumeral, giving reasons.

a

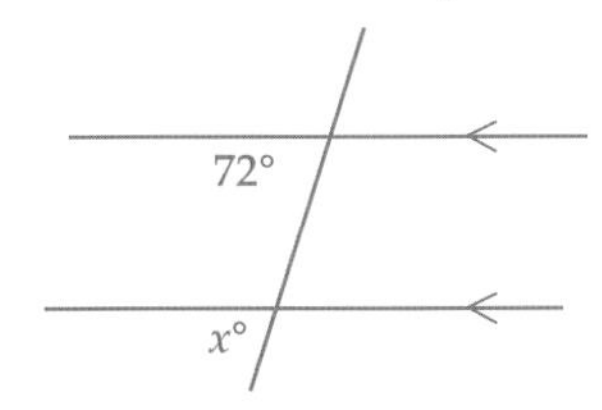

b

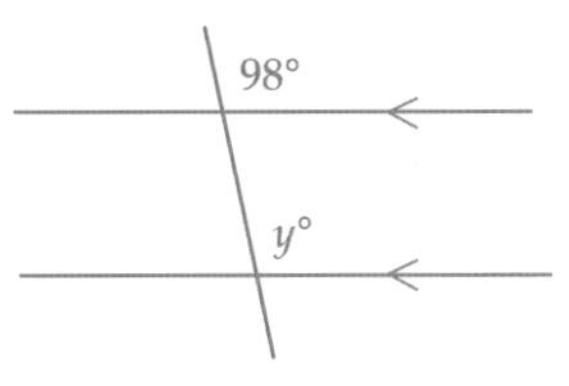

c

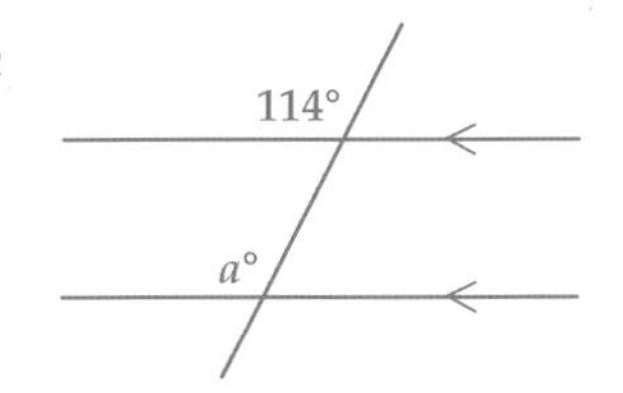

d

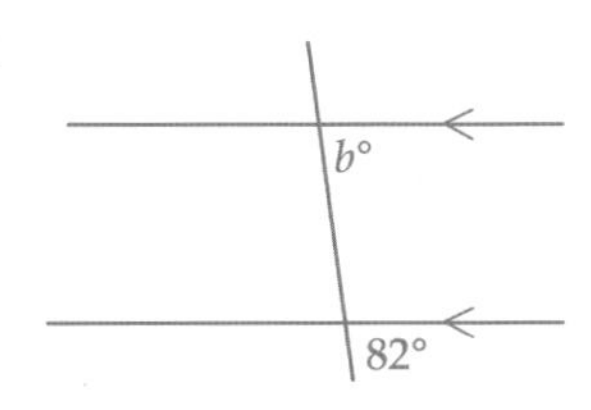

6 Find the value of m.

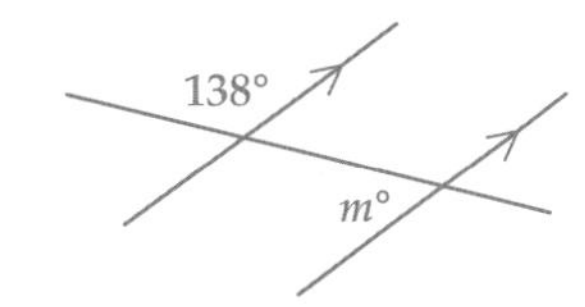

ISBN 9780170350969

2-07 Alternate angles on parallel lines

WORDBANK

alternate angles Pairs of angles formed when a transversal crosses two or more other lines; they are between the lines on opposite sides of the transversal.

Alternate means 'changing direction'.
If a transversal crosses two *parallel* lines, the alternate angles are equal.

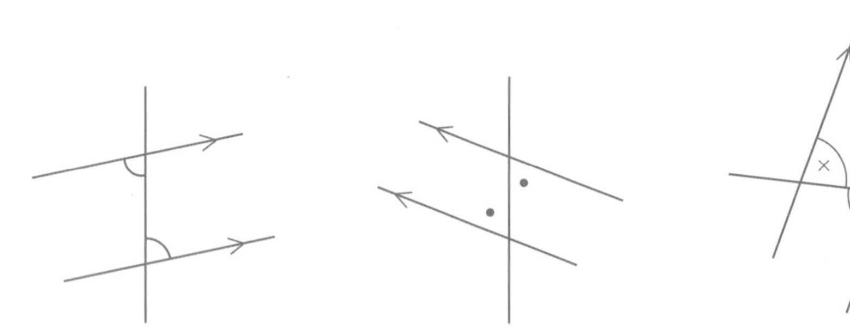

If you turn each diagram of alternate angles upside-down, you will see that the alternate angles fit on each other, so they are equal in size.

Alternate angles on parallel lines are equal.

EXAMPLE 7

Find the value of a, giving a reason.

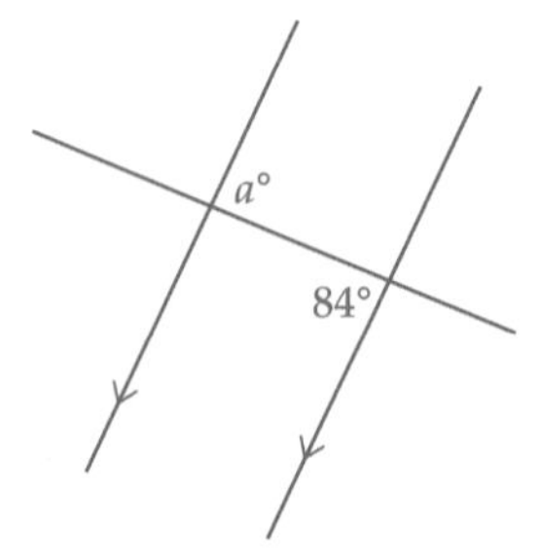

SOLUTION

$a = 84$ (alternate angles on parallel lines)

EXERCISE 2-07

1 Complete: Alternate angles on parallel lines are ________. Select the correct answer A, B, C or D.

A adjacent
B complementary
C equal
D supplementary

 ISBN 9780170350969

EXERCISE 2-07

2 For each diagram, name the angle that is alternate to the angle marked.

a

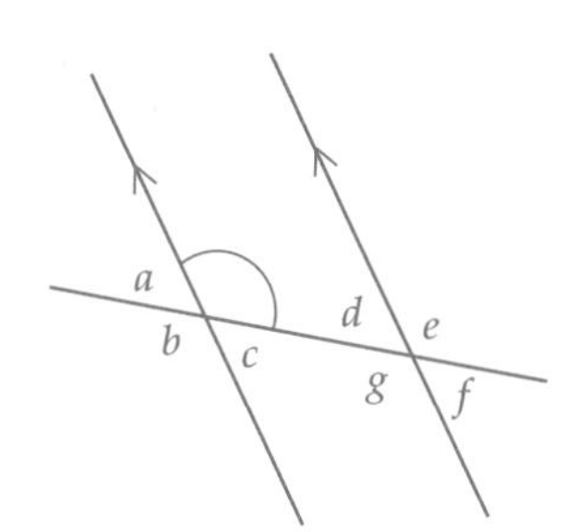

b

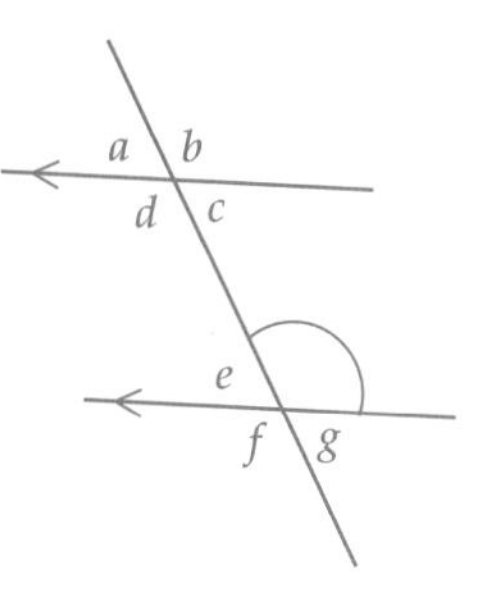

c

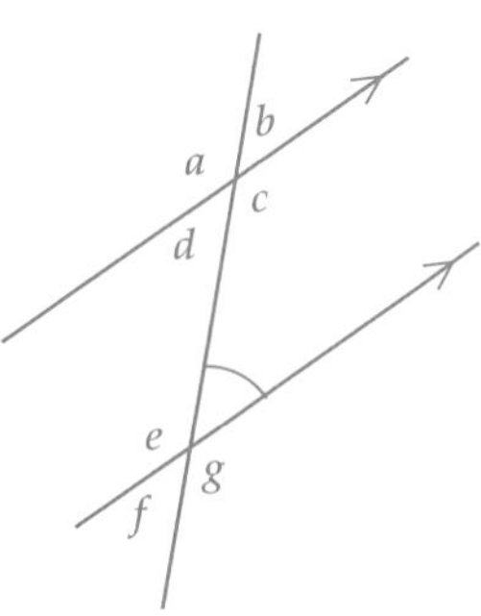

3 Is each statement true or false?

a A transversal is a line that crosses two or more other lines.

b Parallel lines never meet.

c Alternate angles are on the same side of the transversal.

d Alternate angles are equal if the lines are parallel.

4 Copy each diagram and mark a pair of alternate angles on each one.

a

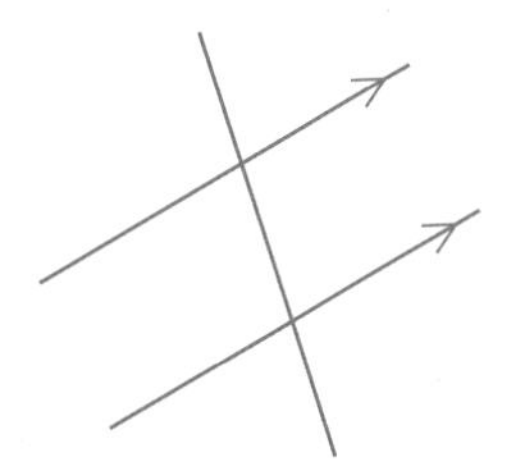

b

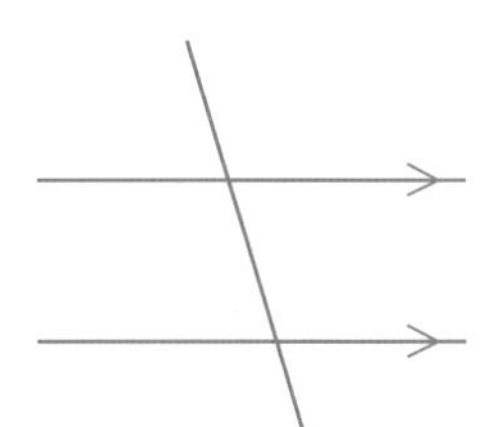

c

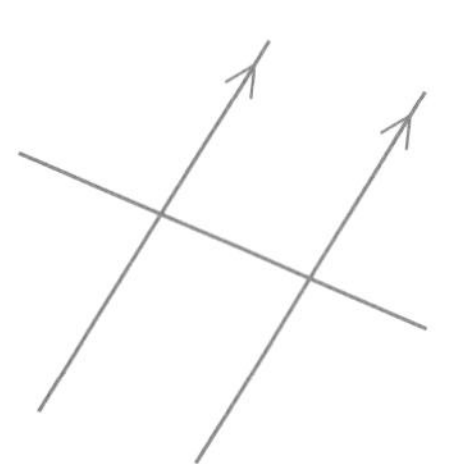

5 Find the value of each pronumeral, giving reasons.

a

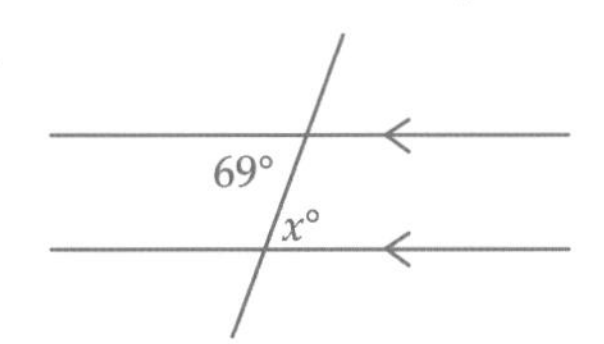

b

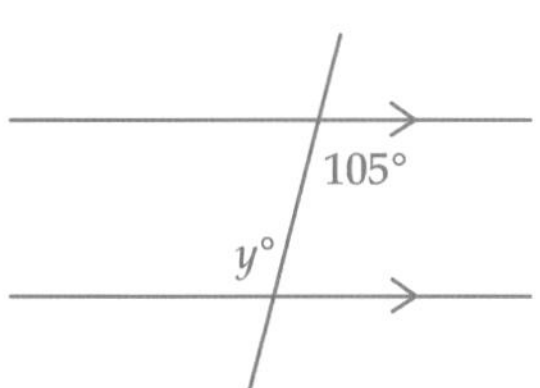

c

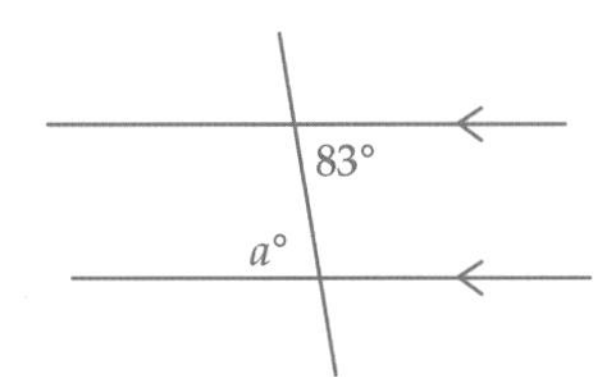

d

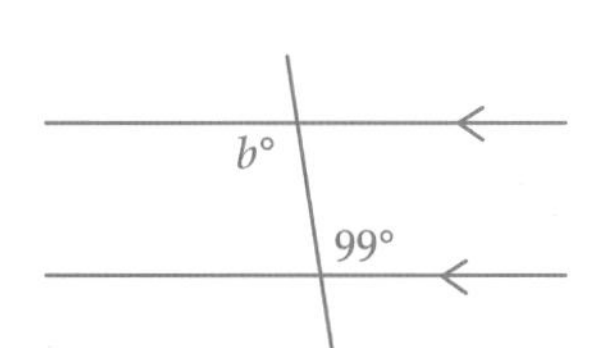

6 Find the value of n.

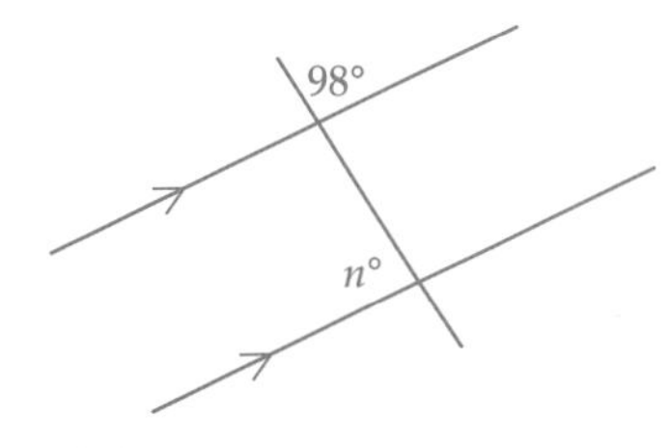

2-08 Co-interior angles on parallel lines

WORDBANK

co-interior angles Pairs of angles formed when a transversal crosses two or more other lines, they are between the lines and on the same side of the transversal.

Co-interior means 'together inside'.
If a transversal crosses two *parallel* lines, the co-inteior angles are supplementary (add to 180°).

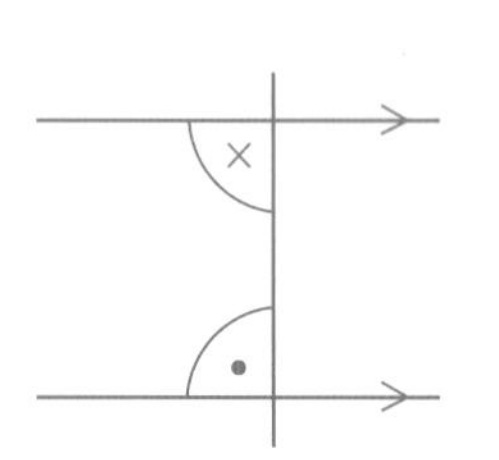

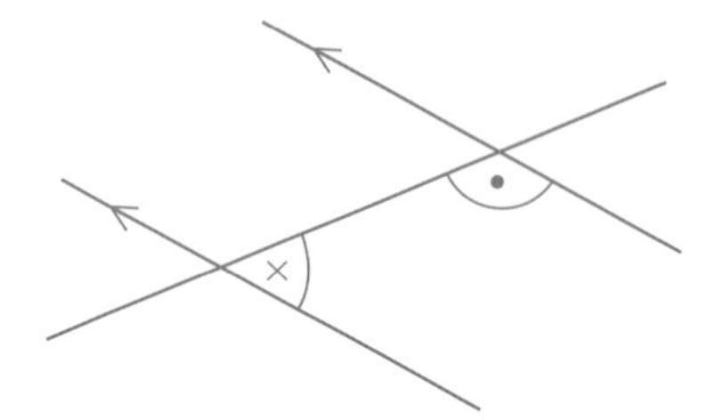

If you move one of the co-interior angles to its 'corresponding angle' position, you will have two angles on a straight line, so they are supplementary.

Co-interior angles on parallel lines are supplementary (add up to 180°).

EXAMPLE 8

Find the value of m, giving a reason.

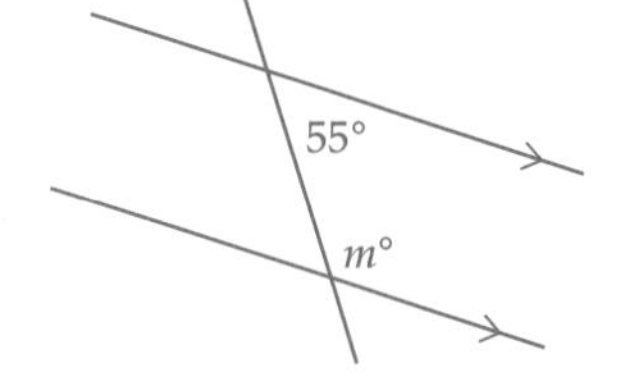

SOLUTION

$m = 180 - 55$ (co-interior angles on parallel lines)
$= 125$

EXERCISE 2-08

1 Complete: Co-interior angles on parallel lines are ____________. Select the correct answer A, B, C or D.

A complementary B supplementary C equal D vertically opposite

2 For each diagram, name the angle that is co-interior to the angle marked.

a

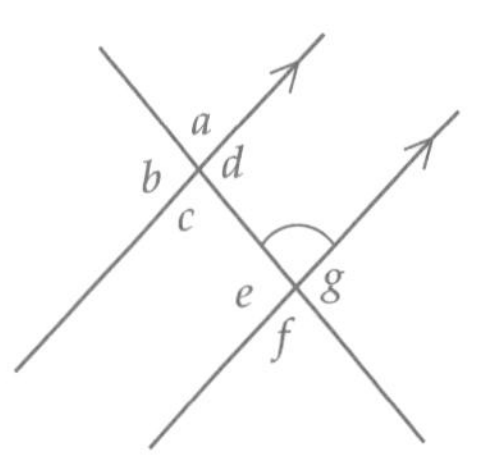

b

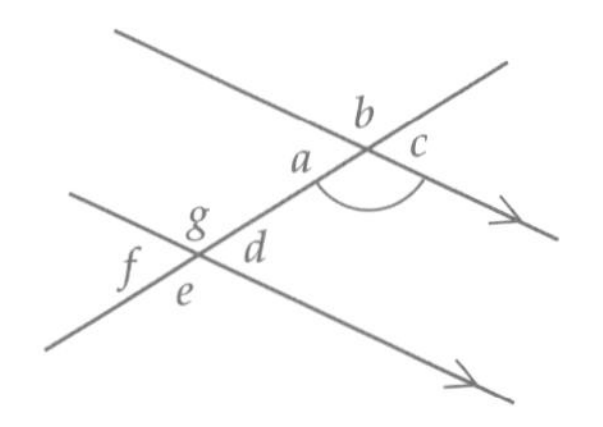

c

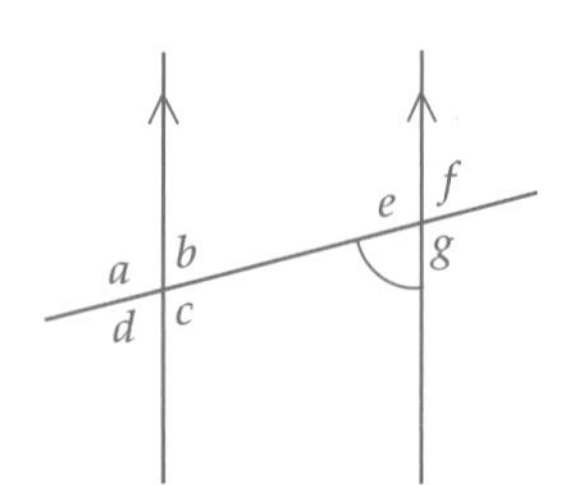

ISBN 9780170350969

EXERCISE 2–08

3 Is each statement true or false?

a A transversal is perpendicular (at 90°) to the other lines.

b Parallel lines cross eventually.

c Co-interior angles are on the same side of the transversal.

d Co-interior angles are equal if the lines are parallel.

4 Copy each diagram and mark a pair of co-interior angles on each one.

a

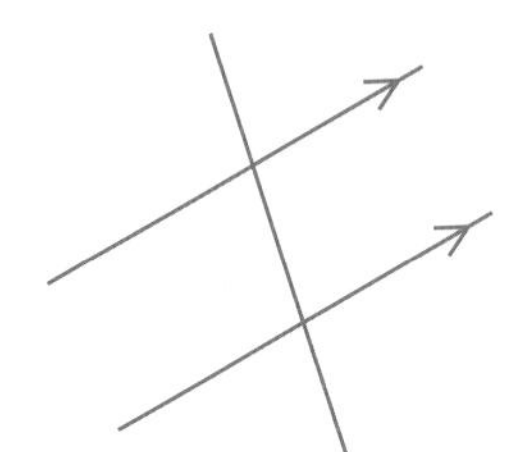

b

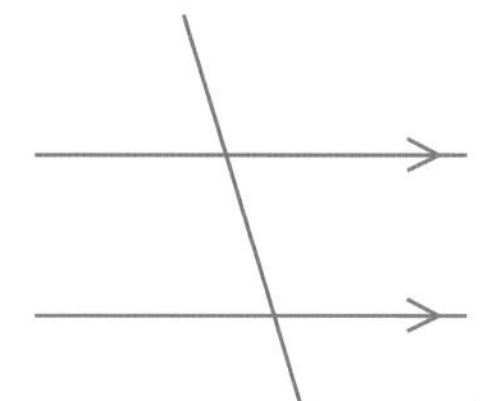

c

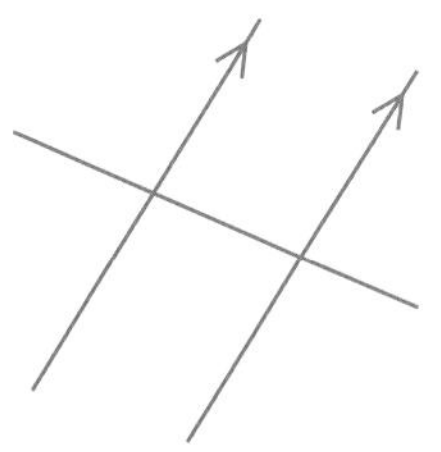

5 Find the value of each pronumeral, giving reasons.

a

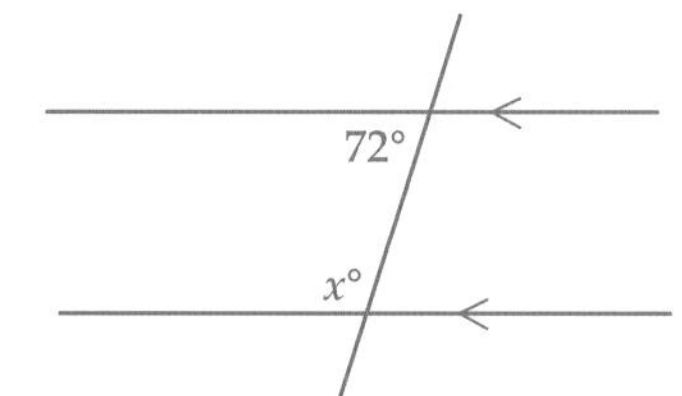

b

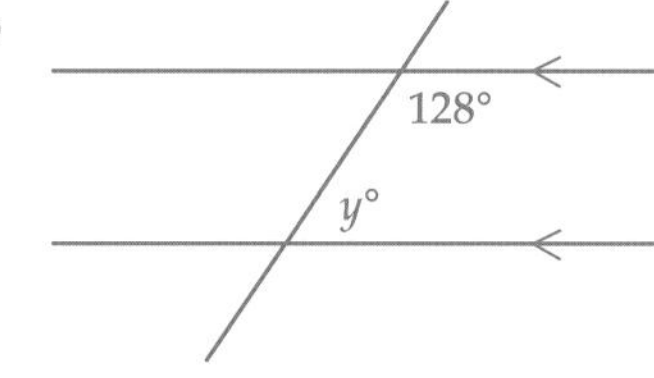

c

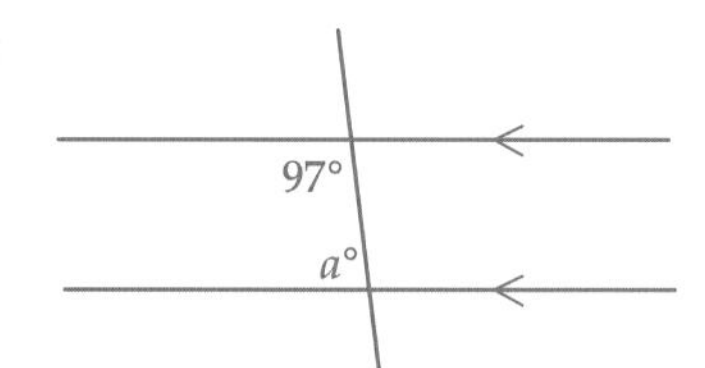

d

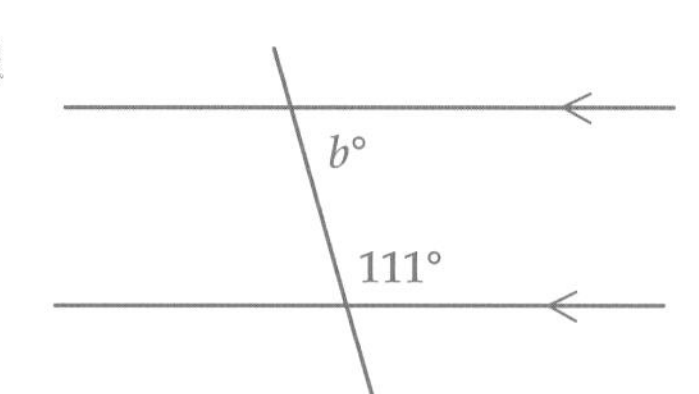

6 Find the value of each pronumeral, giving reasons.

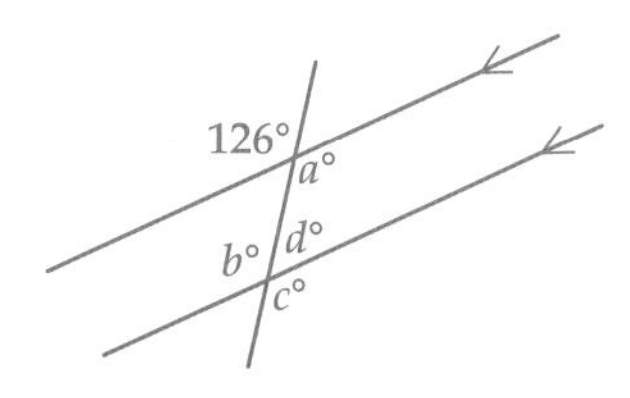

FIND-A-WORD PUZZLE

Make a copy of this page, then find the words below in this grid of letters.

P	A	R	A	L	L	E	L	A	B	G
E	X	E	L	A	I	V	I	C	D	N
R	O	L	T	S	N	C	C	O	V	I
P	R	U	E	R	E	O	N	M	I	D
E	S	R	R	E	S	I	E	P	I	N
N	O	E	N	V	I	N	P	L	D	O
D	I	S	A	S	E	T	O	E	R	P
I	T	S	T	N	O	E	A	M	I	S
C	Q	E	E	A	P	R	E	E	V	E
U	V	E	X	R	Y	I	V	N	O	R
L	A	R	O	T	S	O	I	T	A	R
A	N	G	P	E	V	R	O	A	P	O
R	U	E	I	R	E	T	N	R	I	C
Y	E	D	A	N	G	L	E	Y	O	T

ALTERNATE	ANGLE	COINTERIOR	COMPLEMENTARY
CORRESPONDING	DEGREES	LINES	PARALLEL
PENCIL	PERPENDICULAR	RULER	TRANSVERSAL

ISBN 9780170350969

Part A General topics

Calculators are not allowed.

1 Find the perimeter of this rectangle.

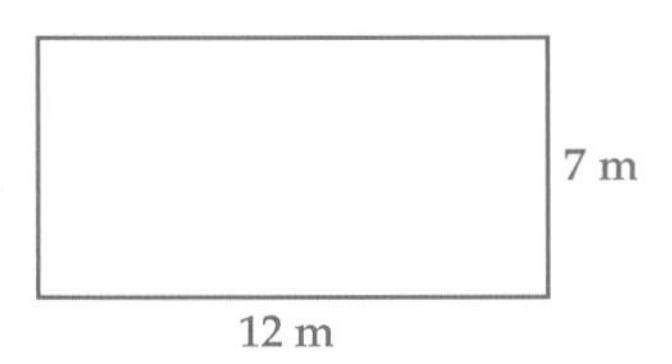

2 Simplify $\frac{16}{24}$.

3 Find 70% of $2000.

4 Name this solid shape.

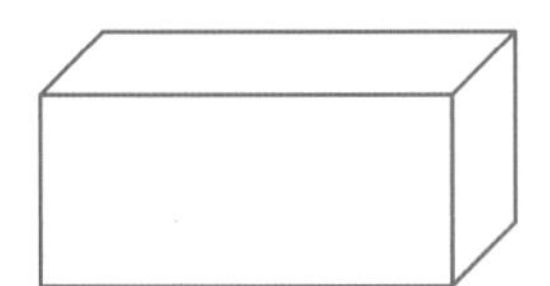

5 Write the decimal halfway between 0.6 and 0.7.

6 The area of a square is 25 m^2. What is the length of one side?

7 Evaluate $4.60 ÷ 4.

8 The time in Sydney is 10 hours ahead of London. If it is 8.30 a.m. in London, what time is it in Sydney?

9 Arrange these decimals in ascending order: 0.4, 0.25, 0.92, 0.46, 0.2.

10 Draw a trapezium.

Part B Angles

Calculators are allowed.

2-01 Naming angles

11 Name the marked angle. Select the correct answer A, B, C or D.

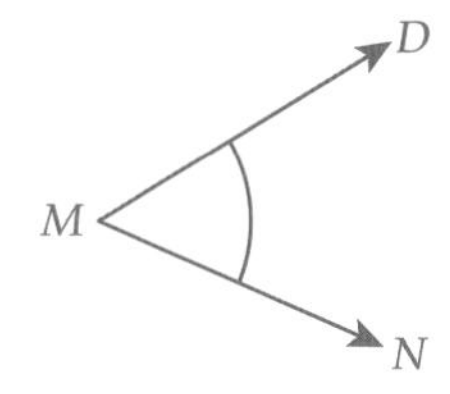

A $\angle MDN$ B $\angle NDM$
C $\angle MND$ D $\angle DMN$

2-02 Types of angles

12 What type of angle is marked? Select A, B, C or D.

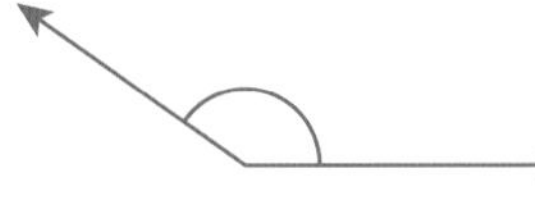

A right B acute
C reflex D obtuse

13 What type of angle is formed by the hands of a clock at 3 o'clock? Select A, B, C or D.

A acute B obtuse C right D straight

2-03 Measuring and drawing angles

14 Draw an angle measuring:

a 72° b 151° c 282°

ISBN 9780170350969

2-04 Complementary and supplementary angles

15 What is the complement of:

a 28°? b 54°?

16 What is the supplement of:

a 32°? b 126°?

17 Find the value of n.

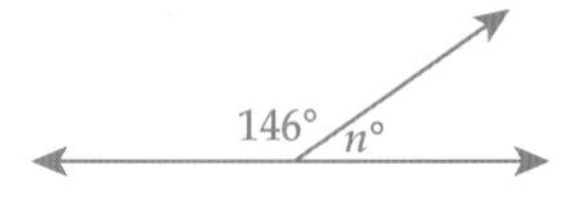

2-05 Angles at a point and vertically opposite angles

18 Find the value of c.

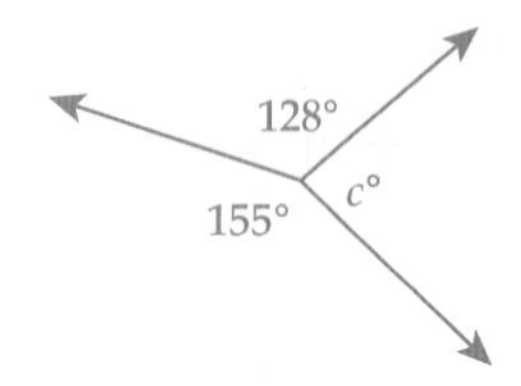

19 Find the value of m.

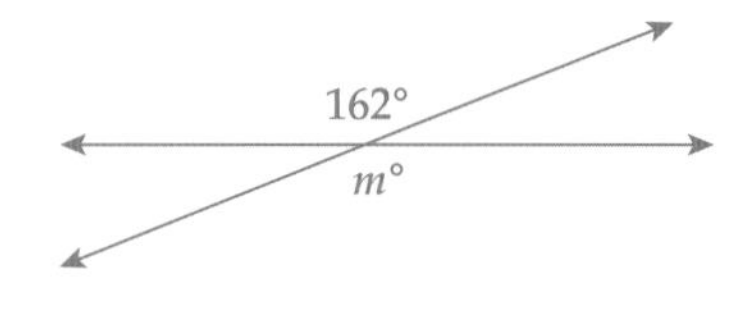

2-06 Corresponding angles on parallel lines

20 Find the value of a.

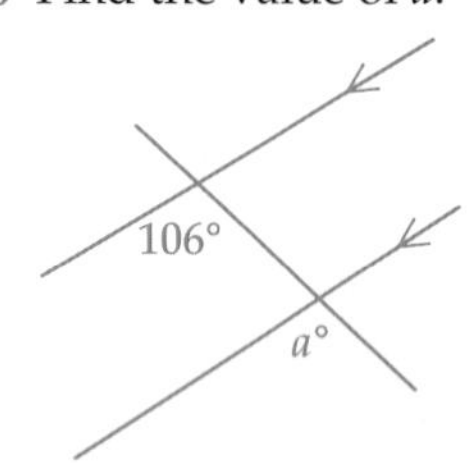

2-07 Alternate angles on parallel lines

21 Find the value of x.

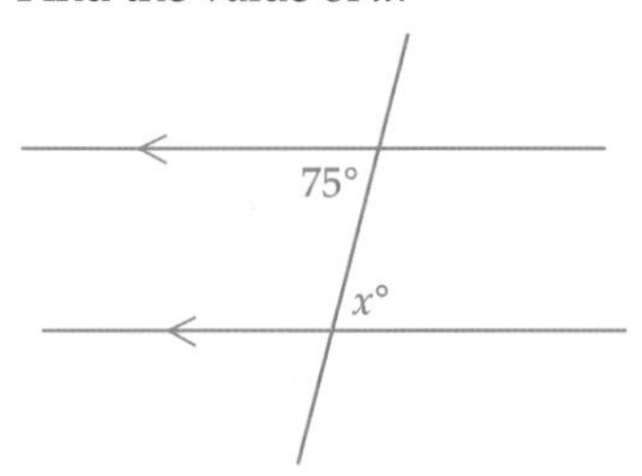

2-08 Co-interior angles on parallel lines

22 Find the values of a, b and c.

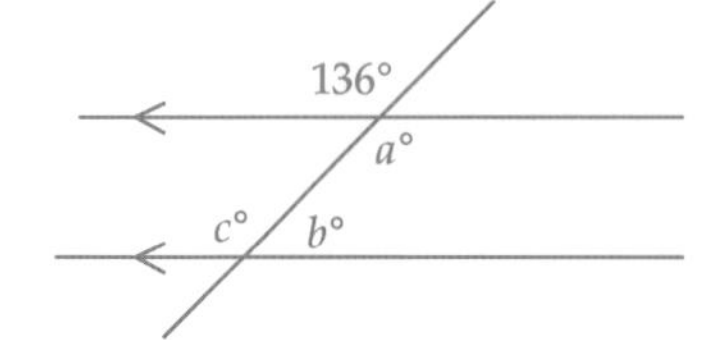

ISBN 9780170350969

WORKING WITH NUMBERS 3

WHAT'S IN CHAPTER 3?

3-01 Rounding and estimating
3-02 Adding numbers
3-03 Subtracting numbers
3-04 Multiplication facts
3-05 Lowest common multiple (LCM)
3-06 Multiplying numbers

IN THIS CHAPTER YOU WILL:

- round numbers and check answers by rounding and estimating
- add and subtract large numbers and solve problems involving sums and differences
- learn the multiplication facts up to 10×10 (or 12×12)
- find the lowest common multiple (LCM) of two or more numbers
- multiply large numbers and solve problems involving products

Shutterstock.com/koosen

ISBN 9780170350969

3-01 Rounding and estimating

WORDBANK

rounding To write a simpler number with roughly the same value.

estimate To make a good (educated) guess of the answer to a problem in round figures.

We often round numbers to estimate an answer to a calculation, or if an exact value is not required. For example, if the number of people attending a music concert was 31 469, we could round it to 31 500 (to the nearest hundred) or 31 000 (to the nearest thousand).

iStockphoto/jjwithers

To **round a number**, 'cut' it at the required place and look at the digit in the next place:

- if the digit is less than 5 (that is, 0, 1, 2, 3 or 4), **round down**
- if the digit is 5 or more (that is, 5, 6, 7, 8 or 9), **round up**.

EXAMPLE 1

Round:

a 1872 to the nearest hundred

b 542 381 to the nearest thousand

SOLUTION

a Counting by hundreds, 1872 is between 1800 and 1900.

- In 1872, the hundreds digit is 8.
- The digit in the next (tens) place is 7, which is more than 5, so **round up** to 1900.

So 1872 ≈ 1900 (rounded to the nearest hundred).

✱ The symbol '≈' means 'approximately equal to'.

b Counting by thousands, 542 381 is between 542 000 and 543 000.

- In 542 381, the thousands digit is 2.
- The digit in the next (hundreds) place is 3, which is less than 5, so **round down** to 542 000.

So 542 381 ≈ 542 000 (rounded to the nearest thousand).

EXAMPLE 2

Estimate the answer to each expression.

a 75 + 33

b 18 × 12

SOLUTION

a $75 + 33 \approx 80 + 30$ ← Rounding
$= 110$ ← Exact answer = 108

b $18 \times 12 \approx 20 \times 12$ ← Rounding
$= 240$ ← Exact answer = 216

ISBN 9780170350969

EXERCISE 3-01

1 Round 26 548 to the nearest hundred. Select the correct answer A, B, C or D.

A 26 600 B 26 550 C 26 540 D 26 500

2 Which is the best estimate of 198 × 12? Select A, B, C or D.

A 2000 B 2400 C 3000 D 2800

3 Round each number to the nearest ten.

a 52 b 78 c 163 d 2587

4 Round each number to the nearest hundred.

a 682 b 249 c 3890 d 5418

5 Round each number to the nearest thousand.

a 1456 b 6832 c 24 863 d 328 491

6 Estimate the value of each expression, then use a calculator to find the exact answer.

a 48 + 72 b 33 – 18 c 672 + 2421 d 3844 – 236

e 5 × 23 f 79 ÷ 4 g 362 × 8 h 644 ÷ 19

7 The extensions on Nina's house are quoted as costing $17 464. Write this amount correct to the nearest $100.

8 The road distance between Perth and Sydney is 3938 km. Round this distance to the nearest:

a 10 km b 100 km c 1000 km

Shutterstock.com/Robyn Mackenzie

9 Round $64 218 to the nearest:

a thousand b hundred c ten d ten thousand

10 a Holly is delivering leaflets to letterboxes. She is told there are 79 284 homes in the local area. She has to round her print order to the nearest ten thousand. How many leaflets should she order?

b The price of the leaflets was $5 per hundred. How much did the printing cost Holly?

c If only one in every 10 leaflets is read, how many leaflets is this?

11 The population of Australia in 2014 was 23 581 800. Round this figure to the nearest thousand.

3-02 Adding numbers

WORDBANK

sum The answer to an addition (+) of two or more numbers.

evaluate To find the value or amount.

To add large numbers:

- write them underneath each other in their place value columns: units, tens, hundreds, and so on
- add the digits in columns: units first, then tens, and so on
- some additions will involve carrying from one column to its left column.

EXAMPLE 3

Evaluate each sum.

a 682 + 65 b 1478 + 303

SOLUTION

✱ Set out putting units under units, tens under tens, and so on.

a
$$\begin{array}{r} {}^{1}6\ 8\ 2\ + \\ 6\ 5\ \ \ \\ \hline 7\ 4\ 7\ \ \ \\ \hline \end{array}$$

682 + 65 = 747.

(Estimating: 682 + 65 ≈ 680 + 70 = 750)

b
$$\begin{array}{r} 1\ 4\ {}^{1}7\ 8\ + \\ 3\ 0\ 3\ \ \ \\ \hline 1\ 7\ 8\ 1\ \ \ \\ \hline \end{array}$$

1478 + 303 = 1781

(Estimating: 1478 + 303 ≈ 1500 + 300 = 1800)

Shutterstock.com/Ljupco Smokovski

ISBN 9780170350969

EXERCISE 3-02

1 What is the sum of 16, 28, 4 and 11? Select the correct answer A, B, C or D.

A 57 B 49 C 59 D 47

2 Increase $258 by $15. Select A, B, C or D.

A $273 B $263 C $264 D $283

3 Evaluate each sum.

a 23 + 784 b 457 + 86 c 64 + 39 d 82 + 58

e 3486 + 563 f 78 + 4732 g 128 + 74 h 483 + 296

i 690 + 2379 j 789 + 2356 k 742 + 387 l 862 + 98

4 What is 26 more than 599?

5 A bakery sells 56 pies on Monday, 38 on Tuesday, 49 on Wednesday, 58 on Thursday and 156 on Friday. Find the total number of pies sold.

Alamy/Simon Evans

6 Gemma saves $65 one week, $86 the next and $128 in the third week. How much did she save altogether?

7 a Increase 939 cm by 82 cm.

b Increase $2876 by $458.

8 Joe went to the supermarket. He bought:

a packet of biscuits for $2.50 3 kg of apples for $12

a bag of tomatoes for $6.50 a box of chocolates for $15

What was the total cost of his purchases?

9 Find the sum of 3582, 796, 1207, 12 450 and 658.

10 a Estimate the total of $45.60, $186.20, 56c, $2.16 and $432.50.

b Find the exact total of the amounts in part a.

3-03 Subtracting numbers

WORDBANK

difference The result of subtracting two numbers.

To subtract large numbers:

- write them underneath each other in their place value columns: units, tens, hundreds, and so on
- subtract the digits in columns: units first, then tens, and so on
- some subtractions will involve trading from one column to its left column

EXAMPLE 4

Evaluate each subtraction.

a 74 – 26 b 809 – 458

SOLUTION

a $^{6}\not{7}\ {}^{1}4\ -$ ⟵ In the units column, change 4 – 6 to 14 – 6 by taking 10 from the tens column

$2\ 6$ ⟵ So 7 in the tens column becomes 6

$4\ 8$

74 – 26 = 48 (Estimating: $74 - 26 \approx 70 - 30 = 40$)

b $^{7}\not{8}\ {}^{1}0\ 9\ -$ ⟵ In the tens column, change 0 – 5 to 10 – 5 by taking 10 from the hundreds column

$4\ 5\ 8$ ⟵ So 8 in the hundreds column becomes 7

$3\ 5\ 1$

809 – 458 = 351 (Estimating: $809 - 458 \approx 800 - 500 = 300$)

Alamy/Peter Titmuss

ISBN 9780170350969

EXERCISE 3-03

1 What is the difference between 459 and 68? Select the correct answer A, B, C or D.

A 371 B 381 C 401 D 391

2 Decrease $425 by $30. Select A, B, C or D.

A $395 B $405 C $396 D $385

3 Evaluate each difference.

a 65 – 37 b 107 – 86 c 64 – 39 d 82 – 58

e 3486 – 563 f 7812 – 473 g 128 – 74 h 483 – 296

i 6908 – 2379 j 4789 – 235 k 3742 – 387 l 4862 – 98

4 What is the difference between 7049 and 4192?

5 Jenny saved $235 and then spent $68 on a present. How much did she have left?

6 Riley set out on a road trip from Bourke to Ballarat, a distance of 568 km. He travelled 186 km in the first 2 hours before taking a break.

a Estimate how far he still had to travel.

b Calculate the exact number of kilometres he still had to travel.

Shutterstock.com/Marcella Miriello

7 Decrease 5000 by 48.

8 A bus was carrying 45 passengers. At the first stop 12 passengers got off the bus. At the second stop 5 passengers got off but 8 got on the bus. At the third stop another 6 passengers got off the bus. How many passengers are now on the bus?

Alamy/Kumar Sriskandan

9 Julia had a $50 note when she went to the supermarket. She bought:

a box of chocolates for $8.75 2 kg of oranges for $6.50

a bag of potatoes for $3.85 a packet of biscuits for $2.90

a What was the total cost of her purchases?

b How much change would Julia receive from $50?

10 Year 7 are asked to set up the hall for assembly with 986 chairs. They already have 328 chairs set up.

a Estimate how many chairs still need to be set up.

b Calculate exactly how many chairs need to be set up.

3-04 Multiplication facts

WORDBANK

product The answer to a multiplication (×) of two or more numbers.

These **multiplication facts** or **times tables** are very useful to learn for multiplying and dividing big numbers. You should know these up to 10 × 10 (or 12 × 12).

1 TIMES TABLE	2 TIMES TABLE	3 TIMES TABLE	4 TIMES TABLE	5 TIMES TABLE	6 TIMES TABLE
1 × 1 = 1	1 × 2 = 2	1 × 3 = 3	1 × 4 = 4	1 × 5 = 5	1 × 6 = 6
2 × 1 = 2	2 × 2 = 4	2 × 3 = 6	2 × 4 = 8	2 × 5 = 10	2 × 6 = 12
3 × 1 = 3	3 × 2 = 6	3 × 3 = 9	3 × 4 = 12	3 × 5 = 15	3 × 6 = 18
4 × 1 = 4	4 × 2 = 8	4 × 3 = 12	4 × 4 = 16	4 × 5 = 20	4 × 6 = 24
5 × 1 = 5	5 × 2 = 10	5 × 3 = 15	5 × 4 = 20	5 × 5 = 25	5 × 6 = 30
6 × 1 = 6	6 × 2 = 12	6 × 3 = 18	6 × 4 = 24	6 × 5 = 30	6 × 6 = 36
7 × 1 = 7	7 × 2 = 14	7 × 3 = 21	7 × 4 = 28	7 × 5 = 35	7 × 6 = 42
8 × 1 = 8	8 × 2 = 16	8 × 3 = 24	8 × 4 = 32	8 × 5 = 40	8 × 6 = 48
9 × 1 = 9	9 × 2 = 18	9 × 3 = 27	9 × 4 = 36	9 × 5 = 45	9 × 6 = 54
10 × 1 = 10	10 × 2 = 20	10 × 3 = 30	10 × 4 = 40	10 × 5 = 50	10 × 6 = 60
11 × 1 = 11	11 × 2 = 22	11 × 3 = 33	11 × 4 = 44	11 × 5 = 55	11 × 6 = 66
12 × 1 = 12	12 × 2 = 24	12 × 3 = 36	12 × 4 = 48	12 × 5 = 60	12 × 6 = 72

7 TIMES TABLE	8 TIMES TABLE	9 TIMES TABLE	10 TIMES TABLE	11 TIMES TABLE	12 TIMES TABLE
1 × 7 = 7	1 × 8 = 8	1 × 9 = 9	1 × 10 = 10	1 × 11 = 11	1 × 12 = 12
2 × 7 = 14	2 × 8 = 16	2 × 9 = 18	2 × 10 = 20	2 × 11 = 22	2 × 12 = 24
3 × 7 = 21	3 × 8 = 24	3 × 9 = 27	3 × 10 = 30	3 × 11 = 33	3 × 12 = 36
4 × 7 = 28	4 × 8 = 32	4 × 9 = 36	4 × 10 = 40	4 × 11 = 44	4 × 12 = 48
5 × 7 = 35	5 × 8 = 40	5 × 9 = 45	5 × 10 = 50	5 × 11 = 55	5 × 12 = 60
6 × 7 = 42	6 × 8 = 48	6 × 9 = 54	6 × 10 = 60	6 × 11 = 66	6 × 12 = 72
7 × 7 = 49	7 × 8 = 56	7 × 9 = 63	7 × 10 = 70	7 × 11 = 77	7 × 12 = 84
8 × 7 = 56	8 × 8 = 64	8 × 9 = 72	8 × 10 = 80	8 × 11 = 88	8 × 12 = 96
9 × 7 = 63	9 × 8 = 72	9 × 9 = 81	9 × 10 = 90	9 × 11 = 99	9 × 12 = 108
10 × 7 = 70	10 × 8 = 80	10 × 9 = 90	10 × 10 = 100	10 × 11 = 110	10 × 12 = 120
11 × 7 = 77	11 × 8 = 88	11 × 9 = 99	11 × 10 = 110	11 × 11 = 121	11 × 12 = 132
12 × 7 = 84	12 × 8 = 96	12 × 9 = 108	12 × 10 = 120	12 × 11 = 132	12 × 12 = 144

ISBN 9780170350969

EXERCISE 3–04

1 Evaluate 8×6. Select the correct answer A, B, C or D.

A 56 B 64 C 48 D 72

2 Evaluate 7×9. Select A, B, C or D.

A 49 B 63 C 56 D 54

3 Evaluate each product.

a 4×3	b 5×5	c 6×7	d 7×8
e 8×10	f 9×12	g 10×2	h 11×3
i 12×5	j 1×6	k 2×7	l 3×8
m 4×9	n 5×10	o 6×11	p 7×12
q 8×1	r 2×9	s 10×3	t 11×4
u 12×4	v 1×5	w 4×8	x 6×9

4 Evaluate each product.

a 6×3	b 7×5	c 6×4	d 8×8
e 9×10	f 6×12	g 10×5	h 11×7
i 12×8	j 1×9	k 2×12	l 5×8
m 3×9	n 4×10	o 8×11	p 11×12
q 3×1	r 2×5	s 10×6	t 11×11
u 12×12	v 6×5	w 9×8	x 6×3

5 Copy and complete this table. Look for any patterns in each row or column.

×	1	2	3	4	5	6	7	8	9	10
1	1	2	3		5		7	8	9	
2	2		6			12		16		20
3		6		12	15		21		27	
4				16		24		32		
5	5		15			30				50
6				24		36		48		
7		14	21				49		63	
8	8		24		40		56		72	
9	9	18	27					72		90
10	10		30		50		70		90	

6 Search for 'times tables games' on the internet and play some of the games to make your multiplication skills quicker and more accurate.

3-05 Lowest common multiple (LCM)

The **multiples of 3** are 3, 6, 9, 12, 15, … and so on.
The **multiples** of a number are the products when the number is multiplied by the whole numbers 1, 2, 3, and so on.
The **multiples of 7** are 7, 14, 21, 28, … and so on.

The **lowest common multiple** (**LCM**) of two (or more) numbers is the smallest number that is a multiple of **both** (or **all**) of these numbers.
To **find the lowest common multiple** (**LCM**) of two or more numbers:
- list the multiples of each number
- underline the common multiples
- select the lowest multiple that is underlined in both lists.

EXAMPLE 5

Find the lowest common multiple of 6 and 8.

SOLUTION

Multiples of 6: 6, 12, 18, <u>24</u>, 30, 36, 42, <u>48</u>, …

Multiples of 8: 8, 16, <u>24</u>, 32, 40, <u>48</u>, …

Underline the common multiples in both lists.

The lowest common multiple of 6 and 8 is 24. ← The lowest multiple underlined

EXAMPLE 6

Every 3rd student on the class list of 30 students receives a red lolly and every 5th student receives a yellow lolly. Which students receive both a red and a yellow lolly?

Shutterstock.com/Jiri Vaclavek

SOLUTION

Multiples of 3: 3, 6, 9, 12, <u>15</u>, 18, 21, 24, 27, <u>30</u> red lollies

Multiples of 5: 5, 10, <u>15</u>, 20, 25, <u>30</u> yellow lollies

So the 15th and 30th students both receive red and yellow lollies.

(15 and 30 are common multiples of 3 and 5.)

 ISBN 9780170350969

EXERCISE 3–05

1 Which number below is a multiple of 9? Select the correct answer A, B, C or D.

A 92 B 81 C 70 D 56

2 Which number below is *not* a multiple of 7? Select A, B, C or D.

A 14 B 35 C 54 D 42

3 a Write the first six multiples of 4.

b Write the first six multiples of 5.

c Underline the common multiples of 4 and 5.

d Hence write down the lowest common multiple of 4 and 5.

4 a List the first seven multiples of 6.

b List the first seven multiples of 9.

c Write down the LCM of 6 and 9.

5 Find the LCM of each pair of numbers.

a 2 and 5 b 3 and 4 c 5 and 9

d 6 and 4 e 20 and 50 f 10 and 12

6 Find the LCM of each pair of numbers.

a 3 and 6 b 4 and 12 c 6 and 18 d 5 and 20

7 In a group of 30 girls, every 2nd girl receives a red sticker and every 3rd girl receives a blue sticker.

a How many girls receive a red sticker?

b How many girls receive a blue sticker?

c How many girls receive one of each colour?

d What position is the first girl to receive one of each colour?

8 In a group of 20 boys, every 5th boy receives a white chocolate and every 4th boy receives a dark chocolate.

a How many boys receive a white chocolate?

b Which boys receive two chocolates?

9 Bill and Samir are racing around a running track. Bill completes one lap every 3 minutes while Samir completes one lap every 4 minutes. How long before they complete a lap at the same time?

AlamyDacorum Gold

ISBN 9780170350969

3-06 Multiplying numbers

WORDBANK

short multiplication A method of multiplying by a number with one digit (2 to 9).

long multiplication A method of multiplying by a number with two or more digits.

EXAMPLE 7

Evaluate each product.

a 230 × 7 b 457 × 68

SOLUTION

a Use short multiplication.

$$\begin{array}{r} {}^{2}2\ 3\ 0\ \times \\ 7 \\ \hline 1\ 6\ 1\ 0 \\ \hline \end{array}$$

← In the units column, 0 × 7 = 0: write down 0

← In the tens column, 3 × 7 = 21: write down 1 and carry 2

← In the hundreds column, 2 × 7 = 14, + 2 = 16

230 × 7 = 1610 (Estimating: 230 × 7 ≈ 200 × 7 = 1400)

b Use long multiplication.

$$\begin{array}{r} 4\ 5\ 7\ \times \\ 6\ 8 \\ \hline 3\ 6\ 5\ 6 \\ 2\ 7\ 4\ 2\ 0 \\ \hline 3\ 1\ 0\ 7\ 6 \\ \hline \end{array}$$

← 457 × 8 = 3656

← Place a 0 in the units column, then 457 × 6 = 2742

← 3656 + 27 420 = 31 076

457 × 68 = 31 076 (Estimating: 457 × 68 ≈ 500 × 70 = 35 000)

Mental multiplication strategies

Multiplying by	Strategy
2	Double
4	Double twice
5	Halve, then multiply by 10
9	Multiply by 10, then subtract the number
10	Add a 0 to the end
100	Add 00 to the end

EXAMPLE 8

Use mental multiplication to evaluate each product.

a 22 × 4 b 36 × 5 c 12 × 9 d 14 × 100

SOLUTION

a $22 \times 4 = 22 \times 2 \times 2$ ← To multiply by 4, double twice

$= 44 \times 2$

$= 88$

b $36 \times 5 = 36 \times \frac{1}{2} \times 10$ ← To multiply by 5, halve, then multiply by 10

$= 18 \times 10$ This is because $\frac{1}{2} \times 10 = 5$

$= 180$ ← To multiply by 10, add a 0 at the end

ISBN 9780170350969

c $12 \times 9 = 12 \times (10 - 1)$ ⟵ This is because $10 - 1 = 9$
$= 12 \times 10 - 12$ ⟵ To multiply by 9, multiply by 10, then subtract the number
$= 120 - 12$
$= 108$

d $14 \times 100 = 1400$ ⟵ To multiply by 100, add 00 at the end

EXERCISE 3-06

1 What is the product of 13 and 11? Select the correct answer A, B, C or D.

A 143 B 133 C 113 D 134

2 Which calculation is the same as 26×9? Select A, B, C or D.

A $26 \times 10 - 1$ B $26 \times 1 - 10$ C $26 \times 10 - 26$ D None of these

3 Evaluate each product by short multiplication.

a 85×4 b 68×7 c 59×8 d 234×6
e 365×8 f 672×5 g 1265×4 h 3478×9

4 Evaluate each product mentally by doubling.

a 17×2 b 43×2 c 55×2 d 28×2

5 Evaluate each product mentally by doubling twice.

a 61×4 b 15×4 c 43×4 d 19×4

6 Evaluate each product by long multiplication.

a 564×16 b 873×27 c 795×34 d 867×19

7 Evaluate each product by multiplying by 10 then subtracting the number.

a 36×9 b 25×9 c 47×9 d 52×9

8 If Brodie sleeps 9 hours each night, how many hours sleep does he get in a fortnight?

9 Evaluate each product mentally by adding 0, 00 or 000 at the end.

a 29×10 b 75×10 c 13×100 d 48×100
e 59×10 f 64×1000 g 20×100 h 47×1000

10 Evaluate each product mentally by halving then multiplying by 10.

a 32×5 b 14×5 c 26×5 d 92×5

11 Lisa plants 46 seeds in each row of her garden. How many seeds does she plant altogether if there are 18 rows?

12 How many minutes are there in:

a one hour? b one day? c one week?

ISBN 9780170350969

CODE PUZZLE

What is the difference between these two fish? Copy and complete the subtractions below to decode the answer. Match each letter of the question to an answer.

Barra

Snapper

A 250 – 146

B 8216 – 4834

C 526 – 22

D 9218 – 418

E 4129 – 3046

G 690 – 360

L 1280 – 430

M 824 – 16

O 1260 – 420

P 696 – 584

R 4290 – 1340

S 2010 – 110

T 1854 – 721

U 5040 – 216

3382-104-2950-2950-104 504-4824-8800-104 330-840-1133

808-840-2950-1083 1900-850-1083-1083-112

ISBN 9780170350969

PRACTICE TEST 3

Part A General topics

Calculators are not allowed.

1 Bryce earns $12 per hour. How much is he paid for 6 hours' work?

2 Write 0.25 as a simple fraction.

3 The area of Australia is 7 692 024 square kilometres. Round 7 692 024 to the nearest thousand.

4 What metric unit is used to measure the length of a football field?

5 Evaluate 36×9.

6 Evaluate $\sqrt{81}$.

7 A bus leaves Sydney at 7:15 a.m. and travels for 8 hours 12 minutes. At what time does the journey end?

8 Find the perimeter of this figure.

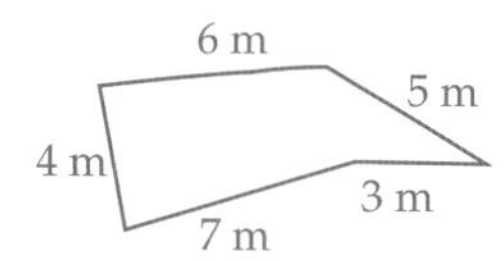

9 Evaluate $160 \div 4 + 8 \times 5$.

10 List the factors of 25.

Part B Working with numbers

Calculators are not allowed.

3–01 Rounding and estimating

11 What is the best estimate for 298×22? Select the correct answer A, B, C or D.

A 5000 B 4000 C 5500 D 6000

12 Round 62 768 correct to the nearest thousand. Select A, B, C or D.

A 60 000 B 62 000 C 63 000 D 6300

3–02 Adding numbers

13 Evaluate 567 + 64 + 128. Select A, B, C or D.

A 759 B 659 C 758 D 769

3–03 Subtracting numbers

14 Evaluate each difference:

a 48 – 29 b 654 – 273 c 2018 – 581

15 Harry was born in 1966. How old will he be in the year 2021?

3–04 Multiplication facts

16 Evaluate each product:

a 8×6 b 7×4 c 9×9 d 12×5

17 When I multiply my favourite number by 4 and subtract 6, the answer is 42. What is my favourite number?

ISBN 9780170350969

3-05 Lowest common multiple (LCM)

18 a Write down the first 6 multiples of 3.

b Write down the first 6 multiples of 4.

c What is the lowest common multiple of 3 and 4?

19 What is the LCM of 16 and 24?

20 Ned and Yumi are swimming laps of a pool. Yumi completes one lap every 5 minutes while Ned completes one lap every 3 minutes. How long before they complete a lap at the same time?

3-06 Multiplying numbers

21 Evaluate each product.

a 27×6

b 368×47

22 Georgia packs 36 chocolates in each box. A carton contains 12 boxes of chocolates. How many chocolates are in 16 cartons?

ISBN 9780170350969

FACTORS AND PRIMES

4

WHAT'S IN CHAPTER 4?

4-01 Division facts
4-02 Dividing numbers
4-03 Division with remainders
4-04 Divisibility tests
4-05 Highest common factor (HCF)
4-06 Prime and composite numbers
4-07 Factor trees

IN THIS CHAPTER YOU WILL:

- learn the division facts based on the multiplication facts up to 10×10 (or 12×12)
- divide by 2 to 10 using short division and solve problems involving quotients
- test whether a number is divisible by 2, 3, 4, 5, 6, 8 or 9
- list the factors of a number
- find the highest common factor (HCF) of two or more numbers
- identify prime and composite numbers
- use factor trees to write a whole number as a product of its prime factors

Shutterstock.com/igor.stevanovic

4-01 Division facts

WORDBANK

quotient The result of dividing (÷) a number by another number. For example, if $12 \div 4 = 3$, the quotient is 3.

Division is the **opposite** of multiplication. Division can be written as: $12 \div 4$ or $\frac{12}{4}$ or $4\overline{)12}$.

The **multiplication facts** or **times tables** can also be used to learn **division facts** and divide numbers. You should know these up to 10×10 (or 12×12).

1 TIMES TABLE	2 TIMES TABLE	3 TIMES TABLE	4 TIMES TABLE	5 TIMES TABLE	6 TIMES TABLE
$1 \times 1 = 1$	$1 \times 2 = 2$	$1 \times 3 = 3$	$1 \times 4 = 4$	$1 \times 5 = 5$	$1 \times 6 = 6$
$2 \times 1 = 2$	$2 \times 2 = 4$	$2 \times 3 = 6$	$2 \times 4 = 8$	$2 \times 5 = 10$	$2 \times 6 = 12$
$3 \times 1 = 3$	$3 \times 2 = 6$	$3 \times 3 = 9$	$3 \times 4 = 12$	$3 \times 5 = 15$	$3 \times 6 = 18$
$4 \times 1 = 4$	$4 \times 2 = 8$	$4 \times 3 = 12$	$4 \times 4 = 16$	$4 \times 5 = 20$	$4 \times 6 = 24$
$5 \times 1 = 5$	$5 \times 2 = 10$	$5 \times 3 = 15$	$5 \times 4 = 20$	$5 \times 5 = 25$	$5 \times 6 = 30$
$6 \times 1 = 6$	$6 \times 2 = 12$	$6 \times 3 = 18$	$6 \times 4 = 24$	$6 \times 5 = 30$	$6 \times 6 = 36$
$7 \times 1 = 7$	$7 \times 2 = 14$	$7 \times 3 = 21$	$7 \times 4 = 28$	$7 \times 5 = 35$	$7 \times 6 = 42$
$8 \times 1 = 8$	$8 \times 2 = 16$	$8 \times 3 = 24$	$8 \times 4 = 32$	$8 \times 5 = 40$	$8 \times 6 = 48$
$9 \times 1 = 9$	$9 \times 2 = 18$	$9 \times 3 = 27$	$9 \times 4 = 36$	$9 \times 5 = 45$	$9 \times 6 = 54$
$10 \times 1 = 10$	$10 \times 2 = 20$	$10 \times 3 = 30$	$10 \times 4 = 40$	$10 \times 5 = 50$	$10 \times 6 = 60$
$11 \times 1 = 11$	$11 \times 2 = 22$	$11 \times 3 = 33$	$11 \times 4 = 44$	$11 \times 5 = 55$	$11 \times 6 = 66$
$12 \times 1 = 12$	$12 \times 2 = 24$	$12 \times 3 = 36$	$12 \times 4 = 48$	$12 \times 5 = 60$	$12 \times 6 = 72$
7 TIMES TABLE	**8 TIMES TABLE**	**9 TIMES TABLE**	**10 TIMES TABLE**	**11 TIMES TABLE**	**12 TIMES TABLE**
$1 \times 7 = 7$	$1 \times 8 = 8$	$1 \times 9 = 9$	$1 \times 10 = 10$	$1 \times 11 = 11$	$1 \times 12 = 12$
$2 \times 7 = 14$	$2 \times 8 = 16$	$2 \times 9 = 18$	$2 \times 10 = 20$	$2 \times 11 = 22$	$2 \times 12 = 24$
$3 \times 7 = 21$	$3 \times 8 = 24$	$3 \times 9 = 27$	$3 \times 10 = 30$	$3 \times 11 = 33$	$3 \times 12 = 36$
$4 \times 7 = 28$	$4 \times 8 = 32$	$4 \times 9 = 36$	$4 \times 10 = 40$	$4 \times 11 = 44$	$4 \times 12 = 48$
$5 \times 7 = 35$	$5 \times 8 = 40$	$5 \times 9 = 45$	$5 \times 10 = 50$	$5 \times 11 = 55$	$5 \times 12 = 60$
$6 \times 7 = 42$	$6 \times 8 = 48$	$6 \times 9 = 54$	$6 \times 10 = 60$	$6 \times 11 = 66$	$6 \times 12 = 72$
$7 \times 7 = 49$	$7 \times 8 = 56$	$7 \times 9 = 63$	$7 \times 10 = 70$	$7 \times 11 = 77$	$7 \times 12 = 84$
$8 \times 7 = 56$	$8 \times 8 = 64$	$8 \times 9 = 72$	$8 \times 10 = 80$	$8 \times 11 = 88$	$8 \times 12 = 96$
$9 \times 7 = 63$	$9 \times 8 = 72$	$9 \times 9 = 81$	$9 \times 10 = 90$	$9 \times 11 = 99$	$9 \times 12 = 108$
$10 \times 7 = 70$	$10 \times 8 = 80$	$10 \times 9 = 90$	$10 \times 10 = 100$	$10 \times 11 = 110$	$10 \times 12 = 120$
$11 \times 7 = 77$	$11 \times 8 = 88$	$11 \times 9 = 99$	$11 \times 10 = 110$	$11 \times 11 = 121$	$11 \times 12 = 132$
$12 \times 7 = 84$	$12 \times 8 = 96$	$12 \times 9 = 108$	$12 \times 10 = 120$	$12 \times 11 = 132$	$12 \times 12 = 144$

EXAMPLE 1

Evaluate each quotient.

a $24 \div 6$ b $30 \div 5$ c $64 \div 8$

SOLUTION

a $4 \times 6 = 24$, so $24 \div 6 = 4$

b $6 \times 5 = 30$, so $30 \div 5 = 6$

c $8 \times 8 = 64$, so $64 \div 8 = 8$

ISBN 9780170350969

EXERCISE 4–01

1 Evaluate $56 \div 8$. Select the correct answer A, B, C or D.

A 6 B 7 C 9 D 8

2 Evaluate $110 \div 10$. Select A, B, C or D.

A 12 B 10 C 9 D 11

3 a Divide these shapes into 4 equal groups.

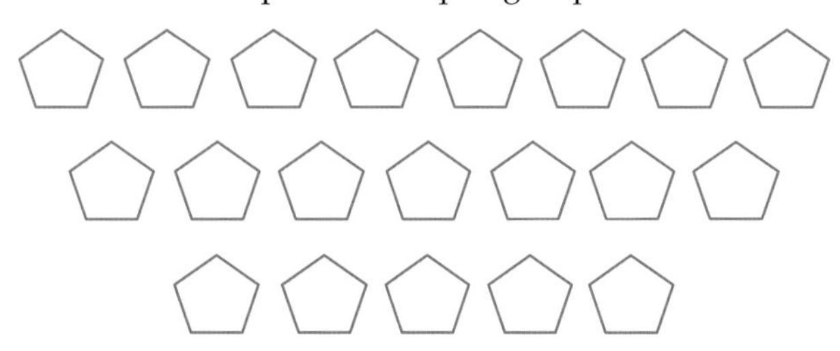

b What division does this illustrate ?

4 Illustrate the division $15 \div 3 = 5$ using triangles.

5 Evaluate each quotient.

a $10 \div 2$	b $12 \div 2$	c $6 \div 3$	d $9 \div 3$
e $15 \div 3$	f $18 \div 6$	g $18 \div 9$	h $20 \div 2$
i $20 \div 4$	j $20 \div 5$	k $32 \div 8$	l $14 \div 7$
m $25 \div 5$	n $28 \div 2$	o $35 \div 7$	p $45 \div 3$
q $63 \div 9$	r $64 \div 8$	s $24 \div 6$	t $50 \div 10$
u $48 \div 12$	v $32 \div 4$	w $56 \div 7$	x $63 \div 7$

6 Evaluate each quotient.

a $14 \div 2$	b $16 \div 2$	c $16 \div 4$	d $18 \div 2$
e $18 \div 3$	f $20 \div 10$	g $21 \div 3$	h $21 \div 7$
i $22 \div 2$	j $22 \div 11$	k $33 \div 3$	l $18 \div 9$
m $15 \div 5$	n $40 \div 4$	o $56 \div 8$	p $44 \div 11$
q $28 \div 7$	r $36 \div 9$	s $42 \div 7$	t $55 \div 5$
u $36 \div 6$	v $60 \div 10$	w $54 \div 6$	x $54 \div 9$

7 Evaluate each quotient.

a $81 \div 9$	b $132 \div 11$	c $144 \div 12$
d $108 \div 9$	e $110 \div 10$	f $121 \div 11$

8 Jill bought 9 hot cross buns and paid \$6.30 for them. How much does one hot cross bun cost?

4-02 Dividing numbers

WORDBANK

short division A method of dividing by a one-digit number (2 to 9).

EXAMPLE 2

Use short division to evaluate each quotient.

a $6312 \div 3$

b $6048 \div 7$

SOLUTION

a
$$\begin{array}{r} 2\,1\,0\,4 \\ 3\overline{)6\,3\,1\,{}^{1}2} \end{array}$$

3 into 6 goes 2

3 into 3 goes 1

3 into 1 goes 0, remainder 1

3 into 12 goes 4

So $6312 \div 3 = 2104$

b
$$\begin{array}{r} 8\;\,6\;\,4 \\ 7\overline{)6^{6}\,0^{4}\,4^{2}\,8} \end{array}$$

7 into 6 goes 0, remainder 6

7 into 60 goes 8, remainder 4

7 into 44 goes 6, remainder 2

7 into 28 goes 4, no remainder

So $6048 \div 7 = 864$

Mental division strategies

DIVIDING BY	STRATEGY
2	Halve
5	Divide by 10, then double
10	For a whole number ending in 0, drop a 0 from the end of the number
100	For a whole number ending in 0s, drop two 0s from the end of the number

EXAMPLE 3

Use mental division to evaluate each quotient.

a $520 \div 10$ b $400 \div 100$ c $260 \div 5$

SOLUTION

a $520 \div 10 = 52$ ← To divide a number ending in 0 by 10, remove the 0

b $400 \div 100 = 4$ ← To divide a number ending in 00 by 100, remove the 00

c $260 \div 5 = 260 \div 10 \times 2$ ← To divide a number ending in 0 by 5, divide by 10 and double

$= 26 \times 2$

$= 52$

ISBN 9780170350969

EXERCISE 4-02

1 Use short division to evaluate 282 ÷ 6. Select the correct answer A, B, C or D.

A 45 B 48 C 46 D 47

2 Evaluate 2016 ÷ 8. Select A, B, C or D.

A 242 B 252 C 262 D 253

3 Evaluate each quotient.

a 88 ÷ 4	b 564 ÷ 6	c 847 ÷ 7	d 968 ÷ 8
e 585 ÷ 5	f 6759 ÷ 9	g 9765 ÷ 3	h 2835 ÷ 5
i 43 284 ÷ 6	j 8973 ÷ 3	k 2056 ÷ 8	l 3438 ÷ 9

4 Jane worked for 8 hours at the local market and was paid $120. How much did she earn per hour?

5 A restaurant bill was $336 for a table of 6. How much should each person pay if they decide to divide the bill equally among them?

Alamy/David Ball

6 Evaluate each quotient mentally.

a 600 ÷ 10	b 3000 ÷ 100	c 1450 ÷ 10	d 8800 ÷ 100
e 2500 ÷ 100	f 7000 ÷ 10	g 7000 ÷ 1000	h 5200 ÷ 100

7 Evaluate each quotient mentally by dividing by 10, then doubling.

a 200 ÷ 5	b 350 ÷ 5	c 80 ÷ 5	d 180 ÷ 5
e 60 ÷ 5	f 400 ÷ 5	g 90 ÷ 5	h 280 ÷ 5

8 A bag of 144 chocolates was shared equally among 9 friends. How many chocolates did each person receive?

9 At Westvale College, there are 135 students in Year 7. If they are placed into five equal-sized classes, how many students are in each class?

10 200 students went on an excursion on 5 buses. If there was the same number of students on each bus, how many students were on each bus?

4-03 Division with remainders

WORDBANK

remainder An amount or number left over from a division.

Sometimes when we are dividing, there is a **remainder** at the end.
When this happens, we can write the remainder as a fraction of the number we are dividing by.

EXAMPLE 4

Use short division to evaluate each quotient.

a $523 \div 5$

b $8644 \div 6$

SOLUTION

a
$$\begin{array}{r} 1\,0\,4\,\text{r}\,3 \\ 5\overline{)\,5\,2^{2}3} \end{array}$$

5 into 5 goes 1

5 into 2 goes 0, remainder 2

5 into 23 goes 4, remainder 3

$523 \div 5 = 104\frac{3}{5}$

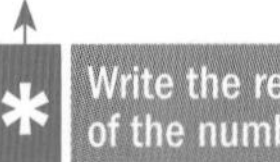

Write the remainder as a fraction of the number we are dividing by

b
$$\begin{array}{r} 1\;4\;4\;0\,\text{r}\,4 \\ 6\overline{)8^{2}6^{2}4\;4} \end{array}$$

6 into 8 goes 1, remainder 2

6 into 26 goes 4, remainder 2

6 into 24 goes 4, no remainder

6 into 4 goes 0, remainder 4

$8644 \div 6 = 1440\frac{4}{6} = 1440\frac{2}{3}$

$\frac{4}{6}$ simplifies to $\frac{2}{3}$

ISBN 9780170350969

EXERCISE 4–03

1 What is the remainder for 239 ÷ 5? Select the correct answer A, B, C or D.

A 1 B 2 C 3 D 4

2 Evaluate 3287 ÷ 8. Select A, B, C or D.

A 411 B 410 C $410\frac{7}{8}$ D $410\frac{1}{8}$

3 Evaluate each quotient, writing any remainder as a simple fraction.

a 89 ÷ 4 b 563 ÷ 6 c 947 ÷ 7 d 548 ÷ 8

e 5816 ÷ 5 f 6059 ÷ 9 g 9745 ÷ 3 h 2839 ÷ 5

i 43 314 ÷ 6 j 8963 ÷ 3 k 2206 ÷ 8 l 3418 ÷ 9

4 Jane worked for 8 hours at the local market and was paid $265. How much did she earn per hour, rounded to the nearest dollar?

5 A restaurant bill was $398 for a table of 7. How much should each person pay if they decide to divide the bill equally among them? Answer to the nearest dollar.

6 a If a 2654 cm length of timber is divided into 6 equal pieces in whole centimetres, find the length of each piece.

b How much is left over?

7 A prize of $5860 was shared equally among 9 friends. How much money did each person receive? Answer to the nearest cent.

8 At Northvale College, there are 146 students in Year 7.

a If they are placed evenly into five classes, how many students are in each class?

b How many students are left over?

Fairfax/Peter Rae

9 Business profits for the first year were $16 550. Sam wanted to share this with his two business partners. How much will each person receive, rounded to the nearest hundred dollars?

ISBN 9780170350969

4-04 Divisibility tests

Is 2016 divisible by 3? Does 3 go into 2016 evenly, with no remainder? How do you know? **Divisibility tests** are rules for deciding whether a number is divisible by any number from 2 to 10.

DIVISIBLE BY:	DIVISIBILITY TEST
2	The number ends with 0, 2, 4, 6, or 8.
3	The sum of the digits in the number is divisible by 3.
4	The last two digits form a number divisible by 4.
5	The number ends with 0 or 5.
6	The number is divisible by *both* 2 and 3.
7	There is no simple divisibility test for 7.
8	The last three digits form a number divisible by 8.
9	The sum of the digits in the number is divisible by 9.
10	The number ends with 0.

Shutterstock.com/nito

Shutterstock.com/iamshutter

Shutterstock.com/Moving Moment

Shutterstock.com/Africa Studio

EXAMPLE 5

Test whether 2016 is divisible by:

a 3   b 4   c 5   d 6   e 8

SOLUTION

a Sum of digits = 2 + 0 + 1 + 6 = 9, which is divisible by 3. ← $9 \div 3 = 3$
So 2016 is divisible by 3. ← $(2016 \div 3 = 672)$

b Last two digits = 16, which is divisible by 4. ← $16 \div 4 = 4$
So 2016 is divisible by 4. ← $(2016 \div 4 = 504)$

c The last digit is not 0 or 5, so 2016 is not divisible by 5.

d The number ends in 6, so it is divisible by 2 (even).
The number is divisible by 3 (from part a).
So 2016 is divisible by 6. ← $(2016 \div 6 = 336)$

e Last three digits = 016, which is divisible by 8. ← $16 \div 8 = 2$
So 2016 is divisible by 8. ← $(2016 \div 8 = 252)$

ISBN 9780170350969

EXERCISE 4–04

1 What is the test for divisibility by 3? Select the correct answer A, B, C or D.

A The last digit must be odd.

B The sum of the digits is 3.

C The sum of the digits is a multiple of 3.

D The number must end with a 3.

2 What is the test for divisibility by 4? Select A, B, C or D.

A The last digit must be 4.

B The last 2 digits must form a number divisible by 4.

C The sum of the digits is a multiple of 4.

D The number must end with an even number.

3 Test whether each number is divisible by 2, 3 and 6.

a 5622 b 878 c 936

d 23 460 e 18 753 f 36 923

4 Test whether each number is divisible by 4 and 8.

a 788 b 968 c 1296

d 45 236 e 32 568 f 234 564

5 Test whether each number is divisible by 5 and 10.

a 560 b 685 c 2580

d 45 656 e 438 950 f 678 900

6 Test whether each number is divisible by 9.

a 684 b 798 c 945

d 12 459 e 65 781 f 324 569

7 There are 186 students in Year 7. Can they be placed into equal groups of:

a 3? b 4? c 5?

8 A jar contains 468 lollies. Test whether they could be shared evenly between:

a 4 people b 6 people c 9 people

9 A tennis club has 256 members. A game of doubles requires 4 players.

a Is 256 divisible by 4?

b How many games of doubles can be played if each member plays only once?

c If only half of the members turn up can they all play doubles?

d How many doubles games could be played if only half of the members turn up?

10 Write down all numbers between 1 and 50 that are divisible by 2, 3 and 4.

ISBN 9780170350969

4-05 Highest common factor (HCF)

WORDBANK

factor A value that divides evenly into a given number. For example, 3 is a factor of 15 as $15 \div 3 = 5$.

highest common factor The largest factor shared by two or more numbers.

To list the factors of a number, write down all the different pairs of numbers that multiply to give that number, then write them in order.

EXAMPLE 6

List the factors of each number.

a 12

b 35

SOLUTION

a $1 \times 12, 2 \times 6, 3 \times 4$
Factors of 12 = 1, 12, 2, 6, 3, 4
= 1, 2, 3, 4, 6, 12

b $1 \times 35, 5 \times 7$
Factors of 35 = 1, 5, 7, 35

The **highest common factor** (**HCF**) of two (or more) numbers is the largest number that is a factor of **both** (or **all**) of these numbers.

To **find the highest common factor** (**HCF**) of two or more numbers:

- list the factors of each number
- underline the common factors
- select the highest factor that is underlined in both lists.

EXAMPLE 7

Find the highest common factor of 24 and 60.

SOLUTION

Factors of 24: $\underline{1}, \underline{2}, \underline{3}, \underline{4}, \underline{6}, 8, \underline{12}, 24$

Factors of 60: $\underline{1}, \underline{2}, \underline{3}, \underline{4}, 5, \underline{6}, 10, \underline{12}, 15, 20, 30, 60$

The common factors are underlined in both lists.

The highest common factor of 24 and 60 is 12. ← This is the largest factor underlined

ISBN 9780170350969

EXERCISE 4-05

1 What are the factors of 8? Select the correct answer A, B, C or D.

A 1, 2, 4, 8 B 1, 2, 8 C 1, 2, 3, 4, 8 D 2, 4, 8

2 What are the factors of 15? Select A, B, C or D.

A 1, 3, 15 B 1, 5, 15 C 1, 3, 5, 15 D 3, 5, 15

3 List the factors of:

a 10 b 16 c 9 d 20 e 48

f 36 g 60 h 24 i 40 j 54

4 What is the highest common factor of 8 and 12? Select A, B, C or D.

A 1 B 2 C 4 D 8

5 What is the HCF of 15 and 25? Select A, B, C or D.

A 15 B 5 C 3 D 1

6 a Write out all the factors of 12.

b Write out all the factors of 15.

c Underline the common factors of 12 and 15.

d Write down the highest common factor of 12 and 15.

7 a List the factors of 30.

b List the factors of 42.

c Underline the common factors of 30 and 42.

d Write down the highest common factor of 30 and 42.

8 Find the highest common factor of each pair of numbers.

a 10 and 18 b 12 and 20 c 15 and 45

d 50 and 80 e 60 and 90 f 36 and 64

9 The highest common factor of two numbers is 6. What could the two numbers be?

10 Find the HCF of 27, 36, 45 and 72.

4-06 Prime and composite numbers

WORDBANK

prime number A number with only two factors, 1 and itself. For example, 7 is a prime number because it has exactly two factors, 1 and 7.

composite number A number with more than two factors. For example, 6 is a composite number because it has four factors: 1, 2, 3 and 6.

- A number is either prime or composite—it cannot be both.
- 2 is the first prime number and the only even prime number. (All other even numbers are composite.)
- 1 is neither prime nor composite because it has only one factor, 1.
- The first 5 prime numbers are 2, 3, 5, 7 and 11.

EXAMPLE 8

State whether each number is prime or composite.

a 23 b 45 c 51 d 37

SOLUTION

a 23 is prime as its only factors are 1 and 23. ← $1 \times 23 = 23$

b 45 is composite as it has more than 2 factors, including 5 and 9. ← $5 \times 9 = 45$

c 51 is composite as it has more than 2 factors, including 3 and 17. ← $3 \times 17 = 51$

d 37 is prime as its only factors are 1 and 37. ← $1 \times 37 = 37$

Shutterstock.com/Natalia Korshunova

ISBN 9780170350969

EXERCISE 4-06

1 Which three numbers are all prime? Select the correct answer A, B, C or D.

A 3, 15, 17 B 5, 13, 17 C 7, 11, 21 D 9, 13, 17

2 Which three numbers are all composite? Select A, B, C or D.

A 33, 51, 63 B 15, 23, 27 C 17, 29, 35 D 31, 39, 42

3 List all the factors of each number and then state whether the number is prime or composite.

a 12 b 23 c 31 d 15 e 17
f 38 g 42 h 57 i 91

4 List all the composite numbers between 60 and 80.

5 The prime numbers can be found by listing the numbers and using the Sieve of Eratosthenes (pronounced 'Siv of Era-tos-the-nees'). Eratosthenes was an ancient Greek mathematician who 'sifted' out the prime numbers by crossing out all of the multiples of numbers (the composite numbers).

a Copy the grid below for 1 to 120 or print out the worksheet 'Sieve of Eratosthenes'.

1	2	3	4	5	6
7	8	9	10	11	12
13	14	15	16	17	18
19	20	21	22	23	24
25	26	27	28	29	30
31	32	33	34	35	36
37	38	39	40	41	42
43	44	45	46	47	48

b Cross out 1. It is neither prime nor composite.

c Except for 2, cross out every multiple of 2: 4, 6, 8, 10, … and notice the pattern.

d Except for 3, cross out every multiple of 3: 6, 9, 12, 15, … and notice the pattern.

e Except for 5, cross out every multiple of 5: 10, 15, 20, 25, … and notice the pattern.

f Except for 7, cross out every multiple of 7: 14, 21, 28, 35, … and notice the pattern.

g Write out the remaining 30 prime numbers between 1 to 120. (The last one is 113.)

6 Look at the prime numbers from Question 5. There are pairs of numbers called twin primes that are only 2 apart. List all 10 pairs of twin primes between 1 and 120.

7 What number am I?

a I have two digits.
I am a twin prime.
I am less than 50 but greater than 25.
The sum of my digits is 5.

b I have three digits.
I am composite.
I am more than 215.
The sum of my digits is 5.

4-07 Factor trees

WORDBANK

factor tree A diagram that lists the prime factors of a number.

Every number can be written as a product of its prime factors.

Remember, **product** means the answer to a multiplication.

The prime factors can be found by using a **factor tree**.

EXAMPLE 9

Write 24 as a product of its prime factors.

SOLUTION

Draw a factor tree for 24.

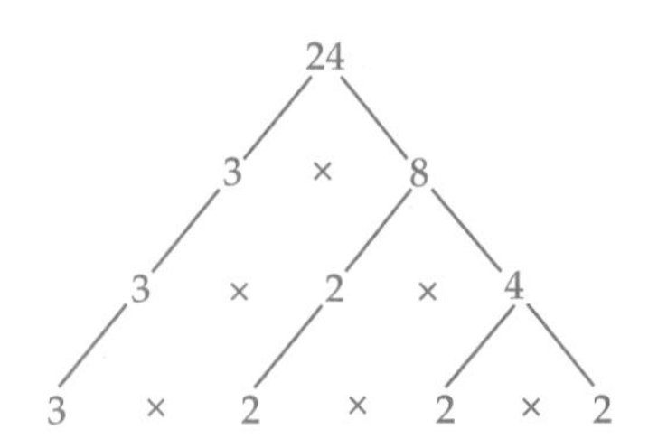

Break 24 into 2 factors: 3 and 8

3 is prime, so leave as 3

Break 8 into 2 factors: 2 and 4

Continue until all factors are prime

As a product of prime factors, $24 = 2 \times 2 \times 2 \times 3$ (or $24 = 2^3 \times 3$)

Note: It is possible to draw different factor trees for the same number, but the final list of prime factors should still be the same. Here is another factor tree for 24:

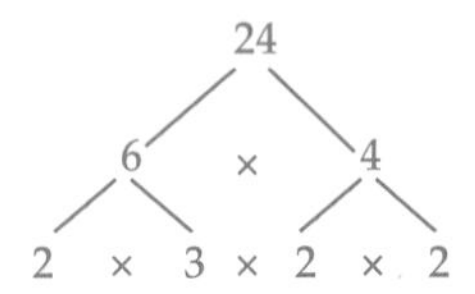

$24 = 2 \times 2 \times 2 \times 3$

ISBN 9780170350969

EXERCISE 4–07

1 If $54 = 3 \times$ ____, what number goes in the blank space? Select the correct answer A, B, C or D.

A 8 B 18 C 28 D 38

2 Copy and complete this factor tree.

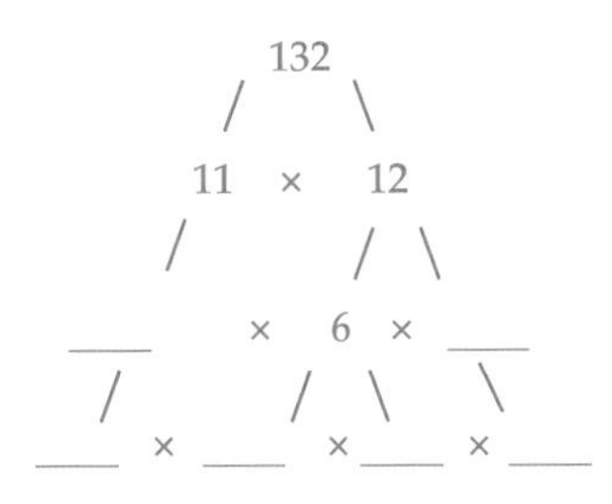

3 Draw two different factor trees for 30 and show that you get the same prime factors for each one.

4 Draw two different factor trees for 48 and show that you get the same prime factors for each one.

5 Use a factor tree to write each number as a product of its prime factors.

a 88 b 63 c 45 d 51

e 132 f 270 g 396 h 218

6 Express 1260 as a product of its prime factors.

7 Use a factor tree to write each number as a product of its prime factors.

a 48 b 200 c 460 d 712

e 98 f 144 g 325 h 135

Shutterstock.com/Photographee.eu

ISBN 9780170350969

CODE PUZZLE

Use the following table to decode the words and phrases below from this chapter.

A	B	C	D	E	F	G	H	I	J	K	L	M
1	2	3	4	5	6	7	8	9	10	11	12	13

N	O	P	Q	R	S	T	U	V	W	X	Y	Z
14	15	16	17	18	19	20	21	22	23	24	25	26

1 4-9-22-9-19-9-15-14

2 17-21-15-20-9-5-14-20

3 19-8-15-18-20

4 20-18-5-5

5 18-5-13-1-9-14-4-5-18

6 4-9-22-9-19-9-2-9-12-9-20-25 20-5-19-20

7 6-1-3-20-15-18

8 8-9-7-8-5-19-20 3-15-13-13-15-14 6-1-3-20-15-18

9 3-15-13-16-15-19-9-20-5

10 16-18-9-13-5

ISBN 9780170350969

PRACTICE TEST 4

Part A General topics

Calculators are not allowed.

1 How many days in March?

2 Find the perimeter of this rhombus.

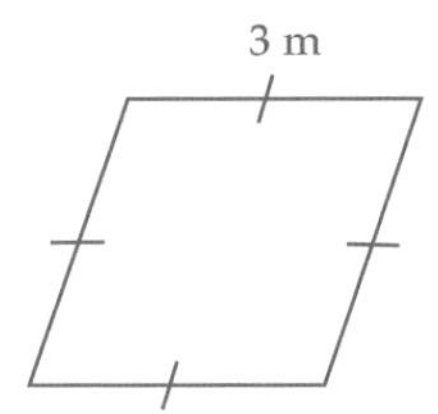

3 What is the name of the shape with 8 sides?

4 How many degrees in a right angle?

5 Find 10% of \$120.

6 A box contains 20 hard-centred chocolates and 12 soft-centred ones. What fraction of the chocolates are soft-centred?

7 How many seconds in 5 minutes?

8 Write these integers in descending order: 8, –3, 0, –1, 4, 1.

9 How many faces does a cube have?

10 Evaluate $120 - 3 \times 8$.

Part B Factors and primes

Calculators are not allowed.

4–01 Division facts

11 Evaluate $48 \div 8$. Select the correct answer A, B, C or D.

A 7　　B 8　　C 6　　D 9

12 Evaluate $72 \div 9$. Select A, B, C or D.

A 7　　B 8　　C 6　　D 9

4–02 Dividing numbers

13 Evaluate $1824 \div 6$. Select A, B, C or D.

A 34　　B 306　　C 304　　D 36

4–03 Division with remainders

14 Evaluate each quotient, showing the remainder as a fraction.

a $165 \div 8$　　b $289 \div 5$　　c $2470 \div 7$

4–04 Divisibility tests

15 Test each number for divisibility by 2, 3 and 6.

a 54　　b 375　　c 1206

16 Test each number for divisibility by 4 or 8.

a 36　　b 288　　c 32 168

ISBN 9780170350969

4-05 Highest common factor (HCF)

17 How many factors does 36 have? Select A, B, C or D.

A 6 B 7 C 8 D 9

18 List the factors of each number.

a 55 b 80

19 What is the highest common factor of 20 and 28?

4-06 Prime and composite numbers

20 How many factors does a prime number have?

21 a List all the prime numbers between 20 and 40.

b List all the composite numbers between 20 and 40.

4-07 Factor trees

22 a Draw a factor tree for 180.

b Hence write 180 as a product of its prime factors.

ISBN 9780170350969

POWERS AND DECIMALS

5

WHAT'S IN CHAPTER 5?

5-01 Powers
5-02 Square and square root
5-03 Cube and cube root
5-04 Order of operations
5-05 Decimals
5-06 Ordering decimals
5-07 Adding and subtracting decimals

IN THIS CHAPTER YOU WILL:

- understand and use index notation
- evaluate expressions involving powers, square root and cube root
- understand and use the order of operations
- understand place value in decimals
- convert decimals to simple fractions
- compare and order decimals
- add and subtract decimals

Getty Images/Fotosearch

ISBN 9780170350969

5-01 Powers

WORDBANK

base The main number that is raised to a power. For example, in 5^3 the base is 5.

power The number at the top right corner of the base that represents repeated multiplication of the base by itself. For example, 5^3 means $5 \times 5 \times 5$ and the power is 3.

index Another word for power.

index notation Using powers to write repeated multiplication. For example, $3^5 = 3 \times 3 \times 3 \times 3 \times 3$.

INDEX NOTATION

- $5^2 = 5 \times 5$, where 5 is called the **base** and 2 is the **power** or **index**.
- For 5^2, we say '5 squared' or '5 to the power of 2'.
- For $6^3 = 6 \times 6 \times 6$, we say '6 cubed' or '6 to the power of 3'.
- The power shows how many times the base appears in the repeated multiplication.

EXAMPLE 1

Write each expression using index notation.

a $4 \times 4 \times 4$

b $7 \times 7 \times 7 \times 7 \times 7$

c $2 \times 2 \times 2 \times 2 \times 2 \times 2 \times 2$

SOLUTION

a $4 \times 4 \times 4 = 4^3$

b $7 \times 7 \times 7 \times 7 \times 7 = 7^5$

c $2 \times 2 \times 2 \times 2 \times 2 \times 2 \times 2 = 2^7$

EXAMPLE 2

Evaluate each expression.

a 3^4

b 8^5

c 20^6

SOLUTION

a $3^4 = 3 \times 3 \times 3 \times 3$
$= 81$

b $8^5 = 8 \times 8 \times 8 \times 8 \times 8$
$= 32\,768$

c $20^6 = 20 \times 20 \times 20 \times 20 \times 20 \times 20$
$= 64\,000\,000$

ISBN 9780170350969

EXERCISE 5-01

1 Write $4 \times 4 \times 4 \times 4 \times 4 \times 4$ in index notation. Select the correct answer A, B, C or D.

A 4^5 B 1024 C 4^6 D 4096

2 Evaluate 7^4. Select A, B, C or D.

A $7 \times 7 \times 7 \times 7 \times 7$ B 2401

C $7 \times 7 \times 7$ D 16 807

3 Write each expression using index notation.

a $3 \times 3 \times 3 \times 3$ b $2 \times 2 \times 2 \times 2 \times 2$

c $5 \times 5 \times 5$ d $4 \times 4 \times 4 \times 4 \times 4 \times 4$

e $7 \times 7 \times 7$ f $6 \times 6 \times 6 \times 6$

g $9 \times 9 \times 9 \times 9 \times 9$ h $12 \times 12 \times 12 \times 12$

i $23 \times 23 \times 23 \times 23 \times 23$

4 Evaluate each expression in Question 3.

5 For each expression, write:

i the base ii the power

a 6^4 b 4^5 c 3^6

d 2^7 e 5^4 f 8^3

g 9^4 h 10^5 i 15^4

6 Write each expression in Question 5 in expanded form, as a repeated multiplication such as $2 \times 2 \times 2 \times 2 \times 2$.

7 Evaluate each expression in Question 6.

8 Evaluate each expression.

a $2^3 \times 3^2$ b $4^2 \times 3^5$ c $2^6 \times 5^3$

d $7^4 \times 3^3$ e $10^3 \times 6^3$ f $9^4 \times 2^7$

9 Evaluate each power of 10.

a 10^3 b 10^5 c 10^2

d 10^6 e 10^8 f 10^7

g 10^9 h 10^4 i 10^1

10 What pattern do you notice in the powers of 10 in Question 9? How are the powers related to the answers?

5-02 Square and square root

WORDBANK

square To square a number means to multiply it by itself. For example, 3 squared = $3^2 = 3 \times 3 = 9$.

square root ($\sqrt{\ }$) The opposite of squaring a number. For example, the square root of $9 = \sqrt{9} = 3$, as $3^2 = 9$.

The **square root** $\sqrt{\ }$ of a number is the positive value which, when squared, equals that number.

EXAMPLE 3

Evaluate each square number.

a 6 squared b 9^2 c $(-2)^2$

SOLUTION

a 6 squared = $6^2 = 6 \times 6 = 36$

b $9^2 = 9 \times 9 = 81$

c $(-2)^2 = (-2) \times (-2) = 4$

EXAMPLE 4

Evaluate each square root.

a $\sqrt{25}$ b $\sqrt{49}$ c $\sqrt{169}$

SOLUTION

a $\sqrt{25} = 5$ ← as $5^2 = 25$

b $\sqrt{49} = 7$ ← as $7^2 = 49$

c $\sqrt{169} = 13$ ← as $13^2 = 169$

Your calculator has keys for calculating squares [x^2] and square roots [$\sqrt{\ }$]. For example,

- to evaluate 6^2, enter 6 [x^2] [=] to get the answer 36
- to evaluate $(-2)^2$, enter [(] [(-)] 2 [)] [x^2] [=] to get the answer 4
- to evaluate $\sqrt{25}$, enter [$\sqrt{\ }$] 25 [=] to get the answer 5

ISBN 9780170350969

EXERCISE 5–02

1 What is the value of 15^2? Select the correct answer A, B, C or D.

A 15×15 B 155 C 115 D 225

2 Evaluate $\sqrt{196}$. Select A, B, C or D.

A 15 B 14 C 16 D 19

3 Evaluate each expression.

a 3 squared b 8 squared c −11 squared

d 7^2 e 5^2 f 9^2

g 10^2 h $(-12)^2$ i $(-4)^2$

4 Use the square x^2 key on your calculator to evaluate each square number.

a 17 squared b −18 squared c 23 squared

d 27^2 e 32^2 f 45^2

g 62^2 h 75^2 i 90^2

5 Evaluate each square root.

a $\sqrt{9}$ b $\sqrt{36}$ c $\sqrt{64}$

d $\sqrt{100}$ e $\sqrt{81}$ f $\sqrt{1}$

g $\sqrt{144}$ h $\sqrt{4}$ i $\sqrt{121}$

6 Use the square root $\sqrt{}$ key on your calculator to evaluate each square root.

a $\sqrt{361}$ b $\sqrt{225}$ c $\sqrt{289}$

d $\sqrt{576}$ e $\sqrt{784}$ f $\sqrt{1024}$

7 Evaluate each square number.

a 20^2 b 200^2 c 2000^2

8 What happened to the number of zeros when you squared each number in Question 7?

9 Evaluate each square root.

a $\sqrt{900}$ b $\sqrt{90\,000}$ c $\sqrt{9\,000\,000}$

10 What happened to the number of zeros when you found the square root of each number in Question 9?

11 Evaluate $\sqrt{64\,000\,000}$.

ISBN 9780170350969

5-03 Cube and cube root

WORDBANK

cube To cube a number means to multiply it by itself to raise it to the power of 3. For example, 2 cubed = $2^3 = 2 \times 2 \times 2 = 8$.

cube root ($\sqrt[3]{\ }$) The opposite of cubing a number. For example, the cube root of 8 = $\sqrt[3]{8} = 2$, as $2 \times 2 \times 2 = 8$.

The **cube root** ($\sqrt[3]{\ }$) of a number is the value which, when cubed, equals that number.

EXAMPLE 5

Evaluate each cube number.

a 4 cubed b 6^3 c $(-10)^3$

SOLUTION

a 4 cubed = $4 \times 4 \times 4 = 64$

b $6^3 = 6 \times 6 \times 6 = 216$

c $(-10)^3 = (-10) \times (-10) \times (-10) = -1000$

EXAMPLE 6

Evaluate each cube root.

a $\sqrt[3]{1}$ b $\sqrt[3]{125}$ c $\sqrt[3]{-729}$

SOLUTION

a $\sqrt[3]{1} = 1$ ⟵ as $1^3 = 1$

b $\sqrt[3]{125} = 5$ ⟵ as $5^3 = 125$

c $\sqrt[3]{-729} = -9$ ⟵ as $(-9)^3 = -729$

Your calculator has keys for calculating cubes [x^3] and cube roots [$\sqrt[3]{\ }$]. For example,

- to evaluate 6^3, enter 6 [x^3] [=] to get the answer 216
- to evaluate $(-10)^3$, enter [(] [(–)] 10 [)] [x^3] [=] to get the answer –1000
- to evaluate $\sqrt{-729}$, enter [$\sqrt[3]{\ }$] [(–)] 729 [=] to get the answer –9

ISBN 9780170350969

EXERCISE 5–03

1 What is the value of 12^3? Select the correct answer A, B, C or D.

A 12×12 B 1728

C 144 D 36

2 Evaluate $\sqrt[3]{-512}$. Select A, B, C or D.

A –8 B –9 C 7 D 8

3 Evaluate each expression.

a 3 cubed b 6 cubed

c 11 cubed d 7^3

e $(-5)^3$ f 9^3

g 13^3 h $(-12)^3$

i $(-4)^3$ j 16^3

k 21^3 l $(-14)^3$

4 Copy and complete:

To cube a number we multiply it by itself to raise it to the ______ of ______. The opposite of cubing a number is ______ ______.

5 Evaluate each cube root.

a $\sqrt[3]{8}$ b $\sqrt[3]{216}$

c $\sqrt[3]{-27}$ d $\sqrt[3]{343}$

e $\sqrt[3]{-64}$ f $\sqrt[3]{512}$

g $\sqrt[3]{-1000}$ h $\sqrt[3]{1331}$

i $\sqrt[3]{-1}$ j $\sqrt[3]{3375}$

k $\sqrt[3]{-216}$ l $\sqrt[3]{9261}$

6 Evaluate each cube number.

a 20^3 b 200^3

7 What happened to the number of zeros when you cubed each number in Question 6?

8 Evaluate each cube root.

a $\sqrt[3]{8000}$ b $\sqrt[3]{8\,000\,000}$

9 What happened to the number of zeros when you found the cube root of each number in Question 8?

10 Evaluate $\sqrt[3]{64\,000\,000}$.

5-04 Order of operations

WORDBANK

operations The four operations in mathematics are addition (+), subtraction (–), multiplication (×) and division (÷).

order of operations The correct order to evaluate a mixed expression involving more than one operation, such as $16 \times 2 - (20 + 4)$.

brackets Grouping symbols around expressions, such as round brackets ( ) or square brackets [ ].

When evaluating **mixed expressions**, calculate using this **order of operations**:

- brackets ( ) first
- then powers (x^y) and square roots ($\sqrt{\ }$)
- then multiplication (×) and division (÷) from left to right
- then addition (+) and subtraction (–) from left to right.

EXAMPLE 7

Evaluate each expression.

a $20 + 32 \div 4$ b $16 \times 2 - (20 + 4)$ c $12 \div 2 \times 6$

d $9 \times 2 - 3 \times 4$ e $16 - 3^2 + (15 - 8)$

SOLUTION

a $20 + \overset{8}{32 \div 4} = 20 + 8$ ← ÷ first

$= 28$

b $16 \times 2 - (\overset{24}{20 + 4}) = \overset{32}{16 \times 2} - 24$ ← ( ) first

$= 32 - 24$ ← × next

$= 8$

c $\overset{6}{12 \div 2} \times 6 = 6 \times 6$ ← Work left to right: do ÷ first

$= 36$

d $\overset{18}{9 \times 2} - \overset{12}{3 \times 4} = 18 - 12$ ← × first

$= 6$

e $16 - 3^2 + (\overset{7}{15 - 8}) = 16 - \overset{9}{3^2} + 7$ ← ( ) first

$= 16 - 9 + 7$ ← Powers next

$= 7 + 7$ ← Work left to right: do – first

$= 14$

ISBN 9780170350969

EXERCISE 5-04

1 Which operation should be done after the brackets in $28 \times 2 - (4 + 8) \div 6$? Select the correct answer A, B, C or D.

A $\div$ B $+$ C $-$ D $\times$

2 Evaluate the expression in Question 1. Select A, B, C or D.

A 54 B 0 C 52 D 10

3 Which operation do you do first if evaluating a mixed expression that involves:

a $+$ and $\times$? b $+$ and $-$? c $-$ and $\div$? d $\times$ and $\div$?

4 Write down which operation you would do first for each expression.

a $6 + 3 \times 7$ b $12 - 9 \div 3$ c $5 + 6 - 4$

d $15 - 6 \times 2$ e $4 \times 6 \div 3$ f $36 \div 3 \times 4$

g $4 \times 2 + 5 \times 3$ h $20 \div 2 - 8 \div 4$ i $64 - 24 \div 6$

5 Evaluate each expression in Question 4.

6 Evaluate each expression.

a $5 + (4 \times 2)$ b $(7 + 4) - 6$ c $(12 - 4) + 7$

d $3 \times (5 + 4)$ e $(45 \div 9) + 8$ f $(18 + 12) - 6$

g $(8 + 4) \times 5$ h $24 \div (3 + 5)$ i $(49 \div 7) + 8$

7 Evaluate each expression.

a $6 \times 3 + 7 \times 2$ b $9 \times 6 - 4 \times 8$

c $5 \times 7 + 4 \times 9$ d $56 \div 8 - 28 \div 7$

e $36 \div 4 + 27 \div 3$ f $55 \div 11 + 48 \div 8$

g $(3 + 4) \times 5^2$ h $(7 + 8) \times 2^3$

i $(9 + 18) \div 3^2$ j $(40 \div 8) + (3 \times 6)$

k $(3 + 4) \times (18 \div 2)$ l $(864 \div 8) - 4^3$

8 Bryony bought 3 packets of biscuits costing \$2.50 per packet and 5 packets of chips costing \$3.20 per packet. How much did she spend altogether?

9 A fishing club charges a \$5 yearly fee and a \$20 boat fee for each fishing trip. How much money does Ray pay over one year if he goes fishing each month?

10 Mel is planting trees in her garden. On Monday she plants 14 trees, Tuesday 18, Wednesday 12, Thursday 8, and Friday 15. Each tree costs \$12.50.

a How many trees does she plant altogether?

b What is the total cost of the trees in her garden?

5-05 Decimals

WORDBANK

decimal A number that uses a decimal point and place value to show tenths $\left(\frac{1}{10}\right)$, hundredths $\left(\frac{1}{100}\right)$, thousandths $\left(\frac{1}{1000}\right)$ and so on.

decimal places The number of digits after the decimal point. For example, 6.24 has 2 decimal places as there are 2 digits after the decimal point.

The size of a decimal such as 6345.284 is shown by its **place value**.

THOUSANDS (1000s)	HUNDREDS (100s)	TENS (10s)	UNITS (1s)	DECIMAL POINT	TENTHS $\left(\frac{1}{10}\text{s}\right)$	HUNDREDTHS $\left(\frac{1}{100}\text{s}\right)$	THOUSANDTHS $\left(\frac{1}{1000}\text{s}\right)$
6	3	4	5	.	2	8	4
6000	300	40	5		$\frac{2}{10}$	$\frac{8}{100}$	$\frac{4}{1000}$

So $6345.284 = (6 \times 1000) + (3 \times 100) + (4 \times 10) + (5 \times 1) + \left(2 \times \frac{1}{10}\right) + \left(8 \times \frac{1}{100}\right) + \left(4 \times \frac{1}{1000}\right)$

TO WRITE A DECIMAL AS A FRACTION, REMEMBER THIS:

1 decimal place is 0.__ $= \frac{\square}{10}$ **1** number after the point, **1** zero in the denominator.

2 decimal places is 0.__ __ $= \frac{\square}{100}$ **2** numbers after the point, **2** zeros in the denominator.

3 decimal places is 0.__ __ __ $= \frac{\square}{1000}$ **3** numbers after the point, **3** zeros in the denominator.

EXAMPLE 8

Convert each decimal to a simple fraction.

a 0.5 b 0.03 c 0.28 d 1.034

SOLUTION

a $0.5 = \frac{5}{10} = \frac{1}{2}$

b $0.03 = \frac{3}{100}$

c $0.28 = \frac{28}{100} = \frac{7}{25}$

d $1.034 = 1\frac{34}{1000} = 1\frac{17}{500}$

✱ Always simplify the fraction if possible.

EXAMPLE 9

Convert each fraction to a decimal.

a $\frac{4}{10}$ b $\frac{32}{100}$ c $\frac{19}{10\,000}$

SOLUTION

a $\frac{4}{10} = 0.4$
1 zero, 1 decimal place

b $\frac{32}{100} = 0.32$
2 zeros, 2 decimal places

c $\frac{19}{10\,000} = 0.0019$
4 zeros, 4 decimal places

ISBN 9780170350969

EXERCISE 5-05

1 Which fraction is the same as 0.39? Select the correct answer A, B, C or D.

A $\frac{39}{10}$ B $\frac{39}{100}$ C $\frac{39}{1000}$ D $\frac{39}{10\,000}$

2 Convert 0.006 to a simplified fraction. Select A, B, C or D.

A $\frac{6}{10}$ B $\frac{3}{50}$ C $\frac{3}{500}$ D $\frac{6}{1000}$

3 For the decimal 3.142, which digit has the place value of:

a tenth $\left(\frac{1}{10}\right)$? b thousandth $\left(\frac{1}{1000}\right)$?

c unit? d hundredth $\left(\frac{1}{100}\right)$?

4 Convert each decimal to a fraction.

a 0.3 b 0.03 c 0.003 d 0.0003

5 Convert each decimal to a fraction with a denominator of 10, 100, 1000 or 10 000.

a 0.6 b 0.08 c 0.0009 d 0.005

e 0.04 f 0.56 g 0.078 h 0.342

i 0.0056 j 1.2 k 2.34 l 4.084

m 3.25 n 1.08 o 5.008 p 2.0762

6 Write each answer to Question 5 as a simple fraction.

7 Convert each fraction to a decimal.

a $\frac{3}{10}$ b $\frac{7}{100}$ c $\frac{4}{1000}$ d $\frac{9}{10}$

e $\frac{23}{100}$ f $\frac{35}{1000}$ g $\frac{451}{1000}$ h $\frac{83}{10\,000}$

i $1\frac{3}{10}$ j $2\frac{5}{100}$ k $4\frac{27}{1000}$ l $\frac{452}{100}$

m $\frac{94}{10}$ n $\frac{456}{1000}$ o $\frac{78}{1000}$ p $3\frac{23}{10\,000}$

Alamy/Sadequl Hussain - Concept Images

5-06 Ordering decimals

Adding 0s to the end of a decimal does not change the size of the decimal.

For example, 3.4 = 3.40 = 3.400 because $3\frac{4}{10} = 3\frac{40}{100} = 3\frac{400}{1000}$.

To compare and order decimals, first add 0s to each decimal where required so that all decimals have the same number of decimal places. This makes them easier to compare.

EXAMPLE 10

Write these decimals in ascending order:

0.7, 0.78, 0.703, 0.75, 0.777, 0.7006

SOLUTION

* **Ascending order** means from smallest to largest.

0.7006 has the greatest number of decimal places. First, write every decimal with 4 decimal places by adding 0s where required.

0.7000, 0.7800, 0.7030, 0.7500, 0.7770, 0.7006

Now order them from smallest to largest, ignoring the decimal points:

0.7000, 0.7006, 0.7030, 0.7500, 0.7770, 0.7800

Write the decimals as they were in the question:

0.7, 0.7006, 0.703, 0.75, 0.777, 0.78

EXAMPLE 11

Complete each statement with a < or > sign.

a 0.3 ____ 0.35 b 0.98 ____ 0.918

SOLUTION

a $0.30 < 0.35$

$0.3 < 0.35$

0.3 is less than 0.35

b $0.980 > 0.918$

$0.98 > 0.918$

0.98 is greater than 0.918

ISBN 9780170350969

EXERCISE 5–06

1 How many decimal places does 5.658 have? Select the correct answer A, B, C or D.

A 1 B 2 C 3 D 4

2 Which decimal is the largest number? Select A, B, C or D.

A 0.85 B 0.8 C 0.08 D 0.825

3 Is each statement true or false?

a 0.45 = 0.450
b 0.706 = 0.7006
c 0.540 = 0.54
d 1.2 = 1.22

4 Copy and complete each statement with a < or > sign.

a 0.6 ____ 0.65
b 0.86 ____ 0.8
c 0.925 ____ 0.95
d 0.87 ____ 0.8
e 0.75 ____ 0.755
f 1.3 ____ 1.35
g 2.14 ____ 2.1
h 0.964 ____ 0.96
i 3.2 ____ 3.25
j 0.75 ____ 0.777
k 0.8 ____ 0.88
l 3.21 ____ 3.2

5 Write each set of decimals in ascending order.

a 0.6, 0.64, 0.632, 0.06, 0.699
b 0.54, 0.05, 0.589, 0.55, 0.505
c 0.989, 0.9, 0.96, 0.99, 0.9999
d 2.1, 2.11, 2.011, 2.0111, 2.111

6 Write each set of decimals in descending order.

a 0.83, 0.899, 0.8, 0.82, 0.888
b 0.3, 0.305, 0.33, 0.003, 0.311
c 1.6, 1.666, 1.006, 1.65, 1.696
d 3.2, 3.002, 3.219, 3.209, 3.2292

7 Copy the number line below and plot the following decimals on the line:

0.5, 0.7, 0.2, 0.4, 0.9, 0.1, 1.2, 0.3, 1.4

8 A group of swimmers swam these times in seconds for one lap of the pool.

Ben 48.6	Bobbie 45.4	Gemma 52.15	Jiva 48.9
Mitch 43.8	Mala 52.65	Sam 49.68	Ebony 43.18

a Arrange their times in order from fastest to slowest.

b Who was the fastest swimmer?

5-07 Adding and subtracting decimals

TO ADD OR SUBTRACT DECIMALS:

- write the decimals underneath each other in columns
- make sure the decimal points line up underneath each other
- fill in the gaps with 0s
- add or subtract the digits in columns
- place the decimal point directly underneath
- check your answer by estimating.

EXAMPLE 12

Find the sums and differences below:

a 0.9 + 1.55 b 2.13 + 3.045 + 0.9

c 82.6 − 3.5 d 478.3 − 59.75

SOLUTION

a
$$\begin{array}{r} {}^{1}0.9\,0 \; + \\ 1.5\,5 \\ \hline 2.4\,5 \\ \hline \end{array}$$

← Gaps filled in with 0s

← Points under points

Check by estimating:

$0.9 + 1.55 \approx 1 + 1.6 = 2.6$

(2.45 is close to 2.6)

b
$$\begin{array}{r} {}^{1}2.1\,3\,0 \; + \\ 3.0\,4\,5 \\ 0.9\,0\,0 \\ \hline 6.0\,7\,5 \\ \hline \end{array}$$

$2.13 + 3.045 + 0.9 \approx 2 + 3 + 1 = 6$

(6.075 is close to 6)

c
$$\begin{array}{r} {}^{7}\not{8}\;{}^{1}2.6 \; - \\ 3.5 \\ \hline 7\;9.1 \\ \hline \end{array}$$

← Use trading to subtract

d
$$\begin{array}{r} 4\;{}^{6}\not{7}\;{}^{17}\not{8}.\;{}^{12}\not{3}\;{}^{1}0 \; - \\ 5\;\;9.\;7\;\;5 \\ \hline 4\;1\;\;8.\;5\;\;5 \\ \hline \end{array}$$

EXAMPLE 13

Amelia went to the supermarket with a $50 note and bought some chicken for $15.25 and some vegetables for $18.60. What change will she receive from her $50?

Shutterstock.com/K2 PhotoStudio

SOLUTION

Cost of goods:
$$\begin{array}{r} 1\,5.2\,5 \; + \\ 1\,8.6\,0 \\ \hline 3\,3.8\,5 \\ \hline \end{array}$$

Change:
$$\begin{array}{r} 5\,0.0\,0 \; - \\ 3\,3.8\,5 \\ \hline 1\,6.1\,5 \\ \hline \end{array}$$

Amelia's change was $16.15.

ISBN 9780170350969

EXERCISE 5-07

1. Evaluate 6.28 + 12.9. Select the correct answer A, B, C or D.

   A 19.18 B 18.18 C 7.57 D 75.7

2. Evaluate 128.4 – 39.25. Select A, B, C or D.

   A 8.915 B 124.475 C 26.41 D 89.15

3. Evaluate each sum.

   a 0.6 + 0.85 b 0.48 + 0.003 c 0.78 + 0.5
   d 2.15 + 0.082 e 1.8 + 0.965 f 0.86 + 1.369
   g 12.42 + 8.947 h 8.06 + 15.843 i 23.9 + 0.7542

4. Evaluate each difference.

   a 25.6 – 0.8 b 18.95 – 0.08 c 12.6 – 0.97
   d 156.4 – 23.68 e 98.4 – 45.83 f 57.9 – 28.65
   g 65.9 – 12.56 h 875.4 – 34.86 i 45.8 – 17.975

5. Evaluate each sum.

   a 23.6 + 568.45 + 1267.3 + 0.985 b 13.95 + 0.763 + 212.08 + 1.34
   c 23.789 + 2.56 + 0.999 d 24 + 0.923 + 1.26 + 45.68

6. Goran bought the following items from the supermarket:

   Packet of chips $2.25; Cake $12.50; 2 kg apples $4.88; Bottle of soft drink $1.95; Packet of serviettes $3.60

   a How much did his groceries cost him altogether?
   b How much change will he receive from a $50 note?

7. Paige cut the following amounts from 15 m of curtain fabric: 1.6 m, 2.45 m and 5.9 m.

   a How much material did she cut off altogether?
   b How much material did she have left?

Alamy/Sindre Ellingsen

8. Evaluate 6.01 + 60.001 + 600.1 – 60.11 – 6.001.

ISBN 9780170350969

FIND-A-WORD PUZZLE

Copy this page, then find all the words listed below in this grid of letters.

O	Y	I	J	M	E	C	A	M	P	D	J	A	N	J
U	P	T	N	R	B	D	G	L	P	E	N	O	E	C
Q	T	E	A	D	D	Z	A	N	M	N	I	E	B	Q
X	F	U	R	I	E	C	K	G	I	T	P	E	U	P
Y	Q	V	T	A	E	X	Q	C	A	R	D	I	C	O
S	G	I	Q	E	T	Z	A	C	S	H	E	B	Y	I
P	O	W	E	R	M	I	I	N	Y	Q	C	D	N	N
N	E	U	L	A	V	L	O	F	P	U	I	L	R	T
B	W	I	Z	S	P	L	T	N	W	N	M	A	W	O
D	R	S	U	I	A	H	J	O	S	N	A	M	D	H
S	U	B	T	R	A	C	T	I	O	N	L	I	J	O
Z	F	L	U	E	V	T	Y	R	W	R	S	C	L	W
B	U	N	O	I	S	I	V	I	D	Z	G	E	F	B
M	H	S	W	B	P	E	M	P	I	G	Q	D	A	N
R	V	I	D	G	E	L	T	C	U	H	G	H	D	W

ADDITION	CUBE	DECIMAL	DECIMALS
DIVISION	INDEX	MULTIPLICATION	OPERATIONS
ORDERING	PLACE	POINT	POWER
ROOT	SQUARE	SUBTRACTION	VALUE

 ISBN 9780170350969

PRACTICE TEST 5

Part A General topics

Calculators are not allowed.

1 Write 70% as a simple fraction.

2 Draw an isosceles triangle.

3 Copy and complete:
4120 mm = ________ cm.

4 Find the area of this rectangle.

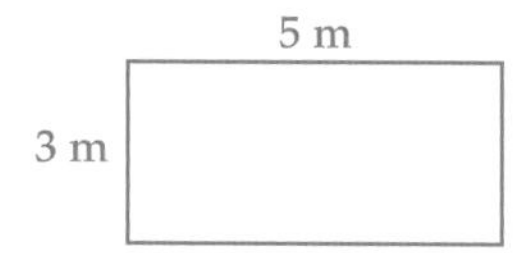

5 What is a quadrilateral?

6 List the first 5 multiples of 8.

7 Evaluate 654 ÷ 6.

8 Write 0415 in a.m./p.m. time.

9 Evaluate $2 \times 2 \times 2 \times 2 \times 2$.

10 Draw in any axes of symmetry for the shape below:

Part B Powers and decimals

Calculators are allowed.

5-01 Powers

11 Evaluate 3^4. Select the correct answer A, B, C or D.

A 12 B 64 C 81 D 3000

5-02 Square and square root

12 Evaluate 13^2. Select A, B, C or D.

A 144 B 169 C 196 D 121

13 Evaluate $\sqrt{225}$. Select A, B, C or D.

A 12 B 13 C 14 D 15

5-03 Cube and cube root

14 Evaluate each expression.

a 5^3 b 7^3 c $(-4)^3$

15 Evaluate each expression.

a $\sqrt[3]{27}$ b $\sqrt[3]{-216}$ c $\sqrt[3]{729}$

5-04 Order of operations

16 Evaluate each expression.

a $28 - 4 \times (6 + 1)$

b $120 - 6^2 + (29 - 13)$

5–05 Decimals

17 Write each decimal as a fraction.

a 0.07   b 0.054   c 0.0008

18 Write each fraction as a decimal.

a $\frac{9}{100}$ b $\frac{17}{1000}$ c $\frac{345}{1000}$

5–06 Ordering decimals

19 Write these decimals in descending order:

0.04, 0.435, 0.0042, 0.004, 0.045

5–07 Adding and subtracting decimals

20 Evaluate 2.34 + 23.045 + 123.08.

21 Evaluate 12.65 – 4.8.

ISBN 9780170350969

MULTIPLYING AND DIVIDING DECIMALS

6

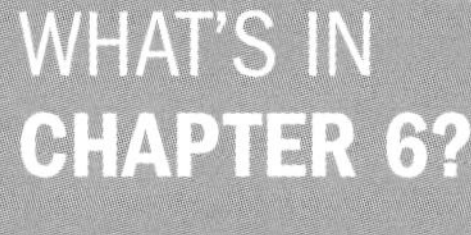

WHAT'S IN CHAPTER 6?

6-01 Multiplying and dividing decimals by powers of 10

6-02 Multiplying decimals by whole numbers

6-03 Multiplying decimals

6-04 Dividing decimals by whole numbers

6-05 Dividing decimals

6-06 Rounding decimals

6-07 Best buys

6-08 Decimal problems

IN THIS CHAPTER YOU WILL:

- multiply and divide decimals by 10, 100, 1000 and so on
- multiply decimals by whole numbers and other decimals
- divide decimals by whole numbers and other decimals
- round decimals to a number of decimal places
- compare the unit cost of items of different sizes and brands to find the 'best buy' (value-for-money)
- solve problems involving decimals

Shutterstock.com/Lex-art

6–01 Multiplying and dividing decimals by powers of 10

$4.3 \times 10 = 43$, so when multiplying a decimal by 10, the decimal point moves one place to the right.

To multiply a decimal by 10, 100 or 1000, move the decimal point to the **right** as the number is getting **larger**.
- To multiply by 10, move the decimal point 1 place to the right.
- To multiply by 100, move the decimal point 2 places to the right.
- To multiply by 1000, move the decimal point 3 places to the right.

EXAMPLE 1

Evaluate each product.

a 3.854×1000 b 0.9×100 c $2.6 \times 10\,000$

SOLUTION

a $3.854 \times 1000 = 3854$

decimal point moves 3 right

b $0.9 \times 100 = 0.90 \times 100 = 90$

adding zeros to the decimal to allow the decimal point to move, decimal point moves 2 right

c $2.6 \times 10\,000 = 2.6000 \times 10\,000 = 26\,000$

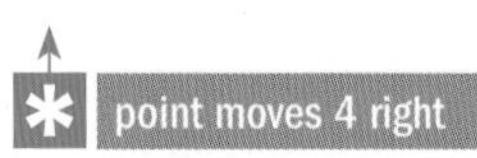

To divide a decimal by 10, 100 or 1000, move the decimal point to the **left** as the number is getting **smaller**.
- To divide by 10, move the decimal point 1 place to the left.
- To divide by 100, move the decimal point 2 places to the left.
- To divide by 1000, move the decimal point 3 places to the left.

EXAMPLE 2

Evaluate each quotient.

a $15.68 \div 10$ b $40.92 \div 1000$

SOLUTION

a $15.68 \div 10 = 1.568$

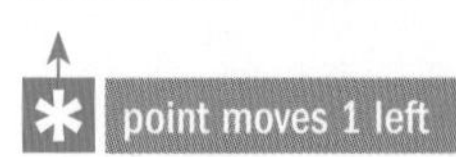

b $40.92 \div 1000 = 0040.92 \div 1000 = 0.040\,92$

adding zeros to the left of the decimal to allow the point to move, point moves 3 left

ISBN 9780170350969

EXERCISE 6-01

1 Evaluate 4.58×1000. Select the correct answer **A**, **B**, **C** or **D**.

A 458 B 45.8 C 45 800 D 4580

2 Evaluate $88.52 \div 1000$. Select **A**, **B**, **C** or **D**.

A 0.088 52 B 0.008 852 C 8.852 D 0.8852

3 Evaluate each product.

a 6.4×100	b 52.48×1000	c 0.96×10
d $23.8 \times 10\,000$	e 0.87×1000	f 5.678×100
g $87.5 \times 100\,000$	h $5.9 \times 1\,000\,000$	i $0.000\,71 \times 10\,000\,000$

4 Evaluate each quotient.

a $5.6 \div 100$	b $8.9 \div 1000$	c $78.95 \div 10\,000$
d $0.85 \div 10$	e $24\,976 \div 100$	f $233 \div 1000$
g $5680 \div 1000$	h $7.6 \div 10\,000$	i $846\,250 \div 1\,000\,000$

5 Copy and complete this sentence.

When dividing by 100 000 we move the decimal _____ 5 places to the _______, and when multiplying by _______ we move the _________ point 7 places to the ________.

6 Georgia was driving at 1.83 km/min on a freeway. She wanted to work out how many metres she had travelled in 1 minute. She was told to multiply 1.83 by 1000.

a What was her distance travelled in metres?

b If she continued driving at the same speed, what would be her distance travelled in metres after 100 minutes?

Shutterstock.com/zhangyang13576997233

7 Henry counted 100 boxes of toothpicks inside a carton.

a How many toothpicks were in the carton if each box held 1000 toothpicks?

b Henry was asked to place all the toothpicks into containers that held 10 toothpicks each. How many containers did Henry have to fill?

8 Evaluate $125\,000 \times 100\,000 \div 1000$.

6–02 Multiplying decimals by whole numbers

To multiply a decimal by a whole number:

- write the numbers underneath each other in columns
- multiply the digits in columns
- place the decimal point directly underneath
- check your answer by estimating.

EXAMPLE 3

Evaluate each product.

a 3.24×7 b 245.6×48

SOLUTION

a

$$\begin{array}{r} 3.24 \times \\ 7 \\ \hline 22.68 \\ \hline \end{array}$$

← 3.24 has 2 decimal places so put 2 decimal places in the answer.

$3.24 \times 7 = 22.68$

Check by estimating:

$3.24 \times 7 \approx 3 \times 7 = 21$ ← 22.68 is close to 21

b

$$\begin{array}{r} 245.6 \times \\ 48 \\ \hline 1964.8 \\ 9824.0 \\ \hline 11788.8 \\ \hline \end{array}$$

1964.8 ← 245.6×8

9824.0 ← place a zero, 245.6×4

11788.8 ← 245.6 has 1 decimal place so put 1 decimal place in the answer.

$245.6 \times 48 = 11\,788.8$

Check by estimating:

$245.6 \times 48 \approx 200 \times 50 = 10\,000$ ← 11 788.88 is close to 10 000

ISBN 9780170350969

EXERCISE 6-02

1 How many decimal places will be in the answer to 62.48×6? Select the correct answer A, B, C or D.

A 1 B 2 C 3 D 4

2 Evaluate 62.48×6. Select A, B, C or D.

A 3748.8 B 37 488 C 374.88 D 37.488

3 How many decimal places are in each decimal below?

a 4.25 b 3.8 c 12.46 d 0.06 e 7.128

4 Evaluate each product.

a 4.25×6 b 3.8×5 c 12.46×7
d 0.06×4 e 7.128×3 f 8.04×9
g 5.42×8 h 638.4×2 i 27.532×11

5 Without multiplying, choose which answer must be correct.

a $2.3 \times 6 =$	138	13.8	1.38
b $45.25 \times 3 =$	13 575	1357.5	135.75
c $0.8 \times 23 =$	184	18.4	1.84
d $12.31 \times 18 =$	22 158	2215.8	221.58
e $5.612 \times 8 =$	4489.6	448.96	44.896

6 Evaluate each product.

a 23.4×24 b 18.63×37 c 24.8×18
d 45.92×35 e 65.9×28 f 126.3×36
g 196.45×84 h 34.76×19 i 32.765×54

7 Roshan went to his local bookstore and bought 5 copies of his favourite book for his friends. The price of each book was \$52.95.

a What did the books cost him altogether?

b If he had a gift voucher for \$150, how much extra did he have to pay?

Getty Images/Hemant Mehta

8 Imogen went to her local supermarket and bought the following items:

2 packets of biscuits for \$2.56 per packet 4 kg bananas at \$2.90 per kg

3 L ice-cream at \$2.75 per litre.

What was her total bill?

ISBN 9780170350969

6-03 Multiplying decimals

To multiply a decimal by another decimal:

- multiply them as whole numbers without the decimal points
- count the total number of decimal places in the question
- write the answer using this number of decimal places
- check your answer by estimating if possible.

EXAMPLE 4

Evaluate each product.

a 0.6×0.02 b 0.04×0.07 c 4.75×6.4

SOLUTION

a $6 \times 2 = 12$

0.6 has 1 decimal place and 0.02 has 2 decimal places. Total = 3 decimal places.

Write 12 with 3 decimal places: 0.$\underline{012}$

$0.6 \times 0.02 = 0.012$

Check by estimating: $0.6 \times 0.02 \approx 0.5 \times 0.02 = 0.01$ (0.012 is close to 0.01)

b $4 \times 7 = 28$

0.04 has 2 decimal places and 0.07 also has 2 decimal places. Total = 4 decimal places.

Write 28 with 4 decimal places: 0.$\underline{0028}$

$0.04 \times 0.07 = 0.0028$

c
$$\begin{array}{r} 475 \quad \times \\ 64 \\ \hline 1900 \\ 28500 \\ \hline 30400 \\ \hline \end{array}$$

4.75 has 2 decimal places and 6.4 has 1 decimal place. Total = 3 decimal places.

Write 30400 with 3 decimal places: 30.$\underline{400}$ = 30.4

$4.75 \times 6.4 = 30.4$

Check by estimating: $4.75 \times 6.4 \approx 5 \times 6 = 30$ (30.4 is close to 30)

ISBN 9780170350969

EXERCISE 6-03

1 How many decimal places will there be in the answer to 3.78×0.4? Select the correct answer A, B, C or D.

A 1 B 2 C 3 D 4

2 Evaluate 3.78×0.4. Select A, B, C or D.

A 1.512 B 15.12 C 1512 D 151.2

3 Write the number of decimal places in each product.

a 0.4×0.2 b 0.03×0.5 c 0.006×0.8 d 0.9×0.007

4 Evaluate each product in Question 3.

5 Evaluate each product.

a 2.1×0.4 b 3.8×0.06 c 2.45×0.8

d 3.652×0.5 e 2.86×0.8 f 6.732×0.9

g 234.8×0.02 h 158.4×0.62 i 12.8×0.04

6 How many decimal places would you place in the answer for each multiplication?

a 4.8×1.6 b 12.6×3.4 c 2.87×4.2

d 86.3×5.7 e 18.65×0.28 f 127.9×1.5

g 28.94×4.8 h 16.8×0.64 i 158.7×2.9

7 Evaluate each product in Question 6.

8 Mai was sewing a dress that needed the following cuts of fabric: 3 pieces each 1.2 m long, 2 pieces each 0.6 m long, 4 trims each 0.2 m long.

a How much fabric did she need altogether?

b If the fabric cost \$12.95 per metre, what was the total cost of the fabric?

9 Jake was building a cubby house and needed the following amounts of timber: 24 palings each 2.6 m long, 12 pieces each 3.2 m long, 8 edges each 1.4 m long.

a How much timber did Jake need altogether, without the palings?

b Calculate the total cost of the timber if the palings cost \$2.50 each and the rest of the timber cost \$4.86 per metre.

c Find the total cost of building the cubby house if Max helped him for 8 hours and charged him \$22.50 per hour.

Getty Images/Andersen Ross

ISBN 9780170350969

6–04 Dividing decimals by whole numbers

To **divide a decimal by a whole number**, use short division, then write the answer with the decimal point in the same column as the original decimal.

EXAMPLE 5

Evaluate each quotient.

a $85.2 \div 6$

b $985.5 \div 9$

SOLUTION

a
$$\begin{array}{r} 1\,4.\,2 \\ 6\overline{)8^{2}5.^{1}2} \end{array}$$

$85.2 \div 6 = 14.2$

Check by estimating:

$85.2 \div 6 \approx 66 \div 6 = 11$

(14.2 is close to 11)

b
$$\begin{array}{r} 1\,0\,9.\,5 \\ 9\overline{)9\,8^{8}5.^{4}5} \end{array}$$

$985.5 \div 9 = 109.5$

$985.5 \div 9 \approx 990 \div 9 = 110$

(109.5 is close to 110)

EXAMPLE 6

Evaluate each quotient.

a $127.5 \div 4$

b $486.25 \div 8$

SOLUTION

a
$$\begin{array}{r} 3\,1.\,8\,7\,5 \\ 4\overline{)1^{1}2\,7.^{3}5^{3}0^{2}0} \end{array}$$

← 2 zeros added for the division to stop

If the division leaves a remainder, we can add 0s after the decimal and keep dividing until there is no remainder.

$127.5 \div 4 = 31.875$

Check by estimating: $127.5 \div 4 \approx 120 \div 4 = 30$ (31.875 is close to 30)

b
$$\begin{array}{r} 6\,0.\,7\,8\,1\,2\,5 \\ 8\overline{)4\,^{4}8\,6.^{6}2^{6}5^{1}0^{2}0^{4}0} \end{array}$$

← 3 zeros added for the division to stop

$486.25 \div 8 = 60.781\ 25$

Check by estimating: $486.25 \div 8 \approx 480 \div 8 = 60$ (60.781 25 is close to 60)

ISBN 9780170350969

EXERCISE 6–04

1 Evaluate 72.48 ÷ 8. Select the correct answer A, B, C or D.

A 9.6 B 8.06 C 9.06 D 906

2 Evaluate 678.5 ÷ 5. Select A, B, C or D.

A 135.7 B 145.7 C 13.57 D 14.57

3 Evaluate each quotient.

a 4.8 ÷ 2 b 3.6 ÷ 3 c 12.8 ÷ 4

d 18.6 ÷ 6 e 21.14 ÷ 7 f 48.55 ÷ 5

g 816.39 ÷ 9 h 624.06 ÷ 6 i 568.24 ÷ 8

4 Each division below has a mistake. Find the correct answer for each one.

a $\begin{array}{r} 9.9 \\ 4\overline{)38.6} \end{array}$ b $\begin{array}{r} 8.9 \\ 8\overline{)647.2} \end{array}$ c $\begin{array}{r} 57.1 \\ 5\overline{)287.5} \end{array}$

d $\begin{array}{r} 651.2 \\ 7\overline{)4557.14} \end{array}$ e $\begin{array}{r} 84.2 \\ 9\overline{)756.18} \end{array}$ f $\begin{array}{r} 81.1 \\ 6\overline{)483.6} \end{array}$

5 Evaluate each quotient. It may be necessary to add some 0s after the decimal point.

a 45.8 ÷ 4 b 789.4 ÷ 5 c 284.12 ÷ 8

d 678.3 ÷ 3 e 984.36 ÷ 8 f 531.9 ÷ 9

g 36.785 ÷ 5 h 42.864 ÷ 4 i 785.6 ÷ 5

6 Jai, Liam and Anita ran a business and made a profit of $12 642.78 this year. If they divide the profit equally, how much will each person receive?

Shutterstock.com/Milosz_M

7 Four friends won a lotto prize and share the prize of $1608.24 equally. How much will each person receive?

8 Divide $486.84 equally between 6 diners.

6–05 Dividing decimals

To **divide a decimal by another decimal**, move the points in both decimals the same number of places to the right so that you are dividing by a **whole number**.

For example, to calculate $2.68 \div 0.2$ we need to make 0.2 into the whole number 2. To do this we will need to multiply by 10.
To keep the question the same, we must multiply both numbers by 10.
$2.68 \div 0.2 = 26.8 \div 2$

EXAMPLE 7

Evaluate each quotient.

a $2.68 \div 0.2$ b $468.168 \div 0.04$

SOLUTION

a $2.68 \div 0.2 = 26.8 \div 2$ ⟵ move both points 1 right so 0.2 is the whole number 2

$$2\overline{)26.8} = 13.4$$

$2.68 \div 0.2 = 13.4$

Check by estimating: $26.8 \div 2 \approx 26 \div 2 = 13$ ⟵ 13.4 is close to 13

b $468.168 \div 0.04 = 46\,816.8 \div 4$ ⟵ move both points 2 right so 0.04 is the whole number 4

$$4\overline{)46\,{}^{2}81{}^{1}6.8} = 11\,704.2$$

$468.168 \div 0.04 = 11\,704.2$

Check by estimating: $46\,816.8 \div 4 \approx 44\,000 \div 4 = 11\,000$ ⟵ 11 704.2 is close to 11 000

ISBN 9780170350969

EXERCISE 6–05

1 How many places would you need to move the decimal point to the right to divide 78.655 by 0.05? Select the correct answer A, B, C or D.

A 1   B 2   C 3   D 4

2 Divide 78.655 by 0.05. Select A, B, C or D.

A 15 731   B 157.31   C 1573.1   D 15.731

3 Write down how many places would you need to move the decimal point to the right to evaluate each quotient.

a $48.6 \div 0.6$ b $568.25 \div 0.05$ c $488.24 \div 0.8$

d $128.564 \div 0.04$ e $756.84 \div 0.08$ f $684.189 \div 0.9$

4 Evaluate each quotient in Question 3.

5 Spot the mistake in each equation. Then correct the mistake and evaluate the quotient.

a $48.68 \div 0.4 = 486.8 \div 40$ b $375.584 \div 0.5 = 37\,558.4 \div 5$

c $34.653 \div 0.03 = 346.53 \div 3$ d $569.64 \div 0.4 = 56\,964 \div 4$

6 Evaluate each quotient.

a $12.118 \div 0.2$ b $368.43 \div 0.3$ c $624.186 \div 0.06$

d $819.27 \div 0.9$ e $492.114 \div 0.07$ f $562.432 \div 0.08$

g $1264.16 \div 0.4$ h $864.358 \div 0.002$ i $876.59 \div 0.5$

7 Josie had 689.36 m of copper pipe which had to be cut into pieces 0.4 m long.

a How many pieces could she make?

b Was there any copper pipe left over?

c How much did her company make from selling each copper piece for $28.50?

8 Julian was cutting up fabric for ballet costumes. He had 12.56 m of satin and needed 0.8 m for each costume.

a How many complete costumes could he make with this fabric?

b What decimal of a metre of fabric was left over?

c What did the costumes cost each if the fabric cost $156 to buy and Julian charged $18 per costume?

Corbis/Glow Images

6–06 Rounding decimals

WORDBANK

rounding To write a simpler number with approximately the same value using fewer digits.

EXAMPLE 8

Round each decimal correct to one decimal place.

a 3.24 b 3.27

SOLUTION

To one decimal place, both decimals are between 3.2 and 3.3.

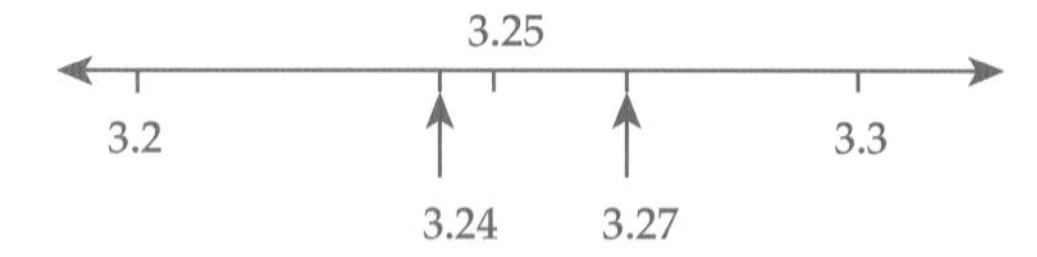

But 3.24 is closer to 3.2 and 3.27 is closer to 3.3. The half-way mark is 3.25.

So rounded to one decimal place,

a $3.24 \approx 3.2$ b $3.27 \approx 3.3$

To **round decimals** to a certain number of decimal places, look at the next digit.
If it 5 or more, round the decimal **up**. If it is less than 5, round the decimal **down**.

EXAMPLE 9

Write each decimal rounded to 2 decimal places.

a 5.274 b 23.64809 c 459.3751

SOLUTION

a Count 2 decimal places, then look at the next digit (4). It is less than 5, so leave the digit 7 as it is.

$\underline{5.27}4 \approx 5.27$

b Look at the 8. It is more than 5, so round the digit 4 up to 5.

$\underline{23.64}809 \approx 23.65$

c Look at the 5. It is 5 or more, so round the digit 7 up to 8.

$\underline{459.37}51 \approx 459.38$

ISBN 9780170350969

EXERCISE 6–06

1 What is 2.863 rounded to 2 decimal places? Select the correct answer A, B, C or D.

A 2.86 B 2.8 C 2.9 D 2.87

2 What is 35.4829 rounded to 1 decimal place? Select A, B, C or D.

A 35.48 B 35.4 C 35.5 D 35.49

3 Copy and complete:

To round a decimal, look at the digit to the left/right of where you want to round. If it is less than ___ round up/down. If it is 5 or more round up/down.

4 Round each decimal to 1 decimal place.

a 4.56 b 8.23 c 12.54 d 18.87

e 7.39 f 23.98 g 124.53 h 7.8246

5 Round each decimal to 2 decimal places.

a 6.284 b 8.357 c 12.652 d 124.568

e 5.873 f 16.659 g 23.495 h 283.2438

6 Round 56.783 928 correct to:

a 2 decimal places b 3 decimal places

c 4 decimal places d 5 decimal places

7 Dani's electricity bill arrived showing the charges listed below:

Peak rate 210.367 kWh @ $0.3670

Off peak 297.542 kWh @ $0.0942

Shoulder 546.372 kWh @ $0.1864

a Find the total amount payable for usage as accurately as possible.

b Round your answer to 2 decimal places. This is the amount she will pay.

Shutterstock.com/Twin Design

8 Round $5687.235829 to 3 decimal places.

6-07 Best buys

WORDBANK

best buy A brand or size of product that gives the best value for money.

unit cost The cost of one item or metric unit, for example, 1 kg.

Unit cost = cost ÷ number of items or units

EXAMPLE 10

Select the best buy among the following sizes of ice-cream.

A 2 L for $3.29 B 5 L for $5.80 C 500 mL for $1.25

SOLUTION

Compare the price of 1 L of ice-cream for each size:

✱ This is called the **unit cost**.

A 1 L costs $\$3.29 \div 2 = \$1.645$

B 1 L costs $\$5.80 \div 5 = \$1.16$

C 1 L costs $\$1.25 \times 2 = \$2.50$ ⟵ $1\text{ L} = 1000\text{ mL} = 2 \times 500\text{ mL}$

The best buy is B with 5 L for $5.80. ⟵ $1.16/L is the cheapest.

EXAMPLE 11

Find the best buy for the following brands of chocolates.

Smoothy: 2 kg box for $15.50

Milky: 1.2 kg for $10.80

Chocky: 875 g for $6.80

Shutterstock.com/Svetlana Foote

SOLUTION

Calculate the price for 1 kg of chocolate:

Smoothy: 1 kg costs $\$15.50 \div 2 = \$7.75$

Milky: 1 kg costs $\$10.80 \div 1.2 = \$9.00$

Chocky: 1 kg costs $\$6.80 \div 0.875 = \$7.771$ ⟵ $875\text{ g} = 0.875\text{ kg}$

The best buy is Smoothy. ⟵ $7.75/kg is the cheapest.

ISBN 9780170350969

EXERCISE 6–07

1 How much for 1 kg if 3.5 kg costs $12.40? Select the correct answer A, B, C or D.

A $3.54 B $4.34 C $43.40 D $2.83

2 How much for 1 kg if 400 g costs $1.62? Select A, B, C or D.

A $2.47 B $4.50 C $4.05 D $4.55

3 At the supermarket, Ashleigh found three sizes of strawberries.

Small 250 g for $2.99 Medium 375 g for $3.50 Large 500 g for $5.20

Which is the best buy?

4 Find the best buy for each set of items.

a 2 kg apples for $3.80, 5 kg for $8.20 or 6.5 kg for $10.50?

b 250 g leg ham for $5.50, 400 g for $8.40 or 1 kg for $17.50?

c 2 L of milk for $3.60, 1.5 L for $2.80 or 4 L for $7.10?

d 1.2 kg steak for $12.80, 800 g for $11.30 or 2.5 kg for $20.65?

e 8 bread rolls for $3.80, 6 bread rolls for $2.60 or 12 bread rolls for $6.20?

f 5 L ice-cream for $8.80, 3 L for $5.25 or 900 mL for $3.40?

Shutterstock.com/s74

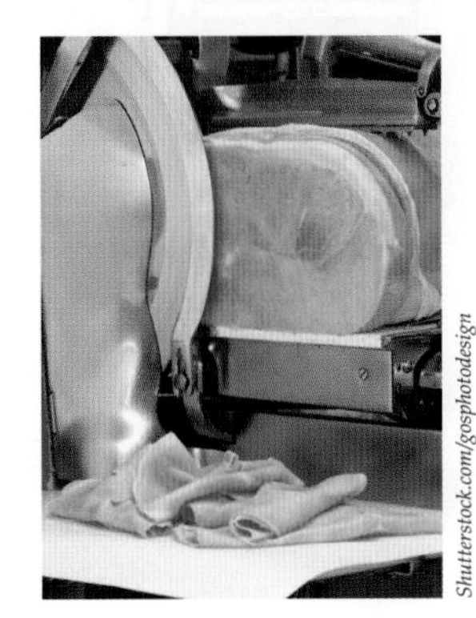
Shutterstock.com/gosphotodesign

Getty Images/Ryan McVay

5 Who found the best buy for the following?

a Tom bought 3 kg prawns for $48 while Bryony bought 2 kg for $31.

b Nouri bought 5 DVDs for $42.50 while Jayden bought 8 DVDs for $58.

c Karina bought 350 mL soft drink for $2.75 while Eli bought 600 mL for $6.10.

d Hugh bought a 420 g hamburger for $6.80 while Philippa bought a 250 g one for $3.15.

e Caitlin bought 1.5 kg oranges for $6.20 while Hashan bought 800 g for $3.80.

6 Victoria wanted to upgrade her mobile phone. She wanted the best buy from the following deals. Which one should she choose?

Modaphone: $29.95/month for 250 calls

Kelstra: $49.95/month for 600 calls

Toptus: $39.95/month for 480 calls

6-08 Decimal problems

- **When adding and subtracting decimals**, keep the decimal points underneath each other.
- **When multiplying decimals**, count the number of decimal places in the question and put the same number of decimal places in your answer.
- **When dividing decimals**, move the points in both decimals the same number of places to the right so that you are dividing by a **whole number**.

EXAMPLE 12

Tom paid the following expenses each week from his wage of \$825.68.

Rent: \$256 Food: \$125.50 Petrol: \$68.95

Clothes: \$135.60 Entertainment: \$165

a Find his total expenses for the week.

b How much does he have left for savings each week?

iStockphoto/segray

iStockphoto/vlynder

iStockphoto/franckreporter

SOLUTION

a Total expenses:

256.00 +
125.50
68.95
135.60
165.00
\$751.05

b Amount left:

825.68 −
751.05
\$74.63

EXAMPLE 13

What is the cost of buying 5 books @ \$35.95 per book?

SOLUTION

Cost = \$35.95 × 5 = \$179.75

ISBN 9780170350969

EXERCISE 6-08

1 Find the difference between 658.45 and 459.50. Select the correct answer A, B, C or D.

A 19.89 B 198.95 C 1117.95 D 189.95

2 Evaluate the product of $56.25 and 8. Select A, B, C or D.

A $45 B $64.25 C $7.03 D $450

3 Find the sum of $45.78, $356.25 and $3096.70.

4 Calculate the difference between 679.546 and 48.93.

5 Find the product of 47.8 and 5.6.

6 Calculate the quotient of 89.4 and 3.

7 a Work out the following supermarket bill.

3 kg beef @ $12.90/kg 2 bottles soft drink @ $1.75/bottle

6 bread rolls @ $0.60 each 3 L ice-cream @ $2.50/L

b How much change would I receive from a $100 note?

8 How many kilometres did Jay's family travel on their recent holiday if the odometer reading was 56 782.9 when they left home and it was 68 543.7 when they returned home later?

9 a Calculate the total cost when 8 shirts priced at $28.90 each are purchased.

b Calculate the change from $300 from this purchase.

10 Lauren was covering a sofa and needed the following cuts of fabric:

2 pieces 1.4 m long 4 pieces 0.3 m long 8 trims 0.2 m long

a How much fabric did she need altogether?

b If the fabric cost $23.95 per metre, what was the total cost of the fabric?

11 Alana was building a sand pit and needed the following amounts of timber:

18 palings 1.8 m long 4 pieces 3.2 m long 6 edges 1.2 m long

a How much timber did Alana need altogether, without the palings?

b Calculate the total cost of the timber if the palings cost $3.50 each and the rest of the timber cost $2.95 per metre.

Getty Images/Juanmonino

CODE PUZZLE

These pictures feature Babe, a talking pig from an Australian film series. What is the difference between the two pictures? Evaluate the expressions below to decode the answer, matching each question letter to an answer value.

A	$12.4 + 8.25$	B	$18.6 - 5.4$	C	3.8×0.6
D	$42.7 \div 7$	E	$35.2 + 6.06$	F	$49.6 - 8.4$
H	2.9×0.8	I	$648.4 \div 4$	M	$20.2 + 1.75$
N	$83.6 - 4.8$	O	8.4×1.2	R	$28.05 \div 0.5$
S	$12.7 - 3.9$	T	$15.4 \div 0.2$	Y	1.9×1.4

13.2 20.65 13.2 41.26 1 162.1 8.8   10.08 78.8   77 2.32 41.26

41.2 20.65 56.1 21.95   20.65 78.8 6.1   13.2 20.65 13.2 41.26 2

162.1 8.8   162.1 78.8   77 2.32 41.26   2.28 162.1 77 2.66

ISBN 9780170350969

PRACTICE TEST 6

Part A General topics

Calculators are not allowed.

1 Evaluate 25×2.

2 Write 6:35 p.m. in 24-hour time.

3 A grasshopper can jump up to 20 times its body length. How far can a grasshopper jump if it is 6.4 cm long?

4 Find the lowest common multiple of 8 and 12.

5 What metric unit would we use to measure water in a cup?

6 Draw a scalene triangle.

7 How many faces does a rectangular prism have?

8 If I am driving north and make a left-hand turn, which direction would I then be travelling?

9 A flight from Sydney to Hong Kong leaves Sydney at 11:10 a.m. and takes 8 hours 40 minutes. What time would I arrive in Hong Kong (Sydney time)?

10 How many weeks in one year?

Part B Multiplying and dividing decimals

Calculators are not allowed.

6-01 Multiplying and dividing decimals by powers of 10

11 How many places right would you move the decimal point if multiplying a decimal by 10 000? Select the correct answer A, B, C or D.

A 1 B 2 C 3 D 4

12 How many places left would you move the decimal point if dividing a decimal by 1 000 000? Select A, B, C or D.

A 3 B 4 C 5 D 6

6-02 Multiplying decimals by whole numbers

13 Evaluate 54.6×8. Select A, B, C or D.

A 436.8 B 43.68 C 4.368 D 4368

6-03 Multiplying decimals

14 Evaluate each product.

a 0.07×0.9 b 21.8×0.4 c 234.9×1.7

6-04 Dividing decimals by whole numbers

15 Evaluate $986.4 \div 3$. Select A, B, C or D.

A 326.6 B 328.8 C 32.88 D 322.1

16 Find the cost of 1 DVD if you bought 6 DVDs for $85.80.

6-05 Dividing decimals

17 Evaluate each quotient.

a 36.85 ÷ 0.5 b 28.665 ÷ 0.09

18 Josh had 84 m of copper pipe which had to be cut into pieces 0.3 m long.

a How many pieces could he make?

b Was there any copper pipe left over?

6-06 Rounding decimals

19 Round each decimal correct to 1 decimal place.

a 5.672 b 12.408 c 156.289

6-07 Best buys

20 Which is the best buy for bread rolls?

8 for $2.40

10 for $3.60

$3.20 per dozen

6 for $2.20

6-08 Decimal problems

21 Breanna went to the supermarket with $50 and bought 3 kg oranges for $2.95 per kg.

a What did it cost her altogether?

b What was Breanna's change from $50?

ISBN 9780170350969

FRACTIONS

7

WHAT'S IN CHAPTER 7?

7-01 Fractions
7-02 Fractions and decimals
7-03 Equivalent fractions
7-04 Simplifying fractions
7-05 Improper fractions and mixed numerals
7-06 Ordering fractions
7-07 Adding and subtracting fractions
7-08 Adding and subtracting mixed numerals

IN THIS CHAPTER YOU WILL:

- revise what a fraction is and the meaning of numerator and denominator
- convert fractions to decimals
- find equivalent fractions
- simplify fractions
- convert between improper fractions and mixed numerals
- order fractions, including on a number line
- add and subtract fractions, including mixed numerals

Shutterstock.com/Skumer

7-01 Fractions

WORDBANK

numerator The number at the top in a fraction.

denominator The number at the bottom in a fraction.

$\frac{2}{5}$ 2 ← numerator, 5 ← denominator

Fractions are made by dividing a quantity into **equal parts**. This shape has been divided into 3 equal parts.

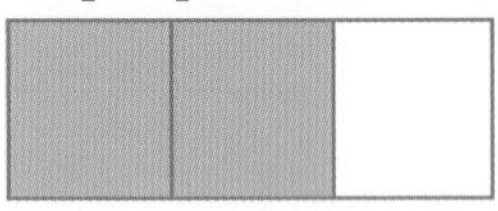

$\frac{2}{3}$ (two-thirds) of the shape is shaded.

This means 2 parts out of 3 are shaded.

EXAMPLE 1

a What fraction of this shape is shaded?

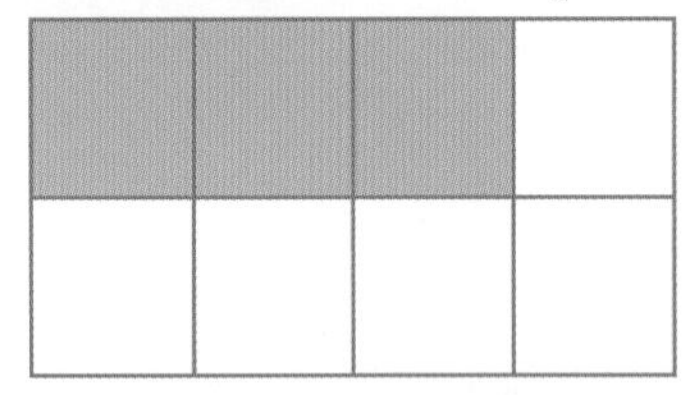

b In the answer to part a, which number is the denominator?

c What fraction of the shape is unshaded?

SOLUTION

a The shaded area is 3 parts out of 8 equal parts. So $\frac{3}{8}$ is shaded.

b The denominator is 8. ← the number at the bottom

c $\frac{5}{8}$ is unshaded. ← 5 parts out of 8 are not shaded.

ISBN 9780170350969

EXERCISE 7-01

1 What is the top number of a fraction called? Select the correct answer A, B, C or D.

A proper B numerator C equal D denominator

2 What fraction of this shape is shaded? Select A, B, C or D.

A $\frac{3}{8}$ B $\frac{2}{8}$ C $\frac{5}{9}$ D $\frac{5}{8}$

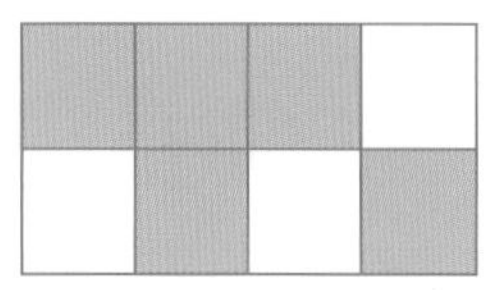

3 What fraction of each shape is shaded?

a

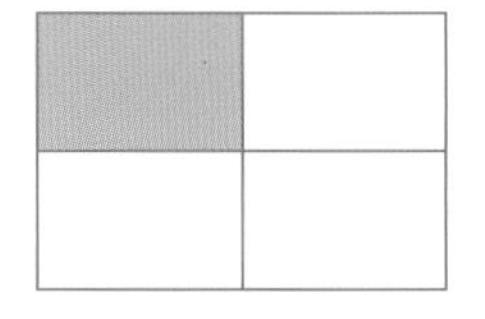

b

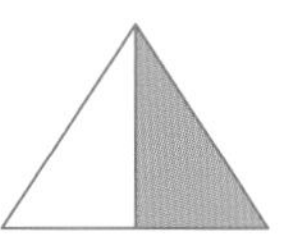

c

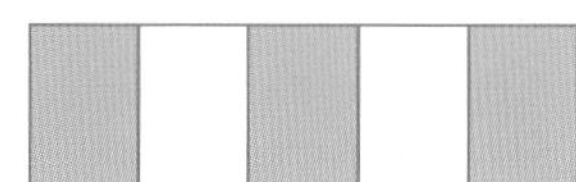

d

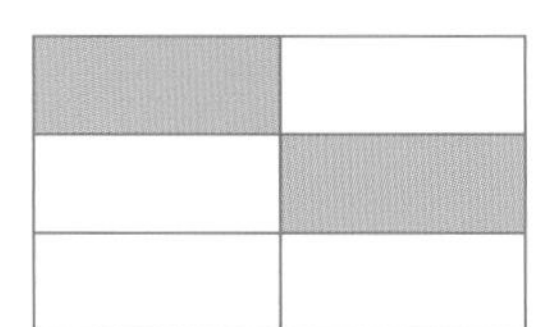

e

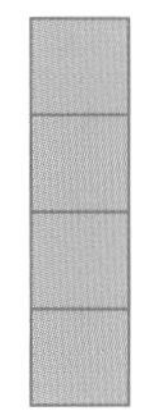

f

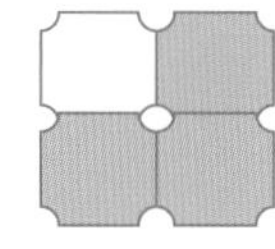

g

h

4 What fraction of each shape in Question 3 is unshaded?

5 Draw a rectangle and divide it into fifths.

6 Approximately what fraction of this glass is filled?

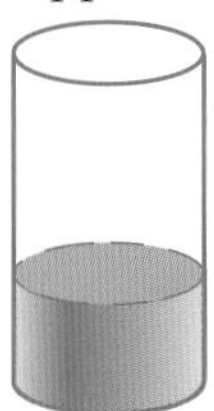

7 Draw a circle and divide it into eighths.

8 What fraction of the month shown on this calendar has been crossed out?

JULY 2015						
Sun	Mon	Tue	Wed	Thu	Fri	Sat
			~~1~~	~~2~~	~~3~~	~~4~~
~~5~~	~~6~~	~~7~~	~~8~~	~~9~~	~~10~~	~~11~~
~~12~~	13	14	15	16	17	18
19	20	21	22	23	24	25
26	27	28	29	30	31	

7–02 Fractions and decimals

To change a fraction to a decimal, use short division to divide the numerator by the denominator.

EXAMPLE 2

Convert to decimals:

a $\frac{3}{4}$ b $\frac{2}{5}$ c $\frac{3}{8}$ d $\frac{5}{4}$

SOLUTION

a $\frac{3}{4} = 3 \div 4$

$4\overline{)3.00}$ = 0.75 ← we need to add 2 zeros

$\frac{3}{4} = 0.75$

b $\frac{2}{5} = 2 \div 5$

$5\overline{)2.0}$ = 0.4

$\frac{2}{5} = 0.4$

c $\frac{3}{8} = 3 \div 8$

$8\overline{)3.000}$ = 0.375

$\frac{3}{8} = 0.375$

d $\frac{5}{4} = 5 \div 4$

$4\overline{)5.00}$ = 1.25

$\frac{5}{4} = 1.25$

This answer is larger than 1 because the fraction is larger than 1.

iStockphoto/DougSchneiderPhoto

ISBN 9780170350969

EXERCISE 7-02

1 Convert $\frac{1}{4}$ to a decimal. Select the correct answer A, B, C or D.

A 2.5 B 0.25 C 1.4 D 0.14

2 Convert $\frac{3}{5}$ to a decimal. Select A, B, C or D.

A 3.5 B 0.35 C 0.6 D 1.66

3 Is each statement true or false?

a $\frac{3}{5} = 3 \div 5$ b $\frac{4}{5} = 4 \div 5$ c $\frac{5}{8} = 8 \div 5$

d $\frac{1}{4} = 4 \div 1$ e $\frac{1}{2} = 1 \div 2$ f $\frac{3}{4} = 4 \div 3$

4 Convert each fraction to a decimal.

a $\frac{1}{2}$ b $\frac{1}{4}$ c $\frac{5}{8}$ d $\frac{3}{5}$

e $\frac{7}{5}$ f $\frac{7}{8}$ g $\frac{4}{5}$ h $\frac{3}{10}$

i $\frac{1}{8}$ j $\frac{3}{2}$ k $\frac{3}{16}$ l $\frac{9}{20}$

m $\frac{7}{4}$ n $\frac{13}{20}$ o $\frac{8}{5}$ p $\frac{11}{8}$

5 Convert the decimals below to fractions by using place value.

Remember: 1 number after the point is tenths, $0.1 = \frac{1}{10}$

2 numbers after the point is hundredths, $0.01 = \frac{1}{100}$, and so on.

a 0.3 b 0.25 c 0.09 d 0.6

e 0.28 f 0.95 g 0.15 h 0.7

7-03 Equivalent fractions

WORDBANK

Equivalent fractions Fractions that are the same size. They have the same value.

For example, $\frac{1}{3}, \frac{2}{6}$ and $\frac{3}{9}$ are all equivalent fractions.

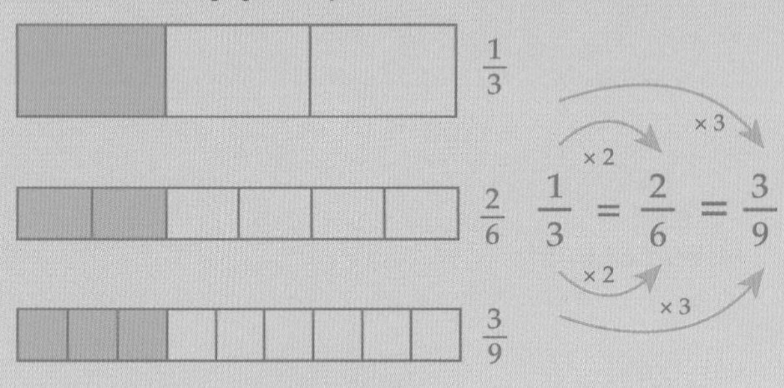

$$\frac{1}{3} = \frac{2}{6} = \frac{3}{9}$$

(numerator: ×2, ×3; denominator: ×2, ×3)

To find an equivalent fraction:

- multiply the numerator and denominator by the same number, or
- divide the numerator and the denominator by the same number.

EXAMPLE 3

Complete each pair of equivalent fractions.

a $\frac{2}{3} = \frac{?}{12}$ b $\frac{4}{5} = \frac{24}{?}$ c $\frac{15}{20} = \frac{?}{4}$

SOLUTION

a To find the missing numerator, look at the two denominators, 3 and 12.

3 is multiplied by 4 to give 12, so do the same thing to the numerator 2.

$\frac{2}{3} = \frac{2 \times 4}{3 \times 4} = \frac{8}{12}$

b To find the missing denominator, look at the two numerators 4 and 24.

4 is multiplied by 6 to give 24, so do the same thing to the denominator 5.

$\frac{4}{5} = \frac{4 \times 6}{5 \times 6} = \frac{24}{30}$

c To find the missing numerator, look at the two denominators 20 and 4.

20 is *divided* by 5 to give 4, so do the same thing to the numerator 15.

$\frac{15}{20} = \frac{15 \div 5}{20 \div 5} = \frac{3}{4}$

ISBN 9780170350969

EXERCISE 7-03

1 Which fraction is equivalent to $\frac{2}{3}$? Select the correct answer A, B, C or D.

A $\frac{4}{9}$ B $\frac{5}{9}$ C $\frac{9}{12}$ D $\frac{6}{9}$

2 Which fraction is equivalent to $\frac{3}{4}$? Select A, B, C or D.

A $\frac{4}{8}$ B $\frac{5}{8}$ C $\frac{9}{12}$ D $\frac{6}{9}$

3 Copy and complete these sentences.

To form an equivalent fraction we multiply both the _________ and __________ by the same number. Or we can divide both the _________ and __________ by the same number.

4 Copy and complete each pair of equivalent fractions.

a $\frac{1}{4} = \frac{3}{?}$ b $\frac{5}{6} = \frac{?}{30}$ c $\frac{2}{3} = \frac{?}{12}$

d $\frac{7}{10} = \frac{?}{80}$ e $\frac{7}{8} = \frac{63}{?}$ f $\frac{3}{4} = \frac{?}{100}$

g $\frac{20}{24} = \frac{5}{?}$ h $\frac{15}{40} = \frac{?}{8}$ i $\frac{18}{32} = \frac{?}{16}$

5 Jared had a bag of chocolates and gave $\frac{1}{4}$ of them to his friend Prem. Jared then ate $\frac{2}{8}$ of the original number of chocolates and gave $\frac{4}{12}$ of the original number to Grace.

a Did they each receive an equivalent number of the chocolates?

b If there were 96 chocolates in the bag, how many did each person receive?

6 Is each statement true or false?

a $\frac{3}{5} = \frac{6}{15}$ b $\frac{5}{8} = \frac{15}{24}$ c $\frac{4}{9} = \frac{20}{45}$

d $\frac{12}{20} = \frac{2}{5}$ e $\frac{18}{30} = \frac{6}{10}$ f $\frac{15}{25} = \frac{3}{5}$

7 Bridie had a pile of 120 beads. She used $\frac{1}{6}$ of them for a wristband and gave Sammy $\frac{3}{18}$ for her wristband.

a Did both girls have the same number of beads?

b How many beads were left after making both wristbands?

Alamy/MBI

7–04 Simplifying fractions

WORDBANK

simplify a fraction To make the numerator and the denominator of a fraction as small as possible dividing by the same factor.

To simplify a fraction, divide the numerator and denominator by the same number, preferably a large number such as their highest common factor (HCF), until the fraction is in lowest form.

EXAMPLE 4

Simplify each fraction.

a $\frac{24}{60}$ b $\frac{25}{75}$

SOLUTION

a $\frac{24 \div 3}{60 \div 3} = \frac{8}{20} = \frac{8 \div 4}{20 \div 4} = \frac{2}{5}$

or $\frac{24 \div 12}{60 \div 12} = \frac{2}{5}$ ← Dividing numerator and denominator by 12, the HCF of 24 and 60

b $\frac{25 \div 5}{75 \div 5} = \frac{5}{15} = \frac{5 \div 5}{15 \div 5} = \frac{1}{3}$

To write an amount as a fraction of a whole, write it as

$$\text{Fraction} = \frac{\text{amount}}{\text{whole amount}}$$

Units must be the same.

EXAMPLE 5

Express:

a 45 minutes as a fraction of 1 hour

b 9 hours as a fraction of 1 day

SOLUTION

a $\text{Fraction} = \frac{45\,\text{min}}{1\,\text{h}}$

$= \frac{45\,\text{min}}{60\,\text{min}}$

$= \frac{3}{4}$

So 45 minutes is $\frac{3}{4}$ of 1 hour.

b $\text{Fraction} = \frac{9\,\text{h}}{1\,\text{day}}$

$= \frac{9\,\text{h}}{24\,\text{h}}$

$= \frac{3}{8}$

So 9 hours is $\frac{3}{8}$ of 1 day.

ISBN 9780170350969

EXERCISE 7-04

1 Simplify $\frac{12}{20}$. Select the correct answer A, B, C or D.

A $\frac{4}{9}$ B $\frac{3}{5}$ C $\frac{9}{12}$ D $\frac{6}{5}$

2 Simplify $\frac{15}{35}$. Select A, B, C or D.

A $\frac{5}{7}$ B $\frac{3}{5}$ C $\frac{9}{12}$ D $\frac{3}{7}$

3 Copy and complete:

To simplify a fraction, ________ both the numerator and __________ by the same number until the fraction is in its ___________ form.

4 Simplify each fraction.

a $\frac{14}{18}$ b $\frac{21}{24}$ c $\frac{25}{60}$ d $\frac{36}{48}$

e $\frac{50}{75}$ f $\frac{32}{56}$ g $\frac{56}{70}$ h $\frac{45}{54}$

5 Is each statement true or false?

a $\frac{6}{15}=\frac{2}{5}$ b $\frac{18}{24}=\frac{9}{8}$ c $\frac{20}{45}=\frac{4}{9}$

d $\frac{16}{20}=\frac{4}{5}$ e $\frac{21}{30}=\frac{3}{10}$ f $\frac{20}{25}=\frac{4}{5}$

6 In a road rules test, Tyson answered 20 questions correctly out of 25. What fraction of the questions did he answer correctly?

Shutterstock.com/hartphotography

7 Convert each test mark to a simplified fraction.

a 74 out of 100

b 28 out of 40

c 50 out of 60

8 Express as a simplified fraction:

a 20 minutes of 1 hour

b 45 cm of 1 m

c 10 hours of 1 day

d 15 seconds of 1 minute

e 8 mm of 1 cm

f 4 weeks of 1 year

7-05 Improper fractions and mixed numerals

WORDBANK

proper fraction A fraction such as $\frac{4}{10}$ where the numerator is smaller than the denominator.

improper fraction A fraction such as $\frac{7}{3}$ where the numerator is larger than or equal to the denominator.

mixed numeral A number such as $3\frac{2}{5}$, made up of a whole number and a fraction.

If a fraction's numerator is larger than its denominator, then the value of the fraction is greater than 1. For example, the improper fraction $\frac{3}{2}$ is represented by the diagram below:

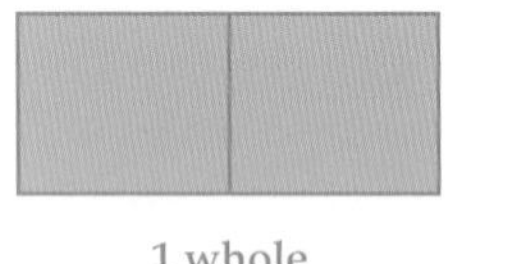

1 whole +

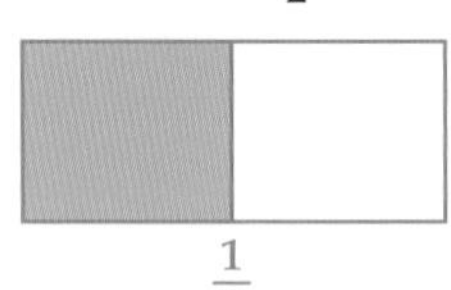

$\frac{1}{2}$

Three parts are shaded and there are two parts in each whole, so $\frac{3}{2} = 1\frac{1}{2}$, a mixed numeral.

To convert an improper fraction to a mixed numeral, divide the numerator by the denominator and write the remainder as a proper fraction.

EXAMPLE 6

Convert each improper fraction to a mixed numeral.

a $\frac{11}{8}$ b $\frac{7}{3}$

SOLUTION

a $\frac{11}{8} = 11 \div 8$

$= 1$ remainder 3

$= 1\frac{3}{8}$

b $\frac{7}{3} = 7 \div 3$

$= 2$ remainder 1

$= 2\frac{1}{3}$

✱ Write the remainder in the numerator of the fraction.

ISBN 9780170350969

7-05 Improper fractions and mixed numerals

EXAMPLE 7

Convert each mixed numeral to an improper fraction.

a $2\frac{1}{2}$ b $5\frac{3}{7}$

SOLUTION

a
$$2\frac{1}{2} = 2 + \frac{1}{2}$$
$$= \frac{2\times 2}{2} + \frac{1}{2}$$
$$= \frac{4}{2} + \frac{1}{2}$$
$$= \frac{5}{2}$$

b
$$5\frac{3}{7} = 5 + \frac{3}{7}$$
$$= \frac{5\times 7}{7} + \frac{3}{7}$$
$$= \frac{35}{7} + \frac{3}{7}$$
$$= \frac{38}{7}$$

A quick shortcut is to work clockwise from the denominator, multiply it by the whole number and add the numerator.

$$5\frac{3}{7} = \frac{7\times 5 + 3}{7} = \frac{38}{7}$$

To convert a mixed numeral to an improper fraction, multiply the denominator by the whole number, then add the numerator. Write the total as the new numerator of the fraction.

Shutterstock.com/ffolas

Shutterstock.com/ffolas

EXERCISE 7-05

1 Convert $\frac{8}{5}$ to a mixed numeral. Select the correct answer A, B, C or D.

A $1\frac{4}{5}$ B $1\frac{3}{8}$ C $1\frac{3}{5}$ D $1\frac{5}{8}$

2 Convert $2\frac{2}{3}$ to an improper fraction. Select A, B, C or D.

A $\frac{8}{3}$ B $\frac{5}{3}$ C $\frac{3}{8}$ D $\frac{7}{3}$

3 Classify each fraction as being proper (**P**), improper (**I**) or a mixed numeral (**M**).

a $1\frac{1}{2}$ b $\frac{3}{100}$ c $3\frac{2}{5}$ d $20\frac{1}{4}$

e $\frac{2}{3}$ f $\frac{100}{45}$ g $\frac{150}{250}$ h $\frac{86}{87}$

i $\frac{3}{2}$ j $\frac{3}{5}$ k $4\frac{1}{20}$ l $\frac{1000}{25}$

4 Describe in words:

a an improper fraction b a mixed numeral

5 What improper fraction is shaded in each diagram?

a 

b   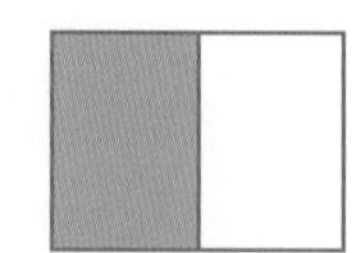

c 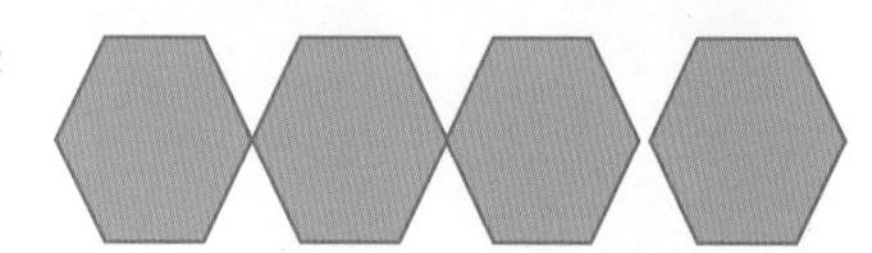 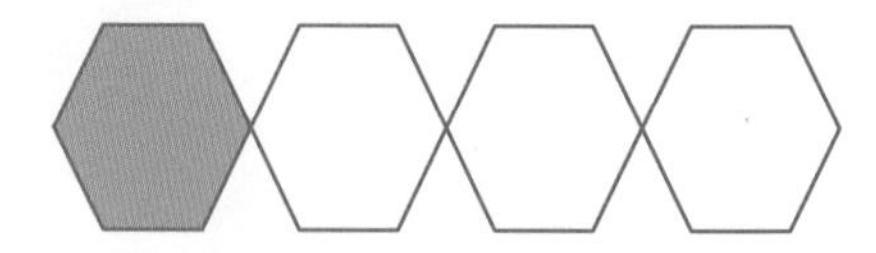

ISBN 9780170350969

EXERCISE 7–05

6 What mixed numeral is shaded in each diagram in Question 5?

7 Convert each improper fraction to a mixed numeral.

a $\frac{6}{5}$ b $\frac{8}{3}$ c $\frac{5}{4}$ d $\frac{9}{4}$

e $\frac{5}{3}$ f $\frac{6}{5}$ g $\frac{10}{9}$ h $\frac{15}{4}$

8 Convert each mixed numeral to an improper fraction.

a $3\frac{1}{3}$ b $2\frac{3}{4}$ c $4\frac{3}{4}$ d $1\frac{7}{8}$

e $6\frac{2}{3}$ f $4\frac{5}{6}$ g $5\frac{2}{5}$ h $7\frac{3}{8}$

Michelle Sole

7-06 Ordering fractions

- **To order fractions**, write them with the same **denominator**, then compare their **numerators**.
- A common denominator can be found by multiplying the denominators of both fractions together or by using the lowest common multiple (LCM) of the denominators.

EXAMPLE 8

Which fraction is larger: $\frac{3}{4}$ or $\frac{5}{6}$?

SOLUTION

Change both fractions to equivalent fractions with the same denominator.

Method 1

Common denominator $= 4 \times 6 = 24$.

$\frac{3}{4} = \frac{3 \times 6}{4 \times 6} = \frac{18}{24}$ $\qquad \frac{5}{6} = \frac{5 \times 4}{6 \times 4} = \frac{20}{24}$

As $20 > 18$, $\frac{20}{24}$ is larger so $\frac{5}{6}$ is larger.

Method 2

The lowest common multiple (LCM) of 4 and 6 is 12.

$\frac{3}{4} = \frac{3 \times 3}{4 \times 3} = \frac{9}{12}$ $\qquad \frac{5}{6} = \frac{5 \times 2}{6 \times 2} = \frac{10}{12}$

As $10 > 9$, $\frac{10}{12}$ is larger so $\frac{5}{6}$ is larger.

EXAMPLE 9

Plot the fractions $\frac{2}{3}$, $-\frac{1}{2}$ and $\frac{1}{6}$ on a number line.

SOLUTION

The lowest common multiple (LCM) of 2, 3 and 6 is 6.

$\frac{2}{3} = \frac{2 \times 2}{3 \times 2} = \frac{4}{6}$ $\qquad -\frac{1}{2} = -\frac{1 \times 3}{2 \times 3} = -\frac{3}{6}$ $\qquad \frac{1}{6} = \frac{1}{6}$

Divide a number line into intervals of $\frac{1}{6}$.

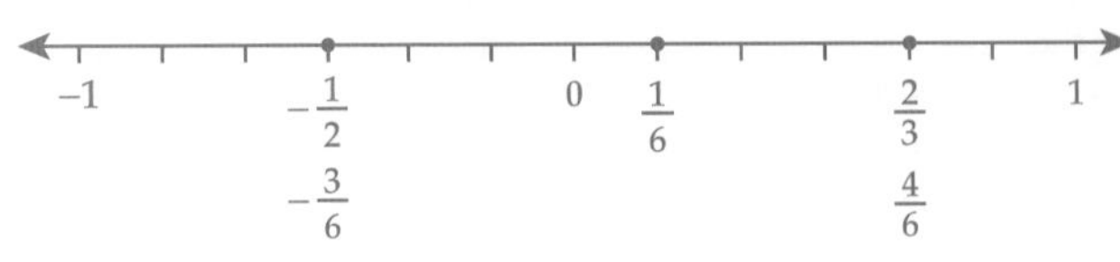

 ISBN 9780170350969

EXERCISE 7–06

1 Which fraction is largest: $\frac{1}{2}, \frac{3}{4}, \frac{2}{3}$ or $\frac{3}{5}$? Select the correct answer A, B, C or D.

A $\frac{1}{2}$ B $\frac{3}{4}$ C $\frac{2}{3}$ D $\frac{3}{5}$

2 Which fraction is smallest: $\frac{1}{2}, \frac{3}{4}, \frac{2}{3}$ or $\frac{3}{5}$? Select A, B, C or D.

A $\frac{1}{2}$ B $\frac{3}{4}$ C $\frac{2}{3}$ D $\frac{3}{5}$

3 Find the larger fraction in each pair.

a $\frac{7}{10}, \frac{3}{10}$ b $\frac{5}{3}, \frac{2}{3}$ c $\frac{3}{4}, \frac{7}{8}$

d $\frac{2}{3}, \frac{3}{10}$ e $2\frac{1}{5}, 2\frac{1}{6}$ f $3\frac{4}{7}, 2\frac{5}{6}$

4 Find the smaller fraction in each pair.

a $\frac{4}{5}, \frac{3}{5}$ b $\frac{11}{8}, \frac{9}{8}$ c $\frac{1}{4}, \frac{3}{10}$

d $\frac{1}{6}, \frac{2}{3}$ e $4\frac{3}{4}, 4\frac{2}{3}$ f $1\frac{2}{7}, 1\frac{1}{5}$

5 Which set of fractions shows $\frac{19}{16}, \frac{7}{6}, \frac{5}{4}, \frac{33}{32}$ in ascending order? Select A, B, C or D.

A $\frac{7}{6}, \frac{5}{4}, \frac{19}{16}, \frac{33}{32}$ B $\frac{5}{4}, \frac{7}{6}, \frac{19}{16}, \frac{33}{32}$

C $\frac{33}{32}, \frac{7}{6}, \frac{19}{16}, \frac{5}{4}$ D $\frac{19}{16}, \frac{33}{32}, \frac{7}{6}, \frac{5}{4}$

6 Write these fractions in ascending order: $\frac{2}{3}, \frac{11}{12}, \frac{1}{4}, \frac{5}{6}$.

7 Write these fractions in descending order: $\frac{3}{4}, \frac{2}{3}, \frac{1}{12}, \frac{5}{6}, \frac{1}{4}$.

8 For each number line, name in simplest form each fraction marked by a dot.

a
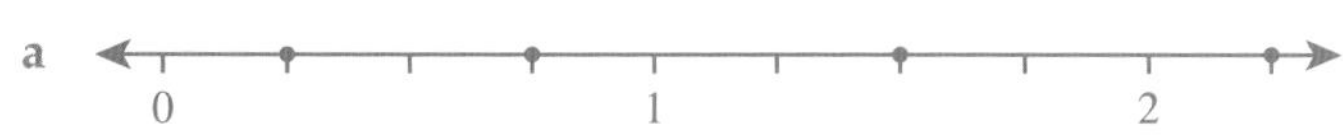

b
0 1 2

c
0 1

9 Plot each set of fractions on a separate number line.

a $\frac{2}{3}, \frac{3}{3}, 1\frac{1}{3}, \frac{7}{3}$ b $\frac{1}{5}, 1\frac{3}{5}, \frac{5}{5}, 2\frac{2}{5}, \frac{3}{5}, \frac{9}{5}$

c $\frac{5}{6}, -\frac{1}{3}, \frac{1}{2}, -1\frac{1}{6}$ d $-\frac{7}{10}, \frac{2}{5}, -\frac{1}{2}, -\frac{1}{4}, \frac{9}{10}$

7-07 Adding and subtracting fractions

To add and subtract fractions:

- **with the same denominator**, simply add or subtract the numerators.
- **with different denominators**, first convert them to equivalent fractions with the same denominator, then add or subtract numerators.

 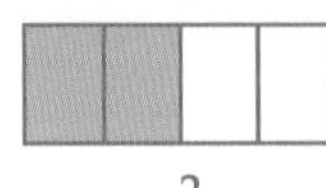 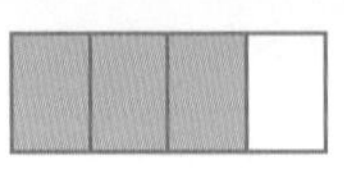

$$\frac{1}{4} \quad + \quad \frac{2}{4} \quad = \quad \frac{3}{4}$$

EXAMPLE 10

Evaluate each expression by first writing both fractions with the same denominator.

a $\frac{1}{8}+\frac{3}{4}$ b $\frac{2}{3}+\frac{3}{4}$ c $\frac{4}{5}-\frac{3}{4}$

SOLUTION

a $\frac{1}{8}+\frac{3}{4}=\frac{1}{8}+\frac{3\times 2}{4\times 2}$ ← Write with a common denominator of 8

$=\frac{1}{8}+\frac{6}{8}$

$=\frac{7}{8}$

b $\frac{2}{3}+\frac{3}{4}=\frac{2\times 4}{3\times 4}+\frac{3\times 3}{4\times 3}$ ← Write with a common denominator of 12

$=\frac{8}{12}+\frac{9}{12}$

$=\frac{17}{12}$

$=1\frac{5}{12}$ ← Change improper fraction to mixed numeral

c $\frac{4}{5}-\frac{3}{4}=\frac{4\times 4}{5\times 4}-\frac{3\times 5}{4\times 5}$ ← Write with a common denominator of 20

$=\frac{16}{20}-\frac{15}{20}$

$=\frac{1}{20}$

ISBN 9780170350969

EXERCISE 7–07

1 Evaluate $\frac{1}{4}+\frac{2}{3}$. Select the correct answer A, B, C or D.

A $\frac{3}{7}$ B $\frac{7}{12}$ C $\frac{11}{12}$ D $\frac{3}{12}$

2 Evaluate $\frac{3}{4}-\frac{2}{5}$. Select A, B, C or D.

A -1 B $\frac{7}{20}$ C $\frac{1}{20}$ D $\frac{5}{20}$

3 Evaluate each expression.

a $\frac{3}{6}+\frac{1}{6}$ b $\frac{7}{10}+\frac{2}{10}$ c $\frac{1}{2}+\frac{1}{2}$ d $\frac{2}{3}+\frac{2}{3}$

e $\frac{3}{5}-\frac{1}{5}$ f $\frac{5}{4}-\frac{3}{4}$ g $\frac{9}{10}-\frac{5}{10}$ h $\frac{7}{12}-\frac{4}{12}$

4 Evaluate each sum by writing both fractions with the same denominator first.

a $\frac{1}{5}+\frac{1}{2}$ b $\frac{1}{3}+\frac{2}{5}$ c $\frac{1}{4}+\frac{2}{5}$ d $\frac{3}{5}+\frac{1}{3}$

5 Evaluate each expression.

a $\frac{3}{4}+\frac{3}{5}$ b $\frac{5}{6}+\frac{3}{4}$ c $\frac{5}{8}+\frac{2}{3}$ d $\frac{8}{9}+\frac{3}{5}$

e $\frac{5}{9}-\frac{1}{3}$ f $\frac{11}{12}-\frac{2}{3}$ g $\frac{14}{15}-\frac{2}{5}$ h $\frac{9}{8}-\frac{3}{4}$

i $\frac{3}{8}+\frac{2}{5}$ j $\frac{7}{9}-\frac{1}{4}$ k $\frac{7}{12}+\frac{3}{5}$ l $\frac{6}{7}-\frac{3}{4}$

6 Quan bought a pizza and ate $\frac{1}{3}$ of it. He gave his friend Reece $\frac{1}{4}$ of it and his flatmate Emad then ate $\frac{1}{6}$ of it. If there were 12 pieces of pizza to start with, find:

a how many pieces each person ate

b how many pieces of the pizza were left

c what fraction of the pizza was left.

Corbis/Stewart Cohen

ISBN 9780170350969

7-08 Adding and subtracting mixed numerals

To add and subtract mixed numerals:

- add or subtract the whole numbers first
- then add or subtract the fractions.

EXAMPLE 11

Evaluate each expression.

a $2\frac{1}{4}+1\frac{2}{4}$ b $3\frac{1}{8}+2\frac{2}{8}$ c $4\frac{3}{8}-2\frac{1}{8}$ d $5\frac{3}{8}-4\frac{5}{8}$

SOLUTION

a $2\frac{1}{4}+1\frac{2}{4}=2+1+\frac{1}{4}+\frac{2}{4}$
$=3+\frac{3}{4}$
$=3\frac{3}{4}$

b $3\frac{1}{8}+2\frac{2}{8}=3+2+\frac{1}{8}+\frac{2}{8}$
$=5+\frac{3}{8}$
$=5\frac{3}{8}$

c $4\frac{3}{8}-2\frac{1}{8}=4-2+\frac{3}{8}-\frac{1}{8}$
$=2+\frac{2}{8}$
$=2\frac{2}{8}$
$=2\frac{1}{4}$ ← Simplifying

d $5\frac{3}{8}-4\frac{5}{8}=5-4+\frac{3}{8}-\frac{5}{8}$
$=1+\left(-\frac{2}{8}\right)$
$=\frac{8}{8}-\frac{2}{8}$ ← $1=\frac{8}{8}$
$=\frac{6}{8}$
$=\frac{3}{4}$ ← Simplifying

EXAMPLE 12

Evaluate each expression using your calculator.

a $1\frac{1}{2}+2\frac{1}{3}$ b $3\frac{7}{8}-2\frac{1}{4}$ c $4\frac{1}{4}-2\frac{2}{5}$

SOLUTION

a $1\frac{1}{2}+2\frac{1}{3}=3\frac{5}{6}$ ← On calculator, enter 1 [a b/c] 1 [a b/c] 2 [+] 2 [a b/c] 1 [a b/c] 3 [=]

b $3\frac{7}{8}-2\frac{1}{4}=1\frac{5}{8}$ ← On calculator, enter 3 [a b/c] 7 [a b/c] 8 [−] 2 [a b/c] 1 [a b/c] 4 [=]

c $4\frac{1}{4}-2\frac{2}{5}=1\frac{17}{20}$ ← On calculator, enter 4 [a b/c] 1 [a b/c] 4 [−] 2 [a b/c] 2 [a b/c] 5 [=]

✱ Ask your teacher if your calculator has different keys for fractions.

ISBN 9780170350969

EXERCISE 7–08

1 Evaluate $1\frac{1}{4}+3\frac{2}{4}$. Select the correct answer A, B, C or D.

A $4\frac{3}{8}$ B $3\frac{3}{4}$ C $4\frac{3}{4}$ D $3\frac{3}{16}$

2 Evaluate $2\frac{3}{5}-\frac{2}{5}$. Select A, B, C or D.

A $1\frac{1}{5}$ B $2\frac{5}{5}$ C 3 D $2\frac{1}{5}$

3 Evaluate each expression.

a $1\frac{1}{4}+1\frac{1}{4}$ b $1\frac{1}{5}+2\frac{2}{5}$ c $2\frac{2}{7}+3\frac{1}{7}$ d $2\frac{2}{5}+3\frac{1}{5}$

e $3\frac{3}{7}-2\frac{2}{7}$ f $4\frac{7}{12}-2\frac{5}{12}$ g $4\frac{8}{11}-2\frac{7}{11}$ h $8\frac{17}{20}-6\frac{11}{20}$

i $3\frac{3}{5}+4\frac{1}{5}$ j $2\frac{4}{10}+4\frac{5}{10}$

4 Evaluate each sum using a calculator.

a $1\frac{1}{2}+1\frac{1}{3}$ b $1\frac{1}{2}+1\frac{2}{3}$ c $1\frac{1}{2}+1\frac{1}{4}$ d $1\frac{1}{2}+2\frac{1}{5}$

e $1\frac{1}{2}+2\frac{3}{4}$ f $2\frac{1}{3}+1\frac{1}{4}$ g $2\frac{1}{4}+1\frac{2}{3}$ h $2\frac{2}{3}+1\frac{3}{4}$

5 Evaluate each difference using a calculator.

a $2\frac{3}{4}-1\frac{1}{3}$ b $4\frac{1}{2}-2\frac{1}{4}$ c $2\frac{5}{8}-1\frac{1}{4}$ d $5\frac{1}{2}-3\frac{1}{3}$

e $2\frac{5}{6}-1\frac{2}{3}$ f $7\frac{1}{3}-3\frac{1}{5}$ g $6\frac{4}{7}-2\frac{1}{2}$ h $5\frac{3}{4}-1\frac{1}{2}$

6 Brad and Sophie went to a family picnic. Brad brought $2\frac{1}{2}$ chickens and Sophie brought $1\frac{3}{4}$ chickens. How many chickens did they bring to the picnic?

Shutterstock.com/wavebreakmedia

CROSSWORD PUZZLE

Make a copy of this page, then complete the crossword using these words:

COMMON DENOMINATOR FRACTION IMPROPER MIXED
NUMERAL NUMERATOR PROPER

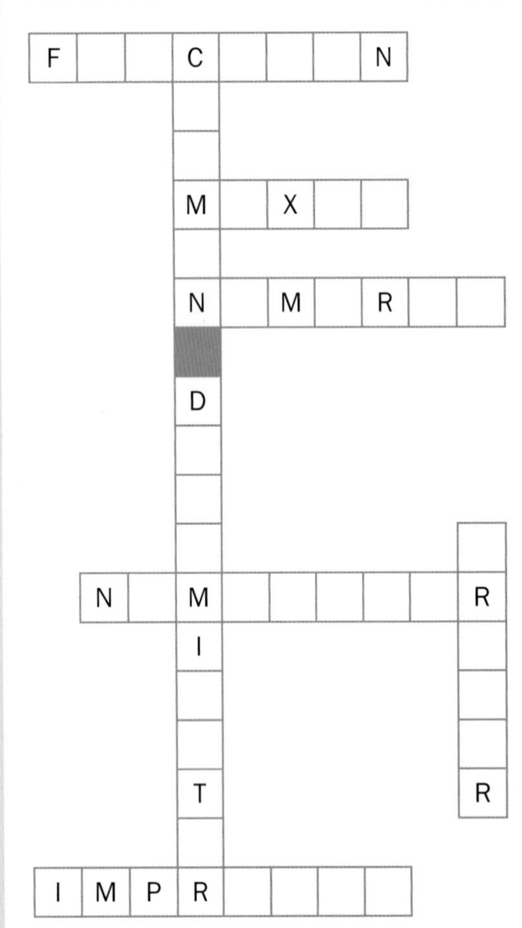

ISBN 9780170350969

PRACTICE TEST 7

Part A General topics

Calculators are not allowed.

1 What type of angle is marked?

2 Evaluate: 0.6×0.03

3 Write in ascending order: 7, –5, 0, –2, 1.

4 Evaluate $84 \div 3$.

5 Copy and complete: 9.41 t = ______ kg.

6 Evaluate $158 + 9 \times 6$.

7 Find the average of 18 and 26.

8 Round 4.6280 to 2 decimal places.

9 Draw a trapezium.

10 List all the prime numbers below 10.

Part B Fractions

Calculators are allowed.

7-01 Fractions

11 What fraction of this shape is shaded? Select the correct answer A, B, C or D.

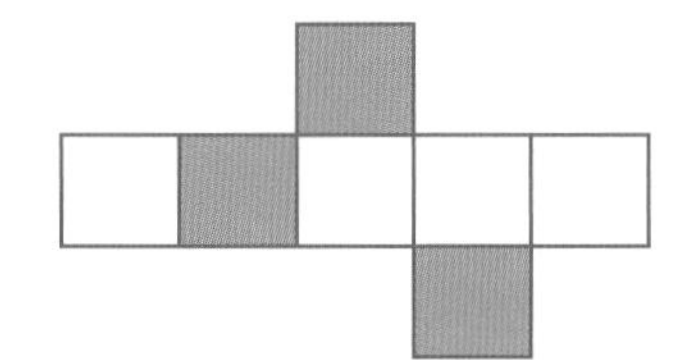

A $\frac{3}{6}$ B $\frac{3}{8}$ C $\frac{2}{7}$ D $\frac{3}{7}$

7-02 Fractions and decimals

12 Convert $\frac{4}{5}$ to a decimal. Select A, B, C or D.

A 0.8 B 0.45 C 0.4 D 0.6

13 Convert $\frac{3}{8}$ to a decimal. Select A, B, C or D.

A 0.38 B 0.83 C 0.375 D 0.3

7-03 Equivalent fractions

14 Is each statement true or false?

a $\frac{1}{3} = \frac{4}{12}$ b $\frac{2}{5} = \frac{6}{18}$ c $\frac{12}{20} = \frac{3}{4}$

7-04 Simplifying fractions

15 Simplify each fraction.

a $\frac{8}{12}$ b $\frac{15}{20}$ c $\frac{22}{40}$

16 Write as a simplified fraction.

a 35 minutes of 1 hour b 120 m of 1 km

PRACTICE TEST 7

7-05 Improper fractions and mixed numerals

17 Write each fraction as a mixed numeral.

a $\frac{13}{8}$ b $\frac{12}{5}$

18 Write each mixed numeral as an improper fraction.

a $1\frac{2}{3}$ b $3\frac{3}{4}$

7-06 Ordering fractions

19 Arrange these fractions in descending order: $\frac{2}{5}, \frac{3}{4}, \frac{5}{12}, \frac{2}{3}$.

7-07 Adding and subtracting fractions

20 Evaluate each expression.

a $\frac{3}{8}+\frac{4}{5}$ b $\frac{5}{6}-\frac{1}{4}$

7-08 Adding and subtracting mixed numerals

21 Evaluate each expression.

a $1\frac{2}{3}+3\frac{3}{5}$ b $5\frac{1}{4}-2\frac{3}{8}$

ISBN 9780170350969

MULTIPLYING AND DIVIDING FRACTIONS

8

WHAT'S IN CHAPTER 8?

8-01 Fraction of a quantity
8-02 Multiplying fractions
8-03 Reciprocals
8-04 Dividing fractions
8-05 Multiplying mixed numerals
8-06 Dividing mixed numerals

IN THIS CHAPTER YOU WILL:

- find a fraction of a number or amount
- multiply fractions, including mixed numerals
- find the reciprocal of a fraction
- divide fractions, including mixed numerals

Shutterstock.com/Jeff Wasserman

8-01 Fraction of a quantity

What is $\frac{2}{3}$ of 12? Here are 12 lollies:

If these lollies were divided into three equal piles, each pile would be $\frac{1}{3}$ of 12.

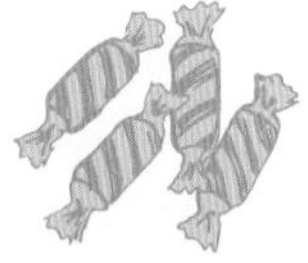

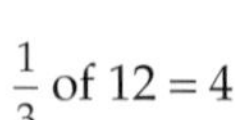

$\frac{1}{3}$ of $12 = 4$ $\quad$ $\frac{1}{3}$ of $12 = 4$ $\quad$ $\frac{1}{3}$ of $12 = 4$

$$\frac{2}{3} \times 12 = 8$$

EXAMPLE 1

Find each amount.

a $\frac{2}{3}$ of \$75 $\qquad$ b $\frac{3}{10}$ of 2 hours (in minutes)

SOLUTION

a $\frac{2}{3}$ of $\$75 = \left(\frac{1}{3} \times \$75\right) \times 2$

$= \$25 \times 2$ ← $75 \div 3 = 25$

$= \$50$

b $\frac{3}{10}$ of 2 hours $= \frac{3}{10}$ of 120 min ← change to min

$= \left(\frac{1}{10} \times 120\right) \times 3$

$= 12 \times 3$ ← $120 \div 10 = 12$

$= 36$ min

✱ The amount is divided by the denominator of the fraction and then multiplied by the numerator.

iStockphoto/BVDC

ISBN 9780170350969

EXERCISE 8-01

1 What is $\frac{1}{4}$ of $180? Select the correct answer A, B, C or D.

A $60 B $80 C $45 D $40

2 What is $\frac{3}{4}$ of $180? Select A, B, C or D.

A $135 B $120 C $150 D $165

3 Find each amount.

a $\frac{1}{4} \times 28$ b $\frac{1}{5} \times 35$ c $\frac{2}{5} \times 50$ d $\frac{1}{2} \times 16$

e $\frac{5}{12}$ of 12 f $\frac{2}{3}$ of 33 g $\frac{1}{9}$ of 27 h $\frac{6}{9}$ of 45

i $\frac{3}{10} \times 50$ j $\frac{5}{6} \times 60$ k $\frac{3}{4} \times 100$ l $\frac{2}{7} \times 42$

4 Find each amount.

a $\frac{1}{3}$ of 2 hours (in min) b $\frac{3}{4}$ of $20 c $\frac{2}{5}$ of 100 km

d $\frac{3}{8}$ of $160 e $\frac{4}{9}$ of 54 m f $\frac{5}{6}$ of 4 min (in seconds)

g $\frac{2}{7}$ of 42 books h $\frac{3}{2}$ of $50 i $\frac{11}{12}$ of 1 day (in hours)

j $\frac{7}{10}$ of 80 mL k $\frac{2}{5}$ of 2 kg (in g) l $\frac{4}{5}$ of $45

5 Michaela was training for a marathon and ran around a training track 8 times. The length of the running track was 250 m.

a How far did she run?

b The next day she ran only $\frac{4}{5}$ of this distance. How far did she run?

c Each day after she ran only $\frac{3}{5}$ of the previous day's distance. How far did she run on the 4th day?

6 Bassam is writing a novel and wrote 120 pages on Monday. On Tuesday he only wrote $\frac{5}{6}$ of this amount. After that, he was able to write only $\frac{4}{5}$ of his previous day's effort. How many pages did Bassam write on:

a Tuesday? b Thursday?

7 At David's party, 12 people ate $\frac{3}{4}$ pizza each and 8 people ate $\frac{5}{8}$ pizza each. How many pizzas were eaten?

ISBN 9780170350969

8-02 Multiplying fractions

What is $\frac{1}{2}$ of $\frac{1}{3}$?

$\frac{1}{3}$ of this diagram is shaded:

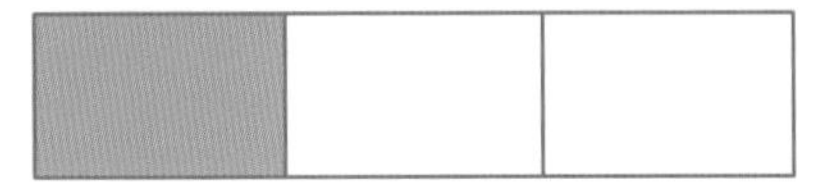

$\frac{1}{2} \times \frac{1}{3} = \frac{1}{2}$ of $\frac{1}{3}$ is shaded.

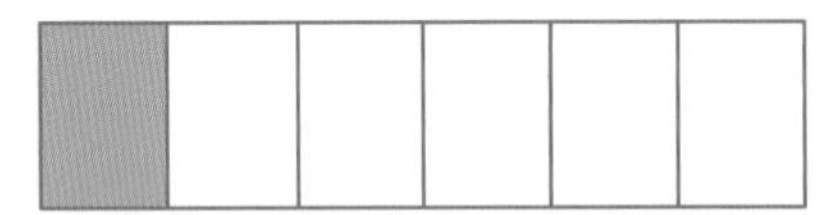

So $\frac{1}{2} \times \frac{1}{3} = \frac{1}{6}$

To multiply fractions:

- simplify numerators with denominators (if possible) by dividing by a common factor
- then **multiply numerators** and **multiply denominators** separately

EXAMPLE 2

Evaluate each product.

a $\frac{3}{4} \times \frac{12}{15}$ b $\frac{5}{6} \times \frac{18}{7}$

SOLUTION

a $\frac{3}{4} \times \frac{12}{15} = \frac{3 \times 12}{4 \times 15} = \frac{36}{60} = \frac{6}{10} = \frac{3}{5}$

or $\frac{3}{4} \times \frac{12}{15} = \frac{3 \times \cancel{12}^{4}}{4 \times \cancel{15}^{5}} = \frac{12}{20} = \frac{3}{5}$ ← simplify first: divide by 3

b $\frac{5}{6} \times \frac{18}{7} = \frac{5 \times 18}{6 \times 7} = \frac{90}{42} = \frac{30}{14} = \frac{15}{7} = 2\frac{1}{7}$ ← change to a mixed numeral

or $\frac{5}{6} \times \frac{18}{7} = \frac{5 \times \cancel{18}^{3}}{\cancel{6}^{1} \times 7} = \frac{15}{7} = 2\frac{1}{7}$ ← simplify first: divide by 6

ISBN 9780170350969

EXERCISE 8-02

1 What is $\frac{1}{3} \times \frac{2}{5}$? Select the correct answer A, B, C or D.

A $\frac{2}{8}$ B $\frac{2}{15}$ C $\frac{3}{15}$ D $\frac{1}{5}$

2 What is $\frac{3}{8} \times \frac{2}{7}$? Select A, B, C or D.

A $\frac{3}{28}$ B $\frac{5}{15}$ C $\frac{5}{56}$ D $\frac{1}{3}$

3 Is each statement true or false?

a $\frac{1}{2} \times \frac{1}{4} = \frac{1}{8}$ b $\frac{1}{4} \times \frac{1}{3} = \frac{1}{7}$

c $\frac{1}{3} \times \frac{1}{6} = \frac{1}{18}$ d $\frac{2}{3} \times \frac{1}{4} = \frac{2}{12}$

4 Evaluate each product.

a $\frac{2}{3} \times \frac{3}{4}$ b $\frac{5}{6} \times \frac{12}{15}$ c $\frac{5}{8} \times \frac{10}{12}$ d $\frac{4}{9} \times \frac{12}{16}$

e $\frac{4}{7} \times \frac{3}{8}$ f $\frac{4}{5} \times \frac{15}{16}$ g $\frac{2}{3} \times \frac{9}{10}$ h $\frac{5}{7} \times \frac{14}{15}$

i $\frac{3}{5} \times \frac{5}{3}$ j $\frac{4}{9} \times \frac{12}{20}$ k $\frac{5}{6} \times \frac{9}{15}$ l $\frac{8}{9} \times \frac{9}{4}$

5 Evaluate $\frac{2}{9} \times \frac{3}{10}$. Select A, B, C or D.

A $\frac{1}{8}$ B $\frac{1}{15}$ C $\frac{6}{19}$ D $\frac{1}{18}$

6 Copy and complete each equation.

a $\frac{1}{4} + \frac{1}{4} + \frac{1}{4} = 3 \times \frac{1}{4}$
$=$

b $\frac{3}{5} + \frac{3}{5} = 2 \times \frac{3}{5}$
$=$
$=$

c $\frac{3}{8} + \frac{3}{8} + \frac{3}{8} + \frac{3}{8} + \frac{3}{8} = 5 \times$
$=$
$=$

d $5 \times \frac{1}{7} =$

e $7 \times \frac{3}{20} =$
$=$

f $8 \times \frac{3}{5} =$
$=$

ISBN 9780170350969

8–03 Reciprocals

WORDBANK

reciprocal of a fraction The fraction 'turned upside down', for example, the reciprocal of $\frac{3}{8}$ is $\frac{8}{3}$.

It is important to find the reciprocal of a fraction when dividing fractions.

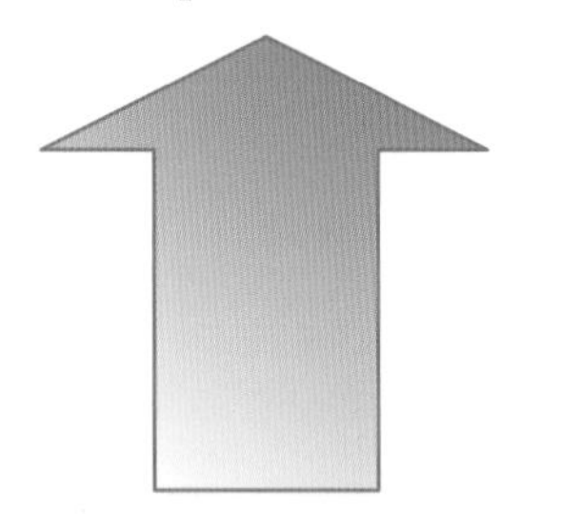

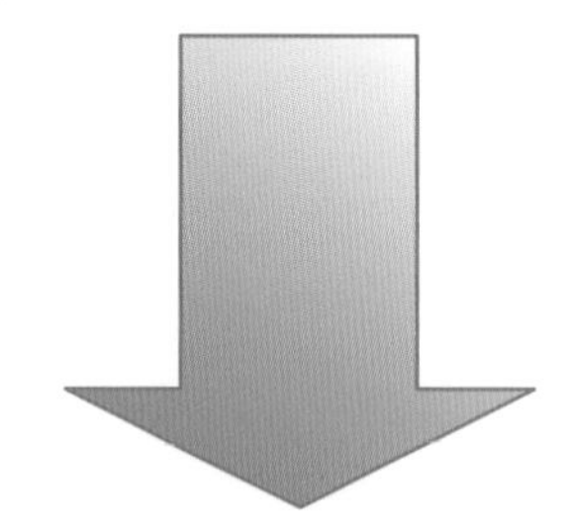

EXAMPLE 3

Find the reciprocal of each number.

a $\frac{2}{5}$ b $\frac{1}{3}$ c $2\frac{1}{2}$

SOLUTION

a The reciprocal of $\frac{2}{5}$ is $\frac{5}{2} = 2\frac{1}{2}$.

b The reciprocal of $\frac{1}{3}$ is $\frac{3}{1} = 3$.

c $2\frac{1}{2} = \frac{5}{2}$ ← Convert $2\frac{1}{2}$ to an improper fraction first.

The reciprocal of $2\frac{1}{2}$ is $\frac{2}{5}$.

Shutterstock.com/Mandy Godbehear

ISBN 9780170350969

EXERCISE 8–03

1 What is the reciprocal of $\frac{3}{4}$? Select the correct answer A, B, C or D.

A $\frac{4}{1}$ B $\frac{3}{1}$ C $1\frac{1}{3}$ D $1\frac{1}{4}$

2 What is the reciprocal of $3\frac{1}{5}$? Select A, B, C or D.

A $\frac{16}{5}$ B $\frac{15}{5}$ C $\frac{5}{16}$ D $\frac{5}{9}$

3 Copy and complete these sentences.

To find the reciprocal of a fraction, simply turn it ______ ______. To find the reciprocal of a mixed numeral, convert it to an ______ fraction first, then turn it ______ ______.

4 Find the reciprocal of each fraction.

a $\frac{2}{3}$ b $\frac{1}{5}$ c $\frac{5}{8}$ d $\frac{1}{10}$

e 7 f $\frac{4}{5}$ g $\frac{5}{6}$ h 9

5 Find the reciprocal of each mixed numeral.

a $1\frac{1}{3}$ b $2\frac{1}{4}$ c $3\frac{2}{3}$

d $1\frac{3}{4}$ e $4\frac{1}{3}$ f $3\frac{5}{8}$

g $5\frac{2}{5}$ h $6\frac{1}{4}$ i $2\frac{4}{5}$

Shutterstock.com/Mandy Godbehear

ISBN 9780170350969

8-04 Dividing fractions

We know that **multiplication** and **division** are the opposite of each other.

So if $\frac{1}{2} \times \frac{1}{3} = \frac{1}{6}$ then $\frac{1}{6} \div \frac{1}{3} = \frac{1}{2}$.

To obtain this result, we must turn the second fraction upside-down and multiply:

$$\begin{aligned}\frac{1}{6} \div \frac{1}{3} &= \frac{1}{6} \times \frac{3}{1} \\ &= \frac{3}{6} \\ &= \frac{1}{2}\end{aligned}$$

To divide by a fraction $\frac{a}{b}$, multiply by its reciprocal $\frac{b}{a}$.

EXAMPLE 4

Evaluate each quotient.

a $\frac{3}{4} \div \frac{5}{8}$

b $\frac{6}{7} \div \frac{12}{5}$

SOLUTION

a $$\begin{aligned}\frac{3}{4} \div \frac{5}{8} &= \frac{3}{4} \times \frac{8}{5} \\ &= \frac{24}{20} \\ &= \frac{6}{5} \\ &= 1\frac{1}{5}\end{aligned}$$

b $$\begin{aligned}\frac{6}{7} \div \frac{12}{5} &= \frac{6}{7} \times \frac{5}{12} \\ &= \frac{\cancel{6}^{1}}{7} \times \frac{5}{\cancel{12}^{2}} \\ &= \frac{5}{14}\end{aligned}$$

* Note that it is the second fraction you are dividing by, so this is the one that turns upside down.

Alamy/Peter Horree

ISBN 9780170350969

EXERCISE 8–04

1 Which product has the same value as $\frac{2}{3} \div \frac{5}{6}$? Select the correct answer A, B, C or D.

A $\frac{3}{2} \times \frac{5}{6}$ B $\frac{2}{3} \times \frac{6}{5}$ C $\frac{3}{2} \times \frac{6}{5}$ D $\frac{2}{3} \times \frac{5}{6}$

2 Evaluate $\frac{2}{3} \div \frac{5}{6}$. Select A, B, C or D.

A $1\frac{1}{4}$ B $1\frac{4}{5}$ C $\frac{4}{5}$ D $\frac{4}{15}$

3 Copy and complete this sentence: To divide by a fraction, ______ by its ______.

4 Is each statement true or false?

a $\frac{5}{6} \div \frac{1}{3} = \frac{5}{3}$ b $\frac{3}{4} \div \frac{6}{7} = \frac{7}{8}$ c $\frac{1}{8} \div \frac{3}{2} = \frac{1}{12}$ d $\frac{2}{5} \div \frac{4}{15} = \frac{2}{3}$

5 Rewrite each division as a multiplication.

a $\frac{3}{8} \div \frac{1}{4}$ b $\frac{4}{7} \div \frac{2}{14}$ c $\frac{5}{8} \div \frac{3}{16}$ d $\frac{4}{5} \div \frac{12}{3}$

e $\frac{7}{8} \div \frac{3}{24}$ f $\frac{2}{3} \div \frac{5}{9}$ g $\frac{5}{7} \div \frac{25}{21}$ h $\frac{5}{8} \div \frac{20}{4}$

i $\frac{5}{7} \div \frac{15}{8}$ j $\frac{8}{3} \div \frac{4}{9}$ k $\frac{6}{5} \div \frac{12}{10}$ l $\frac{6}{9} \div \frac{18}{12}$

6 Find the answer to each division in Question 5.

7 Evaluate each quotient.

a $4 \div \frac{1}{3}$ b $5 \div \frac{1}{2}$ c $7 \div \frac{1}{3}$ d $3 \div \frac{1}{5}$

e $\frac{5}{7} \div 5$ f $\frac{5}{7} \div 7$ g $\frac{1}{4} \div 3$ h $\frac{5}{6} \div 2$

8 A bag of rice weighing $\frac{4}{5}$ kg was divided among 6 friends. How much rice did each person get? Select A, B, C or D.

A 480 g B $\frac{2}{15}$ kg C 800 g D $\frac{4}{15}$ kg

9 Abla had three daughters working with her and wanted to share her annual profit with them. She gave herself $\frac{1}{4}$ of the profit, her oldest daughter Aisha $\frac{2}{5}$ of the profit, and the other two daughters shared the rest equally. How much did each daughter receive if the annual profit was $24 000?

10 Divide $\frac{55}{80}$ by $\frac{25}{40}$.

8–05 Multiplying mixed numerals

To multiply mixed numerals, convert them to improper fractions first.

EXAMPLE 5

Evaluate each product.

a $1\frac{1}{3} \times 1\frac{1}{2}$ b $1\frac{4}{5} \times 2\frac{1}{3}$

SOLUTION

a
$$\begin{aligned} 1\frac{1}{3} \times 1\frac{1}{2} &= \frac{4}{3} \times \frac{3}{2} \\ &= \frac{{}^{2}\cancel{4} \times \cancel{3}^{1}}{{}^{1}\cancel{3} \times \cancel{2}_{1}} \\ &= \frac{2}{1} \\ &= 2 \end{aligned}$$

b
$$\begin{aligned} 1\frac{4}{5} \times 2\frac{1}{3} &= \frac{9}{5} \times \frac{7}{3} \\ &= \frac{{}^{3}\cancel{9} \times 7}{5 \times \cancel{3}_{1}} \\ &= \frac{21}{5} \\ &= 4\frac{1}{5} \end{aligned}$$

These products can also be evaluated on your calculator:

a $1\frac{1}{3} \times 1\frac{1}{2} = 2$ ← On calculator, enter 1 [a b/c] 1 [a b/c] 3 [×] 1 [a b/c] 1 [a b/c] 2 [=] 

b $1\frac{4}{5} \times 2\frac{1}{3} = 4\frac{1}{5}$ ← On calculator, enter 1 [a b/c] 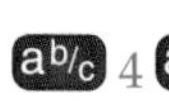 4 [a b/c] 5 [×] 2 [a b/c] 1 [a b/c] 3 [=]

* Ask your teacher if your calculator uses different keys for fractions.

iStockphoto/MachineHeadz

ISBN 9780170350969

EXERCISE 8–05

1 Evaluate $1\frac{1}{4} \times 2\frac{1}{2}$. Select the correct answer A, B, C or D.

A $2\frac{1}{8}$ B $\frac{1}{2}$ C $\frac{8}{25}$ D $3\frac{1}{8}$

2 Evaluate $2\frac{1}{3} \times \frac{6}{5}$. Select A, B, C or D.

A $2\frac{4}{5}$ B $2\frac{1}{15}$ C $2\frac{6}{15}$ D $\frac{42}{5}$

3 Copy and complete:

To multiply mixed numerals, convert them to ______ fractions first, then multiply the ______ and multiply the ______.

4 Evaluate each product.

a $1\frac{1}{2} \times 1\frac{1}{5}$ b $1\frac{1}{2} \times 2\frac{2}{3}$ c $1\frac{1}{2} \times 1\frac{1}{9}$ d $1\frac{1}{3} \times \frac{1}{4}$

e $\frac{3}{4} \times 1\frac{1}{3}$ f $1\frac{1}{3} \times 1\frac{1}{5}$ g $2\frac{2}{3} \times \frac{1}{7}$ h $2\frac{2}{3} \times 1\frac{1}{8}$

i $2\frac{2}{3} \times \frac{6}{7}$ j $2\frac{2}{3} \times 4\frac{1}{2}$

5 Evaluate each product using your calculator.

a $1\frac{1}{4} \times 1\frac{3}{5}$ b $1\frac{1}{4} \times 1\frac{1}{3}$ c $1\frac{1}{4} \times 1\frac{2}{10}$ d $1\frac{1}{4} \times 1\frac{1}{15}$

e $2\frac{1}{2} \times 2\frac{1}{2}$ f $1\frac{2}{3} \times 1\frac{1}{5}$ g $2\frac{1}{4} \times 1\frac{1}{3}$ h $3\frac{3}{4} \times 2\frac{2}{5}$

i $4\frac{1}{2} \times 1\frac{1}{3}$ j $3\frac{3}{4} \times 1\frac{1}{5}$

6 Amy was collecting vintage clothes for a school play and went to one shop and found enough clothes for $3\frac{1}{2}$ outfits. She then visited another shop and they had 6 different sizes of clothes making $2\frac{1}{4}$ outfits each. How many outfits did she have altogether?

8–06 Dividing mixed numerals

To divide mixed numerals, convert them to improper fractions first.

EXAMPLE 6

Evaluate each quotient.

a $1\frac{1}{2} \div \frac{2}{9}$ b $2\frac{1}{4} \div 1\frac{1}{8}$

SOLUTION

a
$$\begin{aligned} 1\frac{1}{2} \div \frac{2}{9} &= \frac{3}{2} \div \frac{2}{9} \\ &= \frac{3}{2} \times \frac{9}{2} \\ &= \frac{27}{4} \\ &= 6\frac{3}{4} \end{aligned}$$

b
$$\begin{aligned} 2\frac{1}{4} \div 1\frac{1}{8} &= \frac{9}{4} \div \frac{9}{8} \\ &= \frac{9}{4} \times \frac{8}{9} \\ &= \frac{{}^{1}\cancel{9} \times \cancel{8}^{2}}{{}_{1}\cancel{4} \times \cancel{9}_{1}} \\ &= \frac{2}{1} \\ &= 2 \end{aligned}$$

These quotients can also be evaluated on your calculator:

a $1\frac{1}{2} \div \frac{2}{9} = 6\frac{3}{4}$ ← On calculator, enter 1 [a b/c] 1 [a b/c] 2 [÷] 2 [a b/c] 9 [=]

b $2\frac{1}{4} \div 1\frac{1}{8} = 2$ ← On calculator, enter 2 [a b/c] 1 [a b/c] 4 [÷] 1 [a b/c] 1 [a b/c] 8 

ISBN 9780170350969

EXERCISE 8-06

1 Evaluate $1\frac{1}{4} \div 2\frac{1}{2}$. Select the correct answer A, B, C or D.

A $2\frac{1}{8}$ B $\frac{1}{2}$ C $\frac{8}{25}$ D $3\frac{1}{8}$

2 Evaluate $2\frac{1}{5} \div 3\frac{2}{3}$. Select A, B, C or D.

A $\frac{3}{5}$ B $8\frac{1}{15}$ C $1\frac{2}{3}$ D $6\frac{2}{15}$

3 Copy and complete:

To divide mixed numerals, convert them to ______ fractions first, then multiply by the ______ of the second fraction.

4 Evaluate each quotient.

a $\frac{5}{7} \div \frac{10}{14}$ b $2\frac{1}{3} \div 4\frac{2}{3}$ c $1\frac{1}{4} \div 2\frac{2}{4}$ d $\frac{3}{8} \div 1\frac{1}{8}$

e $1\frac{1}{5} \div \frac{5}{2}$ f $2\frac{1}{10} \div \frac{8}{3}$ g $1\frac{3}{4} \div \frac{2}{5}$ h $2\frac{1}{5} \div 3\frac{1}{2}$

i $\frac{2}{9} \div \frac{9}{10}$ j $\frac{3}{5} \div 1\frac{4}{6}$

5 Evaluate each quotient using your calculator.

a $1\frac{2}{3} \div \frac{5}{6}$ b $3\frac{1}{3} \div 2\frac{2}{9}$ c $3\frac{1}{5} \div 1\frac{3}{5}$ d $1\frac{5}{7} \div 2\frac{2}{14}$

e $2\frac{2}{9} \div 1\frac{1}{9}$ f $1\frac{1}{2} \div 1\frac{1}{9}$ g $2\frac{1}{3} \div 1\frac{5}{7}$ h $5\frac{1}{4} \div 2\frac{1}{3}$

i $3\frac{3}{4} \div \frac{5}{16}$ j $2\frac{2}{5} \div 3\frac{1}{3}$

6 Tash had $12\frac{5}{6}$ m of timber which had to be cut into pieces $1\frac{1}{12}$ m long.

How many pieces could she make?

Corbis/Jorn

FIND-A-WORD PUZZLE

Make a copy of this page, then find the words listed below in this grid of letters.

N	O	I	S	I	V	I	D	E	F
E	W	S	T	U	A	S	E	V	R
O	M	I	X	E	D	W	N	E	E
N	U	M	B	E	R	A	O	X	C
Y	L	P	I	T	L	U	M	R	I
C	D	V	E	D	I	V	I	D	P
A	I	C	O	M	M	O	N	I	R
N	V	R	E	W	S	N	A	T	O
C	E	I	N	V	E	R	T	O	C
E	S	T	R	E	V	N	O	C	A
L	N	O	I	T	C	A	R	F	L

ANSWER	CANCEL	COMMON	CONVERT
DENOMINATOR	DIVIDE	DIVISION	FRACTION
INVERT	MIXED	MULTIPLY	NUMBER
RECIPROCAL			

ISBN 9780170350969

PRACTICE TEST 8

Part A General topics

Calculators are not allowed.

1 What is an equilateral triangle?

2 Evaluate $3 - (-3)$.

3 List the factors of 20.

4 Evaluate 0.4×0.08.

5 Convert 1000 cents to dollars.

6 Is $2.5 = 2\frac{1}{2}$?

7 Is $\frac{7}{8} > \frac{5}{6}$?

8 How many degrees in a revolution?

9 Evaluate $4 \times 3 \times 5$.

10 Copy this shape and draw its axes of symmetry.

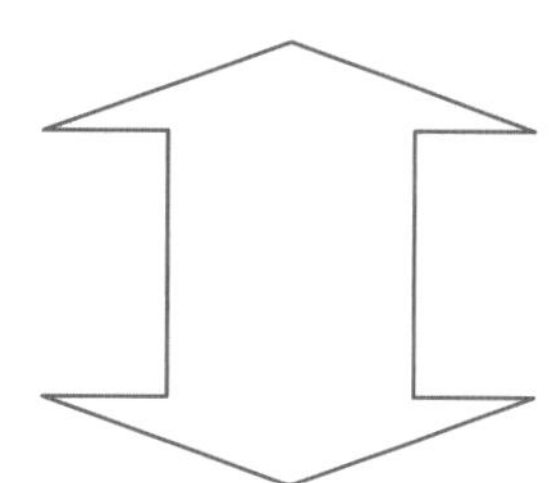

Part B Multiplying and dividing fractions

Calculators are allowed.

8-01 Fraction of a quantity

11 Find $\frac{3}{4}$ of \$60. Select the correct answer A, B, C or D.

A \$15 B \$25 C \$40 D \$45

8-02 Multiplying fractions

12 Evaluate $\frac{3}{7} \times \frac{21}{12}$. Select A, B, C or D.

A $\frac{4}{3}$ B $\frac{5}{4}$ C $\frac{3}{4}$ D $\frac{3}{5}$

8-03 Reciprocals

13 What is the reciprocal of $\frac{4}{6}$? Select A, B, C or D.

A $\frac{2}{3}$ B $\frac{1}{3}$ C $1\frac{1}{2}$ D $\frac{1}{2}$

14 Write the reciprocal of each number.

a $\frac{7}{8}$ b $\frac{1}{5}$ c $2\frac{1}{4}$

PRACTICE TEST 8

8-04 Dividing fractions

15 Is each statement true or false?

a $\frac{2}{5} \div \frac{3}{4} = \frac{5}{2} \times \frac{3}{4}$

b $\frac{5}{8} \div \frac{2}{7} = \frac{5}{8} \times \frac{7}{2}$

16 Evaluate each quotient.

a $\frac{4}{5} \div \frac{16}{3}$

b $\frac{4}{5} \div \frac{12}{25}$

8-05 Multiplying mixed numerals

17 Convert $4\frac{2}{3}$ to an improper fraction. Select **A**, **B**, **C** or **D**.

A $\frac{14}{3}$

B $\frac{3}{14}$

C $\frac{42}{3}$

D $\frac{6}{3}$

18 Evaluate each product.

a $3\frac{1}{2} \times 1\frac{1}{4}$

b $5\frac{1}{3} \times 2\frac{3}{4}$

8-06 Dividing mixed numerals

19 Is each statement true or false?

a $1\frac{1}{8} \div 2\frac{2}{3} = \frac{9}{8} \times \frac{3}{8}$

b $2\frac{1}{6} \div 1\frac{5}{8} = \frac{13}{6} \times \frac{13}{8}$

20 Evaluate each quotient.

a $1\frac{1}{8} \div 2\frac{2}{3}$

b $2\frac{1}{6} \div 1\frac{5}{8}$

ISBN 9780170350969

ALGEBRA AND EQUATIONS

9

WHAT'S IN CHAPTER 9?

9-01 The laws of arithmetic
9-02 Variables
9-03 From words to algebraic expressions
9-04 Substitution
9-05 Equations
9-06 One-step equations
9-07 Two-step equations

IN THIS CHAPTER YOU WILL:

- use the laws of arithmetic to help mental calculation
- use variables to write general rules involving numbers
- use algebraic abbreviations to simplify expressions
- substitute values into algebraic expressions
- solve one-step and two-step equations

Shutterstock.com/Zurijeta

9-01 The laws of arithmetic

WORDBANK

mental Using the mind, not a calculator.

arithmetic The process of adding, subtracting, multiplying and dividing numbers.

The **commutative law** says that any two numbers can be added or multiplied in any order.
For example, $3 + 4 = 4 + 3$ and $3 \times 4 = 4 \times 3$.
The **associative law** says that any three or more numbers can be added or multiplied in any order.
For example, $5 + 1 + 9 = 1 + 9 + 5$ and $2 \times 7 \times 5 = 5 \times 2 \times 7$.

If a, b and c stand for numbers, then:

- $a + b = b + a$
- $(a + b) + c = a + (b + c)$
- $a \times b = b \times a$
- $(a \times b) \times c = a \times (b \times c)$

We can use these laws to help us add and multiply numbers mentally.

EXAMPLE 1

Evaluate each expression.

a $9 + 12$

b $17 + 6 + 3$

c $2 \times 8 \times 5$

SOLUTION

a $9 + 12 = 12 + 9$
$= 21$

b $17 + 6 + 3 = 17 + 3 + 6$
$= 20 + 6$
$= 26$

* group numbers that add to 10, 20, 30, ...

c $2 \times 8 \times 5 = 2 \times 5 \times 8$
$= 10 \times 8$
$= 80$

The **distributive law** says that multiplying can be simplified if you 'split' one of the numbers.
For example, $7 \times 12 = 7 \times (10 + 2) = (7 \times 10) + (7 \times 2)$.

- $a \times (b + c) = (a \times b) + (a \times c)$
- $a \times (b - c) = (a \times b) - (a \times c)$

EXAMPLE 2

Evaluate each product.

a 5×12

b 13×9

c 42×11

SOLUTION

a $5 \times 12 = 5 \times (10 + 2)$
$= (5 \times 10) + (5 \times 2)$
$= 50 + 10$
$= 60$

b $13 \times 9 = 13 \times (10 - 1)$
$= (13 \times 10) - (13 \times 1)$
$= 130 - 13$
$= 117$

c $42 \times 11 = 42 \times (10 + 1)$
$= (42 \times 10) + (42 \times 1)$
$= 420 + 42$
$= 462$

ISBN 9780170350969

EXERCISE 9–01

1 Which equation is true? Select the correct answer A, B, C or D.

A $8 - 5 = 5 - 8$ B $8^5 = 5^8$

C $8 \times 5 = 5 \times 8$ D $8 \div 5 = 5 \div 8$

2 Which expression can be used to evaluate 19×11 mentally? Select A, B, C or D.

A $19 \times (10 - 1)$ B $19 \times (10 - 2)$

C $19 \times (10 + 1)$ D $19 \times (10 + 2)$

3 Copy and complete each equation.

a $6 + 8 = \_ + 6$ b $4 \times 9 = 9 \times \_$

c $a + 7 = 7 + \_$ d $6 \times m = m \times \_$

e $u + v = \_ + u$ f $j \times k = \_ \times j$

4 Evaluate each sum by pairing up numbers that add to 10 or multiples of 10.

a $12 + 23 + 8$ b $34 + 45 + 5$

c $29 + 32 + 11$ d $16 + 5 + 4 + 5$

e $43 + 16 + 7$ f $21 + 12 + 9$

g $53 + 17 + 20$ h $48 + 61 + 12$

5 Is each statement true or false?

a $5 \times 4 \times 2 = 2 \times 5 \times 4$ b $24 \times 9 = 24 \times (10 + 1)$

c $16 \times 11 = 16 \times (10 + 1)$ d $33 \times 12 = 33 \times (10 + 2)$

e $9 + 8 - 3 = 3 - 8 + 9$ f $15 \times 8 = 15 \times (10 - 1)$

6 Copy and complete each line of working.

a $2 \times 9 \times 5 = 2 \times \_\_\_\_ \times 9$
$= \_\_\_\_ \times 9$
$= \_\_\_\_$

b $4 \times 12 \times 25 = 4 \times \_\_\_\_ \times 12$
$= \_\_\_\_ \times 12$
$= \_\_\_\_$

7 Evaluate each product by pairing up numbers that multiply to multiples of 10.

a $2 \times 7 \times 5$ b $5 \times 5 \times 4$

c $2 \times 6 \times 50$ d $5 \times 7 \times 20$

e $3 \times 2 \times 45$ f $50 \times 8 \times 2$

g $4 \times 3 \times 5$ h $15 \times 2 \times 2$

8 Copy and complete each equation.

a $8 \times (10 + 2) = 8 \times \_\_\_\_ + 8 \times 2$ b $7 \times (10 - 1) = \_\_\_\_ \times 10 - 7 \times 1$

c $4 \times (10 + 4) = 4 \times 10 + 4 \times \_\_\_\_$ d $5 \times (10 - 3) = 5 \times \_\_\_\_ - 5 \times 3$

9 Evaluate each product by splitting the second number and using the distributive law.

a 16×11 b 12×9 c 15×8 d 20×12

e 17×9 f 13×8 g 18×11 h 24×12

ISBN 9780170350969

9-02 Variables

A **variable** or **pronumeral** is a symbol or letter of the alphabet used to represent a number. The value of the variable can change or **vary**, which is where the name comes from.

In algebra, we use variables to write general rules for numbers once a pattern has been discovered.

EXAMPLE 3

For each number pattern, write a general rule using a variable.

a $3 \times 1 = 3$
$5 \times 1 = 5$
$8 \times 1 = 8$
$12 \times 1 = 12$

b $3 + 3 = 2 \times 3$
$5 + 5 = 2 \times 5$
$8 + 8 = 2 \times 8$
$12 + 12 = 2 \times 12$

SOLUTION

a $m \times 1 = m$

 For the variable, we can use any letter of the alphabet.

b $w + w = 2 \times w$

We can use abbreviations in algebra when writing variables:

- $1 \times b = b$ — 1 not needed
- $2 \times a = 2a$ — $\times$ not needed
- $3 \times a \times b \times 4 = 12ab$ — numbers multiplied together and written first
- $2 \times a \times a = 2a^2$ — $a \times a = a^2$

EXAMPLE 4

Simplify each expression.

a $n + n + n$

b $3 \times c \times 2$

c $a + a + a + b + b$

d $4 \times d \times d \times 2$

SOLUTION

a $n + n + n = 3n$

b $3 \times c \times 2 = 6c$

c $a + a + a + b + b = 3a + 2b$

d $4 \times d \times d \times 2 = 8d^2$

ISBN 9780170350969

EXERCISE 9-02

1 Simplify $w + w + w + v + v$. Select the correct answer A, B, C or D.

A $2w + 3v$ B $3w + v + v$ C $3w + 2v$ D $w + w + w + 2v$

2 Simplify $a \times 3 \times b \times 5 \times a$. Select A, B, C or D.

A $15ba$ B $35aba$ C $15ab$ D $15a^2b$

3 Write down which of the following could be used as a variable.

4, a, –3, c, 14, m, E, =, s, +, 6, g

4 Simplify:

a $r + r$ b $1e$

c $5 \times n$ d $m + m + m$

e $2 \times s \times t$ f $c \times c$

g $b + b + b + b$ h $3 \times v \times 2$

5 Investigate the number patterns below and then write down a general rule using a variable:

a	b	c	d
$3 + 0 = 3$	$4 \div 1 = 4$	$2 - 2 = 0$	$5 + 5 + 5 = 3 \times 5$
$6 + 0 = 6$	$5 \div 1 = 5$	$4 - 4 = 0$	$2 + 2 + 2 = 3 \times 2$
$8 + 0 = 8$	$7 \div 1 = 7$	$8 - 8 = 0$	$4 + 4 + 4 = 3 \times 4$
$9 + 0 = 9$	$8 \div 1 = 8$	$6 - 6 = 0$	$7 + 7 + 7 = 3 \times 7$

6 Simplify the expressions below:

a $e + e + e$ b $2 \times v \times 5$

c $a + a + b + b$ d $3 \times r \times s \times 4$

e $s + s - s$ f $a \times 6 \times b$

g $d + d + e - e$ h $w + v + w - v - w$

i $6 \times a \times 3$ j $a \times c \times 12$

k $3e - b + e$ l $4t + 2s - t + 2s$

m $6w - 2w$ n $1 \times f \times g \times 3$

o $a + a - a + b$ p $3 \times c \times 7 \times c \times 2$

7 Is each statement true or false?

a $4 - 3 \times a = a$ b $c \times c \times c = 3c$

c $a + a + a = 3a$ d $12 + 5 \times a = 12 + 5a$

8 Write each expression the long way.

a $6b$ b $a + 2b$ c m^2 d $m + 3n$

e $5mn$ f r^3 g $4d$ h $6abc$

9 Use the order of operations to simplify each expression.

a $4 + 2 \times n$ b $7 \times a \times b - a$

c $12 - 3 \times s$ d $a \times a - b \times b$

e $2 \times a - 3 \times b$ f $20 + 3 \times m \times n$

g $8 + 7 \times c$ h $4 + d - d \times d$

9-03 From words to algebraic expressions

To solve problems using algebra we need to be able to convert words into **algebraic expressions** involving variables.

EXAMPLE 5

Write each statement as an algebraic expression. Use a to stand for the number.

a The sum of a number and 5

b The product of 8 and a number

c The difference between a number and 3

d The quotient of a number and 6

SOLUTION

a $a + 5$

b $8 \times a = 8a$

c $a - 3$

d $\frac{a}{6}$

EXAMPLE 6

Translate each statement into an algebraic expression.

a Twice m plus 3

b Decrease k by 8

c Triple the sum of a and b

d Increase 5 by the product of 2 and v

SOLUTION

a $2m + 3$

b $k - 8$

c $3 \times (a + b) = 3(a + b)$

Note that brackets are necessary here as the sum of a and b must be done first.

d $5 + 2 \times v = 5 + 2v$

ISBN 9780170350969

EXERCISE 9-03

1 Write an algebraic expression for twice the sum of m and n. Select the correct answer A, B, C or D.

A $2m + n$ B $m + 2n$ C $2(m + n)$ D $2 + m + n$

2 Write an algebraic expression for the difference between $3x$ and $4y$. Select A, B, C or D.

A $3x + 4y$ B $3x - 4y$ C $3x \times 4y$ D $3x \div 4y$

3 Match each word to a mathematical symbol +, −, ×, ÷.

a difference b increase c product
d sum e quotient f twice
g square h minus i triple

4 Write each statement below as an algebraic expression. Use n for the number.

a the product of 4 and the number
b the difference between the number and 3
c twice the number
d the sum of the number and 7
e triple the number less 9
f the number divided by 4
g decrease the number by 6
h the quotient of the number and 5
i decrease 8 by the number
j the number squared

5 Write as an algebraic expression.

a twice n plus 3
b triple j less 8
c the sum of a and 5
d the product of 7 and b
e twice the sum of s and t
f decrease 10 by the product of 2 and c
g triple the difference of a and 6
h increase d by the sum of a and 4
i decrease a by the sum of b and c
j b squared less the product of a and c

6 Write in words the meaning of each algebraic expression.

a $a + b + c$ b $3b$ c $6 - g$
d $\frac{r}{4}$ e $2a - b$ f $\frac{3b}{c}$
g $5(a + b)$ h $4b + 6$ i $7v - 4$

7 Write each phrase as an algebraic expression.

a Four times the sum of a, b and c
b The product of a, b and c minus 8

9-04 Substitution

WORDBANK

substitution Replacing a variable with a number in an algebraic expression to find the value of the expression.

EXAMPLE 7

If $a = 3$ and $b = -2$, evaluate each expression.

a $2a + b$

b $6a - 5b$

c $4ab$

SOLUTION

a $2a + b = 2 \times 3 + (-2)$
$= 6 - 2$
$= 4$

b $6a - 5b = 6 \times 3 - 5 \times (-2)$
$= 18 + 10$
$= 28$

c $4ab = 4 \times 3 \times (-2)$
$= -24$

Remember the order of operations when substituting:

- brackets ( ) first,
- then powers (x^y) and square roots ($\sqrt{\ }$)
- then multiplication (×) and division (÷) from left to right
- then addition (+) and subtraction (-) from left to right.

EXAMPLE 8

Substitute $m = 3$ and $n = -4$ to evaluate each expression.

a $24 - mn$

b $20 + 3m - n$

c $2mn - 4m^2$

SOLUTION

a $24 - mn = 24 - 3 \times (-4)$
$= 24 + 12$
$= 36$

b $20 + 3m - n = 20 + 3 \times 3 - (-4)$
$= 20 + 9 + 4$
$= 33$

c $2mn - 4m^2 = 2 \times 3 \times (-4) - 4 \times 3^2$
$= -24 - 4 \times 9$
$= -24 - 36$
$= -60$

ISBN 9780170350969

EXERCISE 9–04

1 If $a = -6$ and $b = 5$, evaluate $2a - b$. Select the correct answer A, B, C or D.

A -17 B -7 C 7 D 17

2 Evaluate $3ab - 8a$ if $a = -4$ and $b = -6$. Select A, B, C or D.

A -104 B 34 C -26 D 104

3 Find the value of each algebraic expression if $a = 4$, $b = -2$ and $c = 6$.

a $a + b$ b $2ac$ c $3b - c$

d $2a + 4c$ e abc f $4c - ab$

g $a^2 + 2c$ h $bc - 4$ i $8b^2 - c$

j $\frac{8a}{b}$ k $12 - ab$ l $\frac{c^2}{a}$

4 Use the table below to evaluate each algebraic expression:

m	n	p	q	r
4	–6	2	3	–5

a $2mn$ b pq^2 c $4q - 6r$

d $m^2 + 3n$ e $3p^2r$ f $12 - qr$

g $2m + n - r$ h $pq + n$ i $\sqrt{nr + pq}$

j $\frac{\sqrt{8p}}{m}$ k $-2qr - 8$ l $\frac{np}{m}$

5 Substitute $v = 20$, $w = -5$ and $z = -4$ to evaluate each algebraic expression.

a $2vwz$ b $zw - v$ c $4v - 2w + 3z$

d $wz + 4v$ e $-6vz + w$ f $\frac{3v}{w}$

g $6z - wv$ h $\frac{vw}{z}$ i $8z + 4v - w$

6 Spot the error and find the correct answer.

If $m = 12$ and $n = -4$,

$$\begin{aligned} m^2 - 4n + 6mn &= 122 - 4 - 4 + 6 \times 12 \times -4 \\ &= 144 - 8 - 288 \\ &= -152 \end{aligned}$$

ISBN 9780170350969

9-05 Equations

WORDBANK

equation A number sentence containing algebraic terms, numbers and an equals sign, for example $x - 3 = 8$.

solve an equation To find the value of the variable that makes the equation true.

solution The answer to an equation, the correct value of the variable.

We will first look at solving equations by **guess-and-check**. This involves making an educated guess at the possible value for the variable, then substituting this value into the equation and checking if it is true.

EXAMPLE 9

Solve the equations below using the guess-and-check method:

a $2x - 5 = 17$

b $3a + 6 = 24$

SOLUTION

a Try $x = 8$:

Does $2 \times 8 - 5 = 17$?

$2 \times 8 - 5 = 11$

No, it is too low.

Try $x = 11$:

Does $2 \times 11 - 5 = 17$?

$2 \times 11 - 5 = 17$

YES

So the solution is $x = 11$.

b Try $a = 10$:

Does $3 \times 10 + 6 = 24$?

$3 \times 10 + 6 = 36$

No, it is too high.

Try $a = 6$:

Does $3 \times 6 + 6 = 24$?

$3 \times 6 + 6 = 24$

YES

So the solution is $a = 6$.

ISBN 9780170350969

EXERCISE 9-05

1 Solve $3x + 4 = 19$ by guess-and-check. Select the correct answer A, B, C or D.

A 6 B –5 C 5 D 4

2 Solve $8 - 2x = 12$ by guess-and-check. Select A, B, C or D.

A –2 B 3 C –3 D 2

3 Use guess-and-check to solve each equation.

a $x + 5 = 12$ b $y - 4 = 9$ c $2m = 14$

d $\frac{x}{4} = 6$ e $a - 6 = 15$ f $3w = 27$

g $n + 9 = 22$ h $\frac{b}{5} = 7$ i $t - 6 = 18$

4 Solve each equation.

a $2x - 4 = 6$ b $3a + 4 = 13$ c $5m - 7 = 38$

d $\frac{2x+4}{3} = 8$ e $5w - 4 = 21$ f $\frac{6a-2}{5} = 8$

g $7 - 2x = 13$ h $8 + 3a = 20$ i $\frac{6-2w}{3} = 12$

j $\frac{2x-3}{5} = 9$ k $5 - 3a = 14$ l $10 + 4w = 26$

5 For each problem below, write an equation and solve it to solve the problem.

a Anya thinks of a number. She doubles the number and then subtracts 8. Her answer is 34. What was the number she first thought of?

b Jules thinks of a number. He triples the number and then adds 12. His answer is 39. What was the number he first thought of?

c Tara thinks of a number. She halves it and then adds 15. The result is 24. What was the number Tara first thought of?

Corbis/Laurence Mouton/PhotoAlto

9-06 One-step equations

WORDBANK

inverse operation The 'opposite' process. For example, the inverse operation to adding (+) is subtracting (–).

LHS Left-hand-side of an equation, for example, in $m + 3 = 15$, the LHS is $m + 3$.

RHS Right-hand-side of an equation, for example, in $m + 3 = 15$, the RHS is 15.

These two sides of an equation are equal and when we solve an equation we must keep **LHS = RHS** at all times. The equation must stay **balanced**.

To solve an equation, we use **inverse operations**:

- The inverse of + is –, the inverse of – is +, the inverse of × is ÷ and the inverse of ÷ is ×
- To keep the equation balanced, whatever we do to the LHS, we must do the same to the RHS.

The solution can be checked by substituting the answer into the original equation.

EXAMPLE 10

Solve each equation.

a $x + 3 = 17$ b $w - 5 = 9$ c $4m = 36$ d $\frac{x}{8} = 12$

SOLUTION

a
$x + 3 = 17$ ← opposite of +3 is –3
$x + 3 - 3 = 17 - 3$ ← subtract 3 from both sides
$x = 14$
Check: $14 + 3 = 17$

b
$w - 5 = 9$ ← opposite of –5 is +5
$w - 5 + 5 = 9 + 5$ ← add 5 to both sides
$w = 14$
Check: $14 - 5 = 9$

✱ When setting out, notice we keep the = underneath each other.

c
$4m = 36$ ← opposite of ×4 is ÷4
$\frac{4m}{4} = \frac{36}{4}$ ← divide both sides by 4
$m = 9$
Check: $4 \times 9 = 36$

d
$\frac{x}{8} = 12$ ← opposite of ÷8 is ×8
$\frac{x}{8} \times 8 = 12 \times 8$ ← multiply both sides by 8
$x = 96$
Check: $\frac{96}{8} = 12$

ISBN 9780170350969

EXERCISE 9-06

1 What is the LHS of the equation $x + 8 = 5$? Select the correct answer A, B, C or D.

A 5 B 8 C x D $x + 8$

2 Solve the equation $x + 8 = 5$. Select A, B, C or D.

A $x = 5$ B $x = 3$ C $x = -3$ D $x = 8$

3 Copy and complete:

To solve an equation we use ______ operations. Whatever we do to one side of the equation we must do the ______ to the other. This will keep the equation ______. When solving an equation we must keep the equal signs ______ each other.

4 Write down which operation is the inverse of:

a addition b multiplication c division d subtraction

5 Copy and complete each solution to an equation.

a
$$x + 3 = 8$$
$$x + 3 - 3 = 8 - \_\_\_\_\_\_$$
$$x = \_\_\_\_\_\_$$

b
$$m - 6 = 9$$
$$m - 6 + 6 = \_\_\_\_\_\_ + 6$$
$$m = \_\_\_\_\_\_$$

c
$$3v = 36$$
$$\frac{3v}{3} = \frac{36}{}$$
$$v = \_\_\_\_\_\_$$

d
$$\frac{x}{4} = 5$$
$$\frac{x}{} \times 4 = 5 \times \_\_$$
$$x = \_\_\_\_\_\_$$

6 Solve each equation.

a $w + 6 = 9$ b $m - 4 = 12$ c $b + 14 = 20$

d $v - 18 = 32$ e $7n = 63$ f $9g = 108$

g $\frac{w}{6} = 7$ h $\frac{n}{9} = 5$ i $\frac{x}{8} = 11$

j $r + 15 = 78$ k $12n = 132$ l $8 - y = 24$

7 Solve each problem using an equation.

a Ray thinks of a number, subtracts 13 and ends up with 54. What was the number he first thought of?

b Olivia thinks of a number, divides it by 8 and ends up with 17. What was the number she first thought of?

ISBN 9780170350969

9-07 Two-step equations

To solve an equation:

- do the same to both sides of the equation: this will keep it balanced.
- use inverse operations to simplify the equation (+ and – are inverse operations, × and ÷ are inverse operations)
- write the **solution** (answer) as: x = a number.

EXAMPLE 11

Solve each equation.

a $2a + 5 = 13$ b $\frac{x}{3} - 4 = 8$

SOLUTION

a
$$2a + 5 = 13$$
$$2a + 5 - 5 = 13 - 5 \quad \leftarrow \text{Subtract 5 from both sides}$$
$$2a = 8$$
$$\frac{2a}{2} = \frac{8}{2} \quad \leftarrow \text{Divide both sides by 2.}$$
$$a = 4 \quad \leftarrow \text{This is the solution.}$$

Check: $2 \times 4 + 5 = 13$

b
$$\frac{x}{3} - 4 = 8$$
$$\frac{x}{3} - 4 + 4 = 8 + 4 \quad \leftarrow \text{Add 4 to both sides}$$
$$\frac{x}{3} = 12$$
$$\frac{x}{3} \times 3 = 12 \times 3 \quad \leftarrow \text{Multiply both sides by 3}$$
$$x = 36$$

Check: $\frac{36}{3} - 4 = 8$

Alamy/Leon Swart

ISBN 9780170350969

EXERCISE 9–07

1 Which operation would you do first to solve $3n - 4 = 11$? Select the correct answer A, B, C or D.

A -4 B $\div 3$ C $+4$ D $\times 3$

2 Solve $3n - 4 = 11$. Select A, B, C or D.

A $n = -5$ B $n = 4$ C $n = 3$ D $n = 5$

3 Write down which operation you would do first to solve each equation.

a $3a + 2 = 8$ b $2g - 1 = 5$ c $4m + 3 = 11$

d $5v - 2 = 28$ e $7x + 3 = 24$ f $3c - 6 = 9$

g $8a + 5 = 37$ h $9e - 3 = 24$ i $5v + 6 = 51$

4 Solve each equation in Question 3.

5 Write down which operation you would do first when solving each equation.

a $\frac{x}{2} + 1 = 5$ b $\frac{m}{3} - 2 = 3$ c $\frac{n}{4} + 3 = 6$

d $\frac{s}{8} - 2 = 5$ e $\frac{w}{4} + 3 = 9$ f $\frac{x}{5} + 6 = 8$

g $\frac{x}{7} - 4 = 8$ h $\frac{s}{3} - 8 = 2$ i $\frac{m}{6} + 4 = 7$

6 Solve each equation in Question 5.

7 Solve each problem using an equation.

a Uma thinks of a number. He triples the number and then adds 4 to the number. The result is 49. What is the number?

b Madison thinks of a number. She halves the number and then subtracts 5. The result is 16. What was the number Madison thought of?

iStockphoto/galdzer

CROSSWORD PUZZLE

Make a copy of this page. The numbers from 1 to 26 represent the letters of the alphabet in order, so 1 = A, 2 = B, 3 = C and so on. In the crossword, replace each number with the letter to find the words.

			1						
16			12						
1			7		3				16
20			5		15				18
20			2		14				15
5	24	16	18	5	19	19	9	15	14
18			1		20		14		21
14					1		4		13
					14		5		5
					20		24		18
									1
									12

ISBN 9780170350969

PRACTICE TEST 9

Part A General topics

Calculators are not allowed.

1 Copy and complete: $14 \times 7 = 7 \times$ ______

2 What shape has four right angles?

3 Find the average of 9 and 17.

4 Find $\frac{1}{10}$ of \$250.

5 Find the sum of the first 3 prime numbers.

6 How many degrees in a straight angle?

7 Evaluate $174 \div 3$.

8 Copy and complete: 5, 10, 20, 40, ______.

9 Copy and complete: 2 days = ______ hours.

10 Write these decimals in ascending order: 0.7, 0.07, 0.702, 0.75.

Part B Algebra and equations

Calculators are allowed.

9–01 The laws of arithmetic

11 Which law of arithmetic is shown by $6 \times (10 + 4) = (6 \times 10) + (6 \times 4)$? Select the correct answer A, B, C or D.

A commutative B associative C algebraic D distributive

12 Evaluate 35×11. Select A, B, C or D.

A 353 B 355 C 365 D 385

9–02 Variables

13 Simplify each expression.

a $3 \times a \times b$ b $y \times y$

9–03 From words to algebraic expressions

14 Write as an algebraic expression: triple the sum of w and 5.

9–04 Substitution

15 If $a = 4$ and $b = -2$, find the value of:

a $2a - 3b$ b $6ab + 5b$

16 Find the value of $\frac{3x - y}{6}$ if $x = 6$ and $y = -12$.

9–05 Equations

17 Solve each equation by guess-and-check.

a $5x - 4 = 21$ b $18 - 2a = 10$

9–06 One-step equations

18 Solve each equation.

a $x + 6 = 23$ b $\frac{x}{8} = 12$

9-07 Two-step equations

19 Solve each equation.

a $3w - 6 = 18$ b $\frac{x}{4} + 5 = 8$

20 In which line is an error made when solving the equation below?

$$3a - 8 = 13$$

$3a - 8 + 8 = 13 - 8$	Line 1
$3a = 5$	Line 2
$\frac{3a}{3} = \frac{5}{3}$	Line 3
$a = 1\frac{2}{3}$	Line 4

ISBN 9780170350969

SHAPES AND SYMMETRY

10

WHAT'S IN CHAPTER 10?

10-01 Polygons
10-02 Translation and reflection
10-03 Rotation
10-04 Composite transformations
10-05 Line symmetry
10-06 Rotational symmetry
10-07 Prisms and pyramids

IN THIS CHAPTER YOU WILL:

- name the different types of polygons, including regular polygons
- perform transformations: translate, reflect and rotate shapes
- draw and count the axes of symmetry of different shapes
- identify the rotational symmetry of different shapes
- identify and sketch prisms and pyramids

Shutterstock.com/STILLFX

10–01 Polygons

WORDBANK

polygon A flat shape with straight sides.

regular polygon A polygon with equal sides and equal angles.

A **polygon** is a shape with straight sides. For example, a triangle and square are polygons but a circle and oval are not. Polygons have special names depending on the number of sides they have.

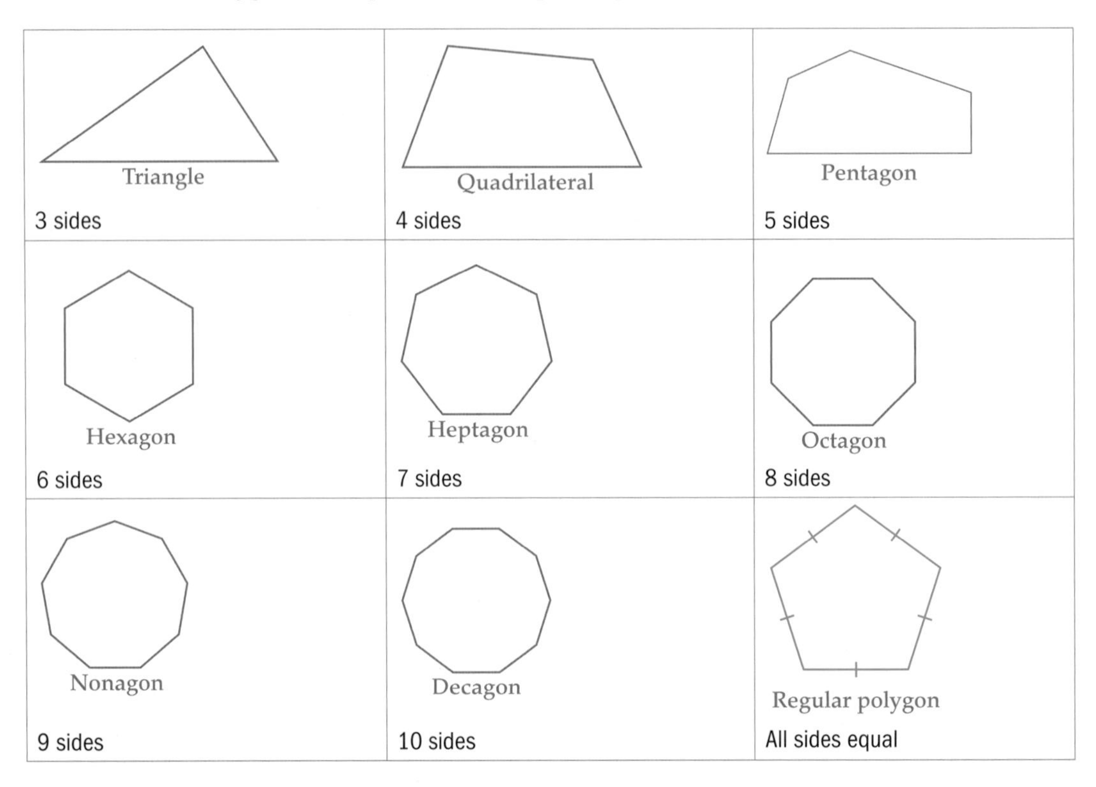

A **regular polygon** has all sides equal and all angles equal. An **irregular** polygon has unequal sides.

EXAMPLE 1

Name each polygon, stating whether they are regular or irregular.

a 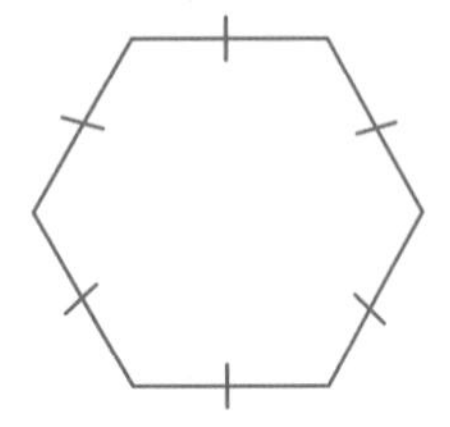

b 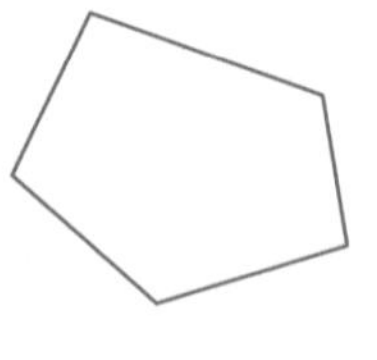

c 

SOLUTION

a hexagon, regular ← 6 sides, marks on sides

b pentagon, irregular ← 5 sides, no marks on sides

c decagon, irregular ← 10 sides, no marks on sides

ISBN 9780170350969

EXERCISE 10–01

1 What is a 4-sided polygon called? Select the correct answer A, B, C or D.

A pentagon B quadrilateral C hexagon D decagon

2 What is the best name for this shape? Select A, B, C or D.

A triangle B quadrilateral

C irregular triangle D regular triangle

3 Name each polygon.

a

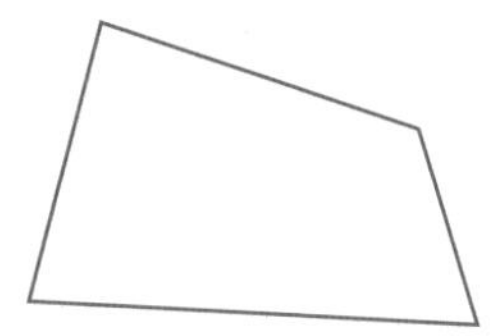

b

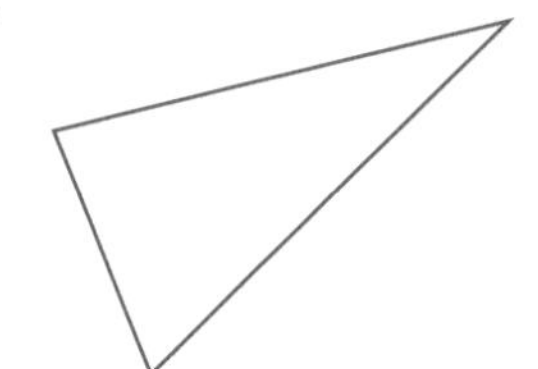

c

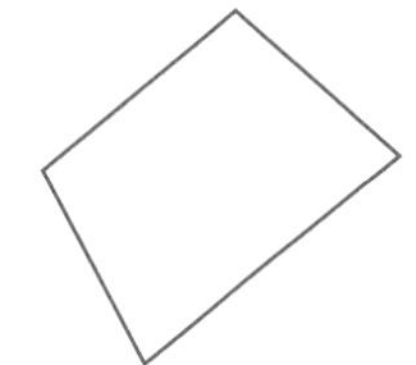

d

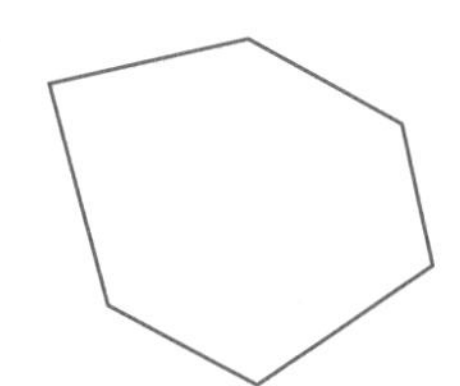

e

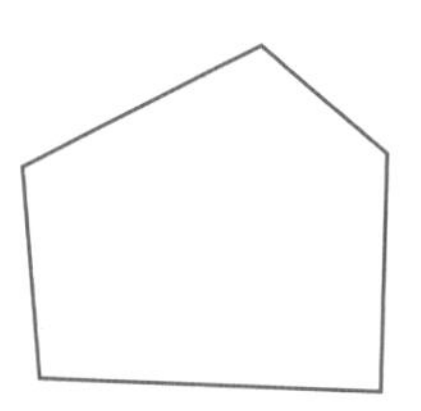

4 Draw a neat sketch of each polygon.

a triangle b hexagon c pentagon d quadrilateral

5 Name the polygon that has:

a 8 sides b 10 sides

6 Name the polygon shaded in this picture.

7 What polygons do you see on this wooden door?

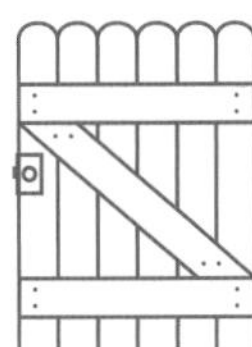

ISBN 9780170350969

8 i What polygons can you see in these road signs?

a

b

c

ii What does each of these road signs mean?

9 Classify these shapes as regular or irregular, and name them.

a

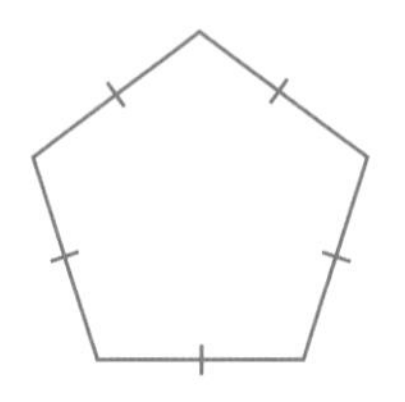

b

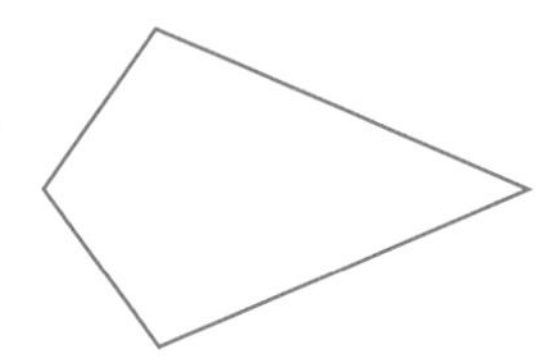

c

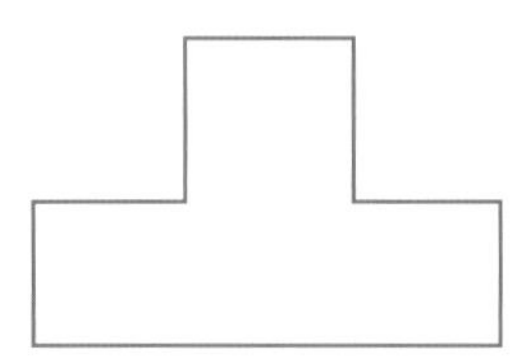

d

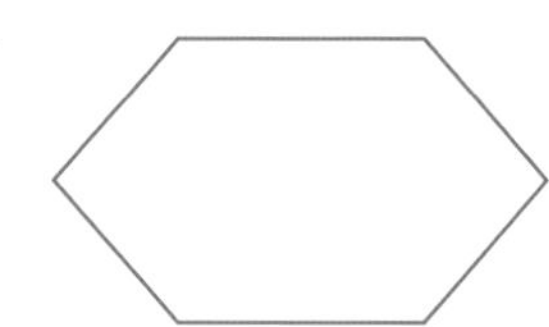

e

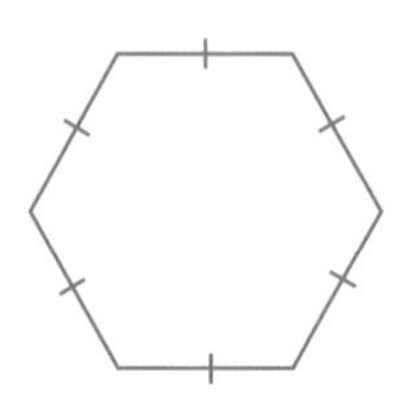

10 Make up a design using at least five different shapes. You can use a template to draw the shapes. Colour in your design.

Shutterstock.com/anaken2012

 ISBN 9780170350969

10–02 Translation and reflection

WORDBANK

transformation The process of moving or changing a shape, by translation, reflection or rotation.

translation The process of 'sliding' a shape, moving it up, down, left or right.

reflection The process of 'flipping' a shape, making it back-to-front like in a mirror.

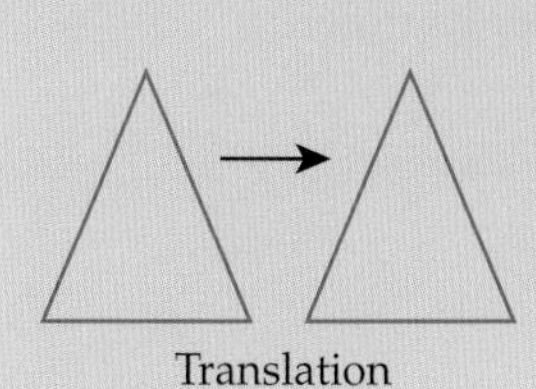
Translation

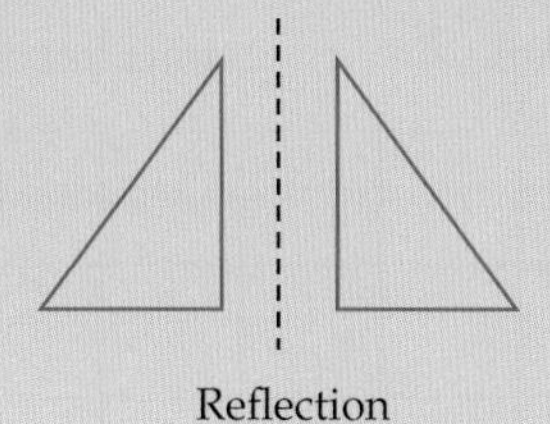
Reflection

EXAMPLE 2

Translate the L shape 3 units right.

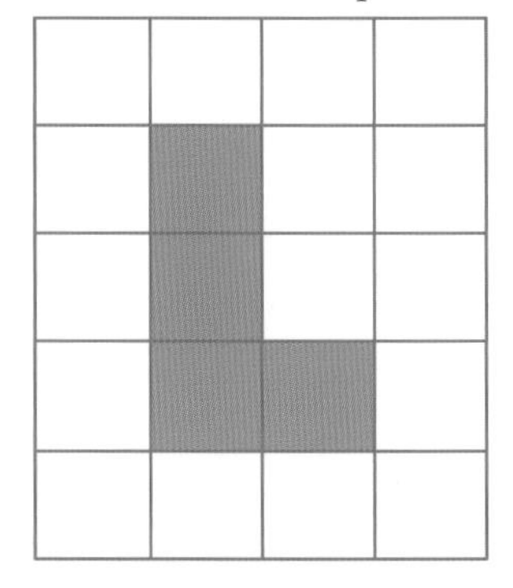

SOLUTION

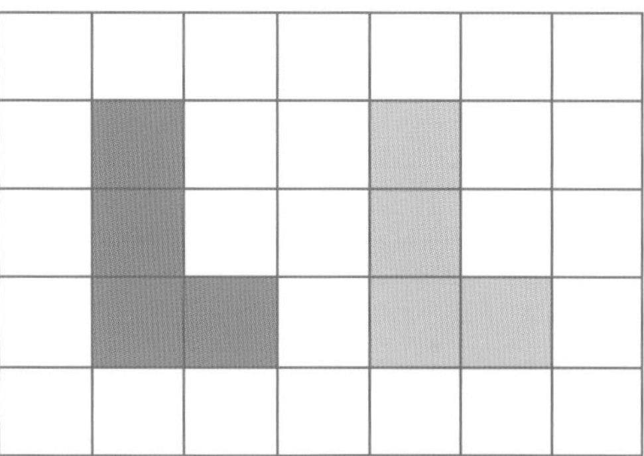

EXAMPLE 3

Reflect the L shape in the line AB.

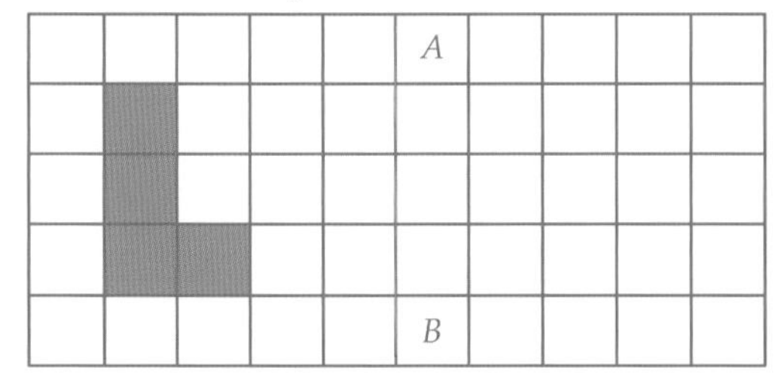

SOLUTION

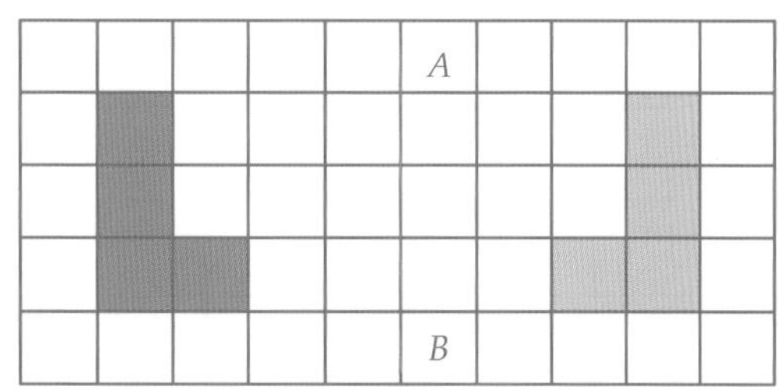

EXERCISE 10–02

1 What type of transformation involves a shape being flipped over a line? Select the correct answer A, B, C or D.

A translation B reflection C transformation D enlargement

2 What type of transformation involves a shape being slid 4 units down? Select A, B, C or D.

A translation B reflection C transformation D enlargement

3 Copy this rhombus on grid paper and perform each translation on it.

a 4 units right b 2 units up c 2 units left d 3 units down

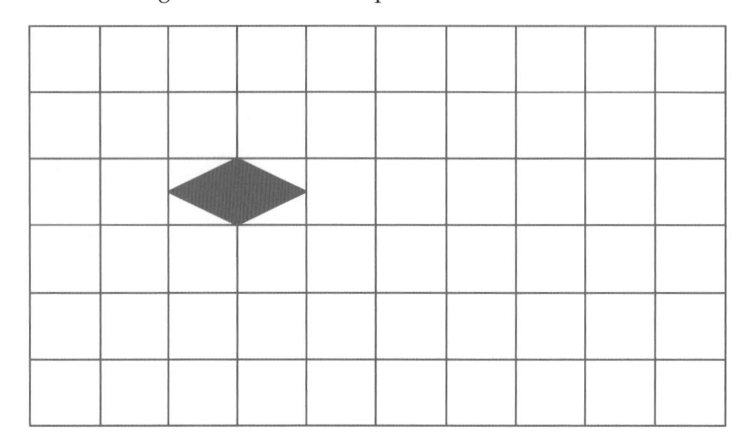

4 Copy each shape onto grid paper and reflect it in the line drawn.

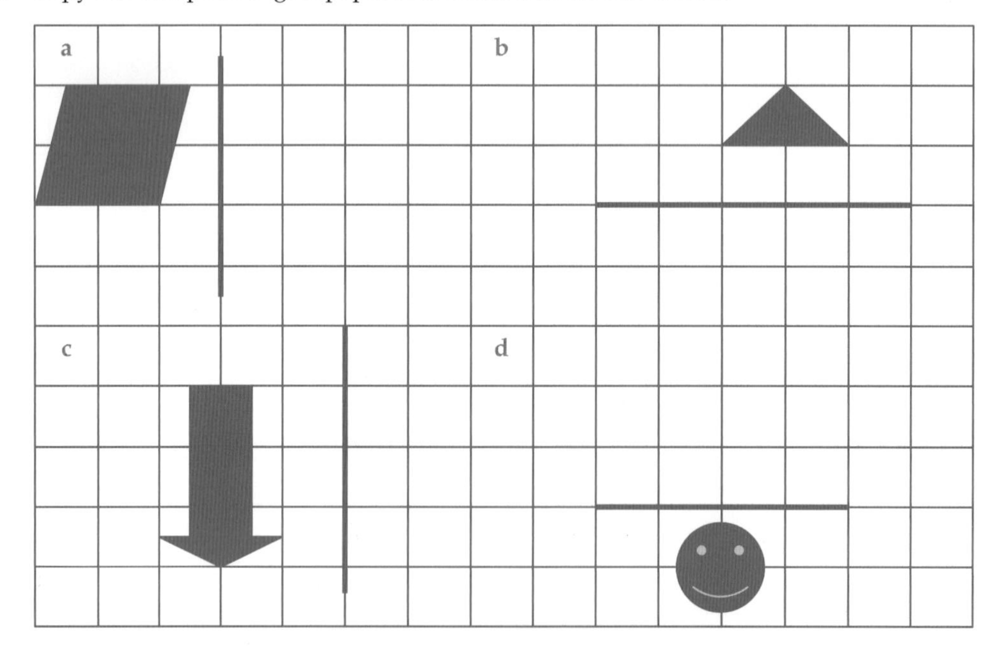

 ISBN 9780170350969

5 Copy each shape onto square dot paper and translate it in the direction and distance indicated by the arrow.

a

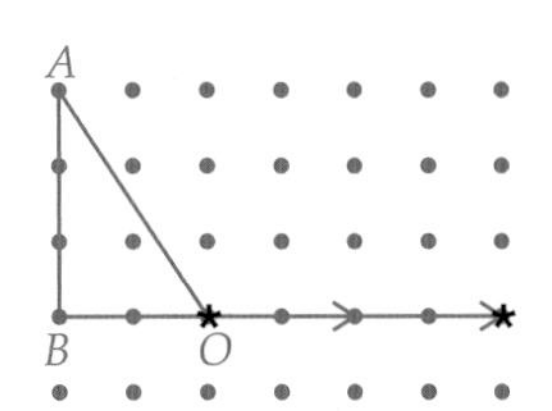

b

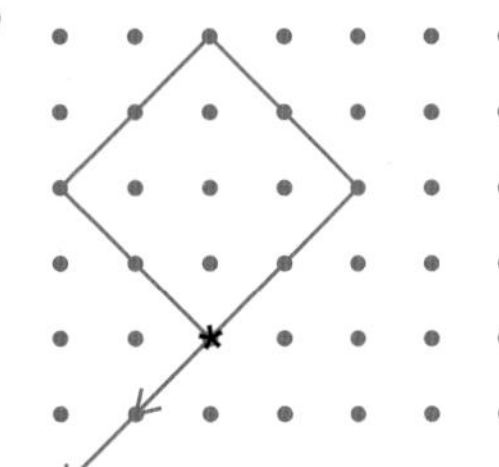

c

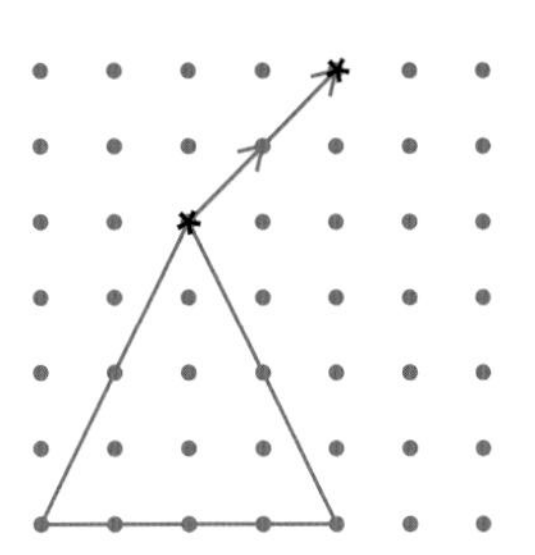

d

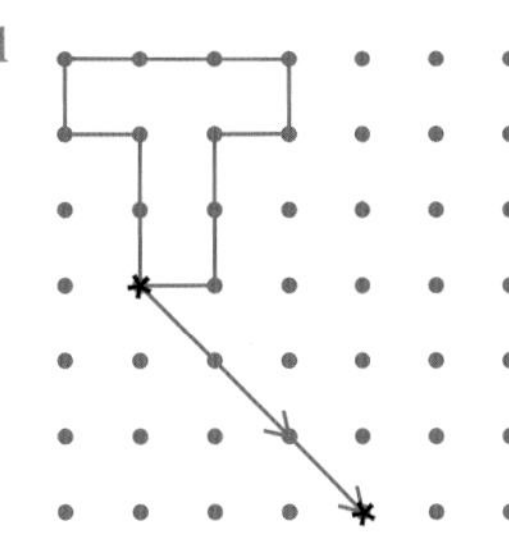

6 Copy each shape onto grid paper and reflect it across the dotted line.

a

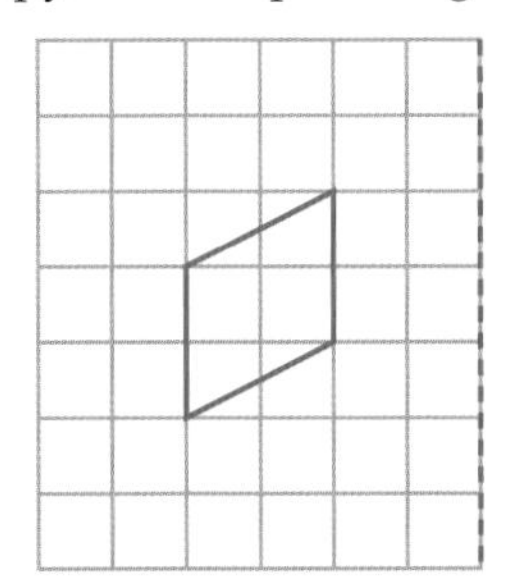

b

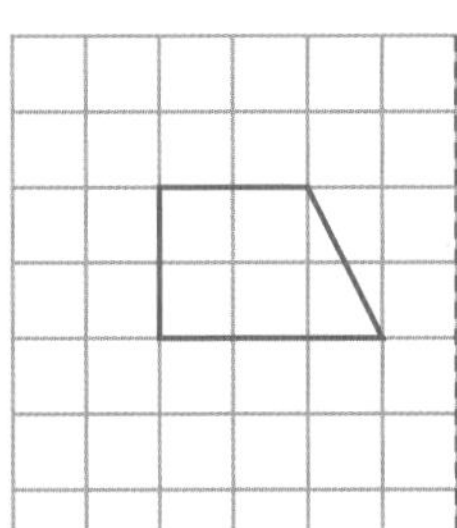

c

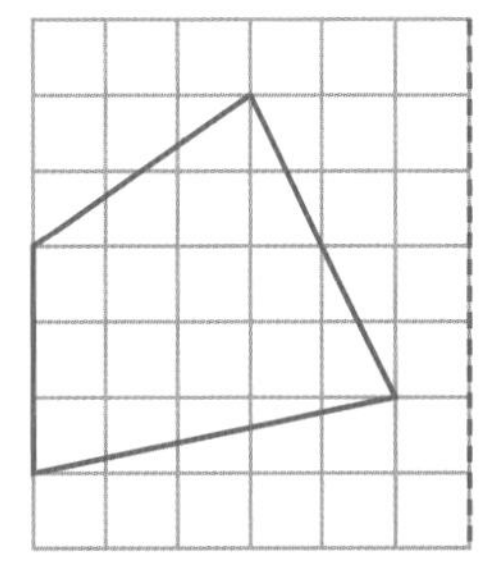

7 For each transformation shown state if it is a translation or reflection.

a

b

ISBN 9780170350969

WORDBANK

rotation The process of 'spinning' a shape, around a point, tilting it sideways or upside-down

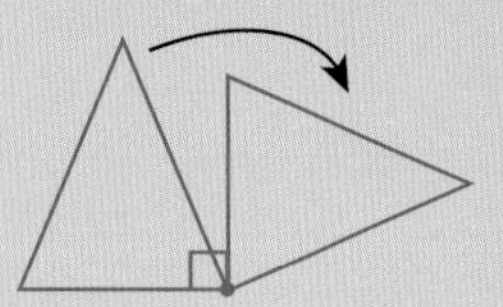

Rotation of a triangle about a vertex 90° clockwise

Clockwise rotation is the same direction as the hands on a clock:

Anti-clockwise rotation is in the opposite direction:

EXAMPLE 4

Rotate the L shape clockwise 90° about point O.

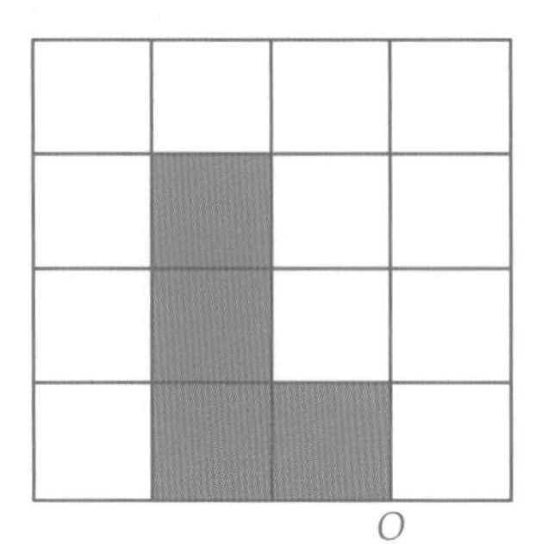

SOLUTION

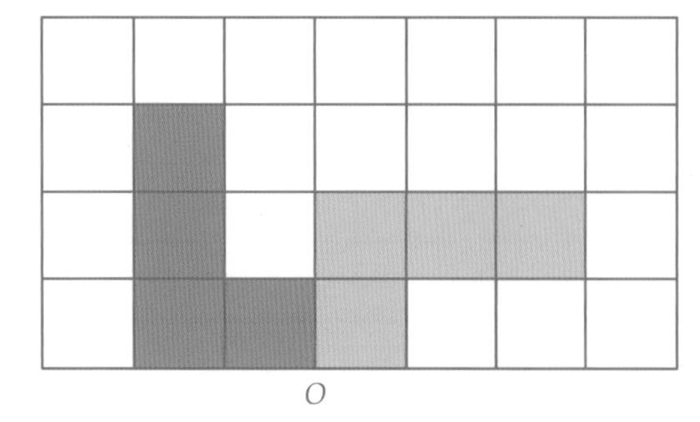

 ISBN 9780170350969

EXERCISE 10–03

1 What type of transformation involves a shape being turned 45° clockwise? Select the correct answer A, B, C or D.

A translation B reflection C transformation D rotation

2 Describe this rotation. Select A, B, C or D.

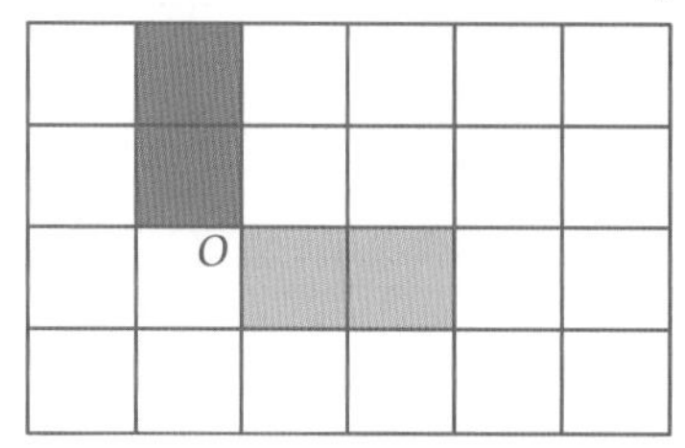

A clockwise 90°
B anticlockwise 180°
C clockwise 180°
D anticlockwise 90°

3 Copy this diagram onto grid paper and perform each rotation about point O.

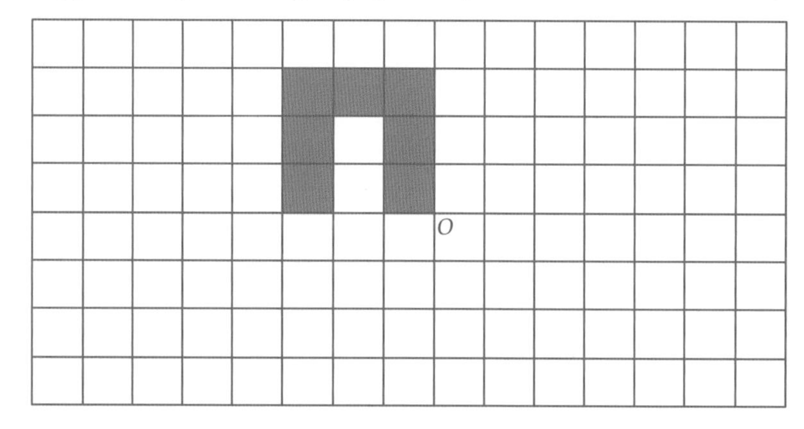

a 90° clockwise
b 180° anticlockwise
c 90° anticlockwise
d 270° clockwise

4 Copy each diagram onto grid paper and perform each rotation about point O.

a

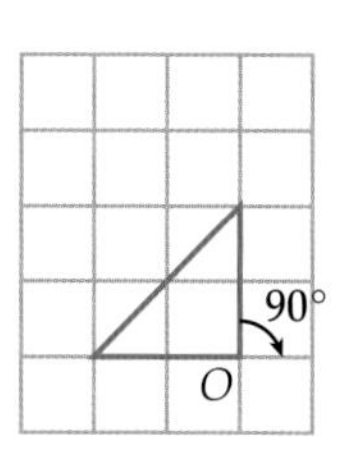

b

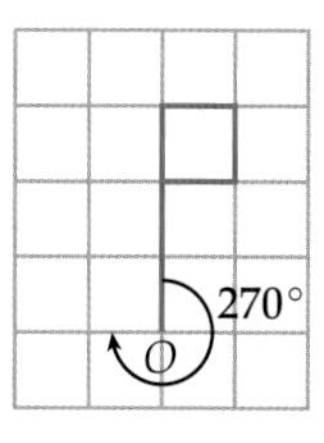

c

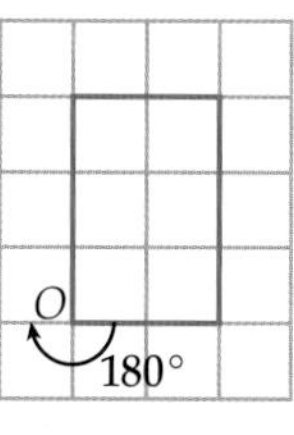

d

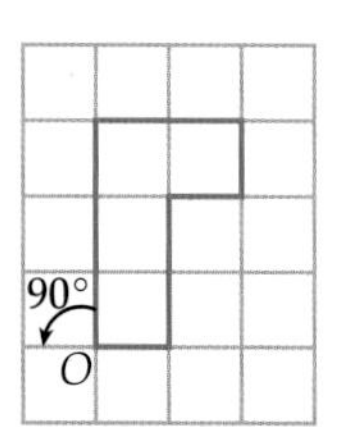

e

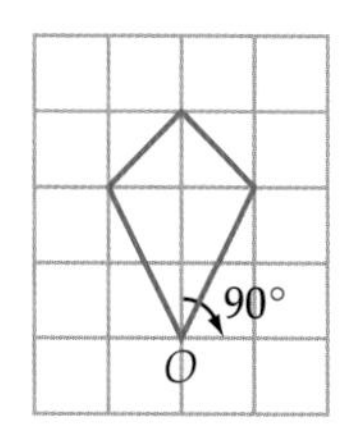

f

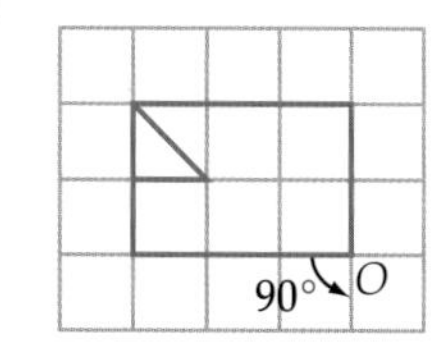

10–04 Composite transformations

A **composite transformation** is a combination of two or more transformations on the one shape, such as a reflection followed by a rotation.
This hexagon has been translated 6 units right and then reflected in the line AB.

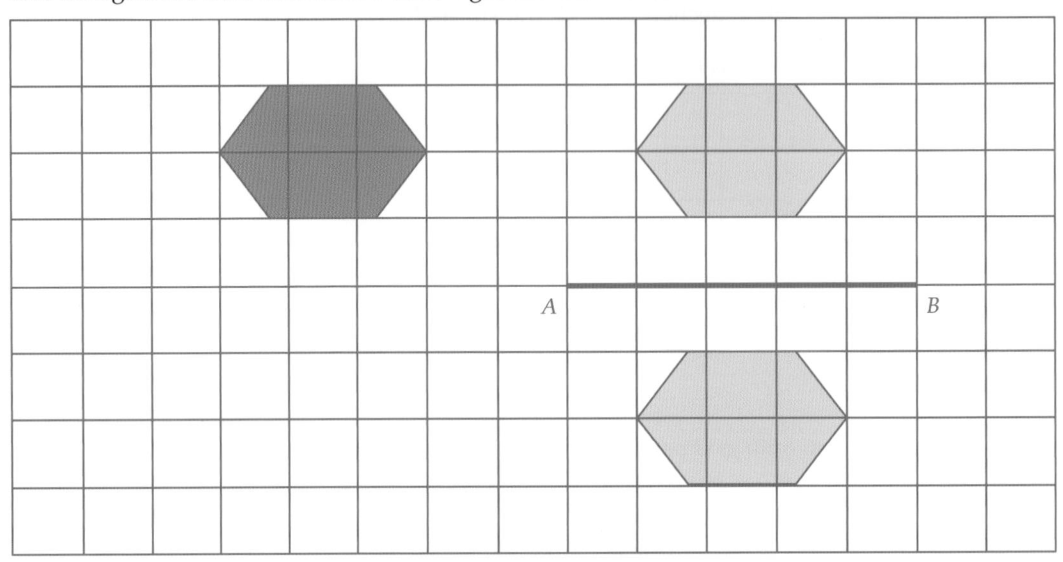

This shape has been rotated 180° about O and then translated 5 units left.

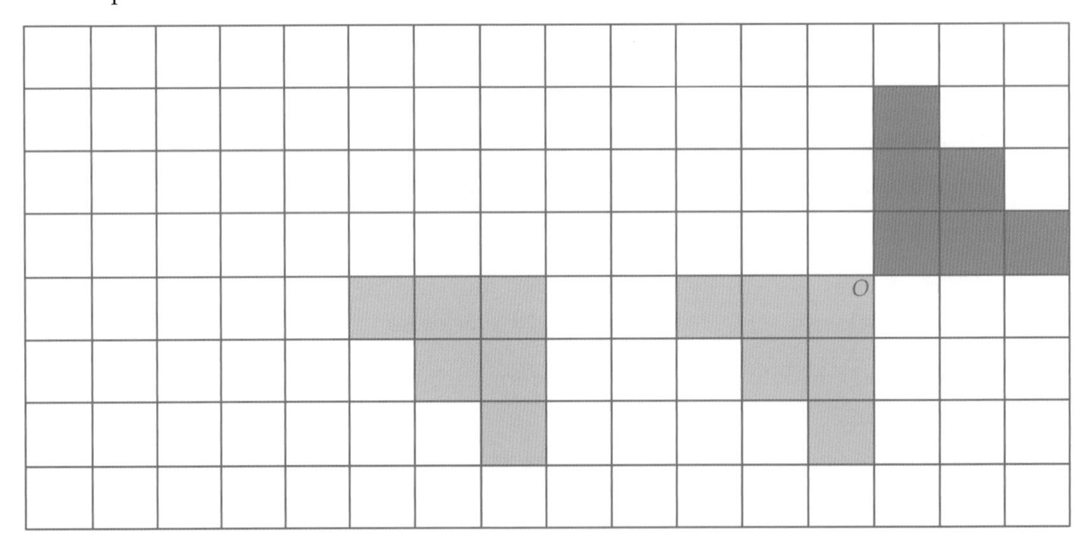

ISBN 9780170350969

EXERCISE 10–04

1 Which transformation is a slide? Select the correct answer A, B, C or D.

A translation B reflection C enlargement D rotation

2 Which transformation is a turn? Select A, B, C or D.

A translation B reflection C enlargement D rotation

3 Copy each diagram onto grid paper and perform the composite transformations stated.

a Translate 4 units right and then reflect in line AB

b Rotate 180° clockwise about O and then reflect in line CD

c Translate 2 units right and 1 unit down and then rotate 270° anticlockwise about point P

d Reflect in line EF and then rotate 90° clockwise about Q

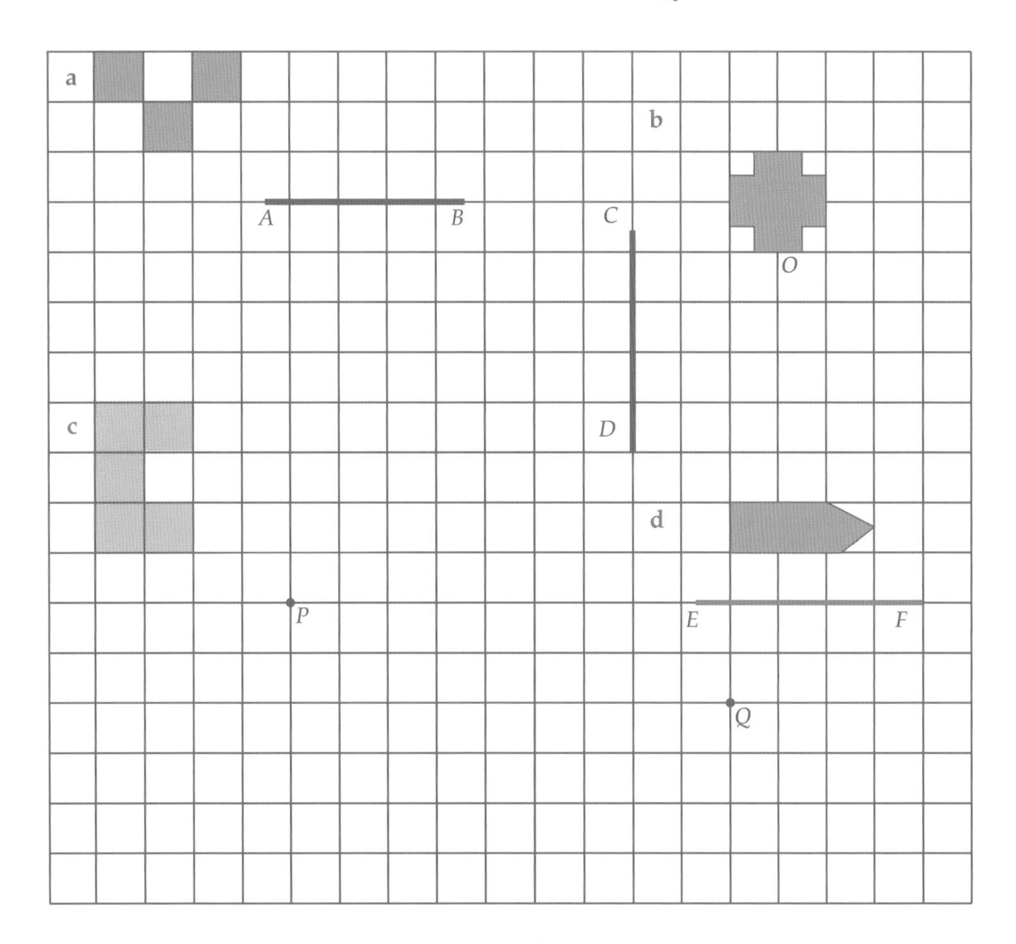

4 Describe each composite transformation.

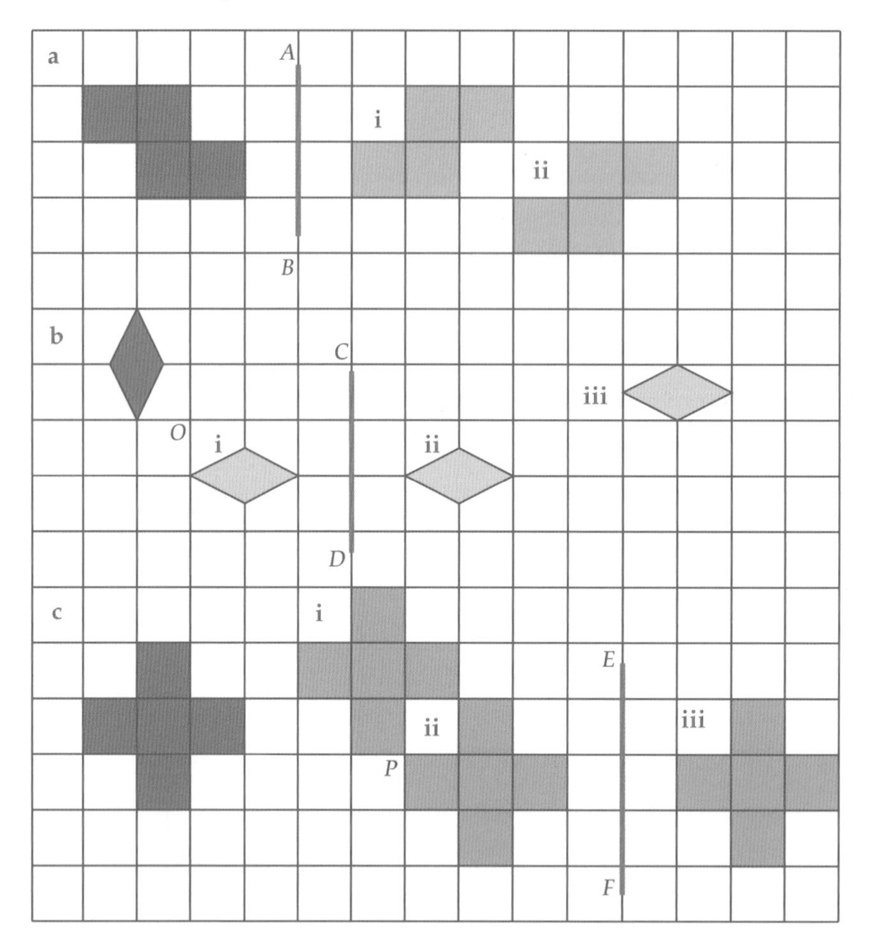

5 Copy the pentagon, rotate it 180° about *O*, then translate 5 units right and then reflect in the line *AB*.

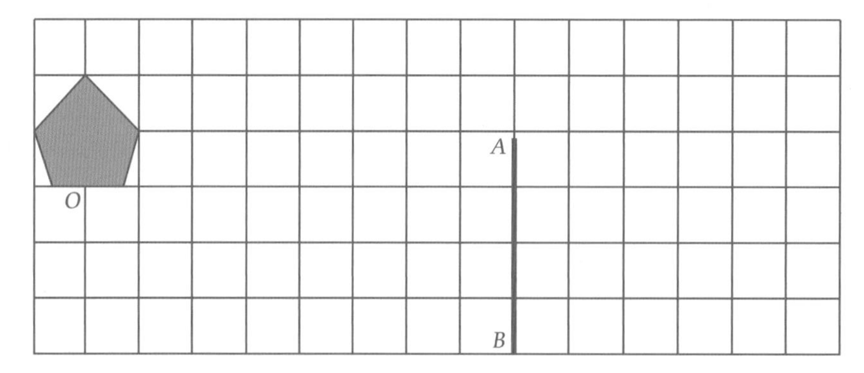

 ISBN 9780170350969

10–05 Line symmetry

WORDBANK

line symmetry A shape has line symmetry if one half exactly folds onto the other half. One half is the mirror-image of the other half.

axis of symmetry The line that divides a symmetrical shape in half. 'Axis' means line (plural is 'axes').

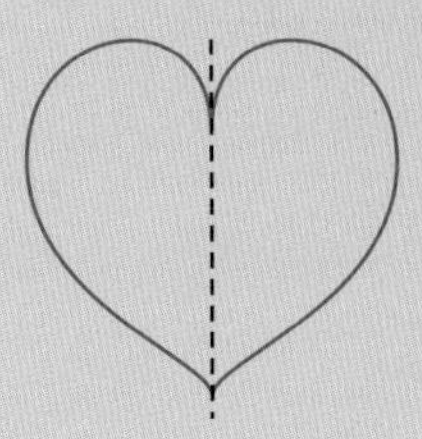

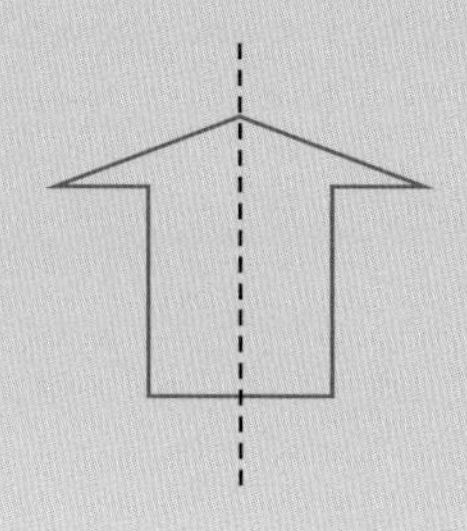

iStockphoto/zxcynosure

Shutterstock.com/Hatchapong Palurtchaivong

Shutterstock.com/Radu Bercan

EXAMPLE 5

For each symmetrical shape, draw the axes of symmetry.

a

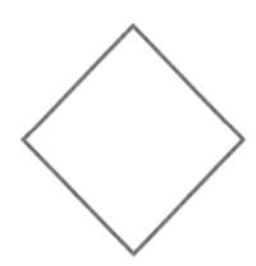

b

c

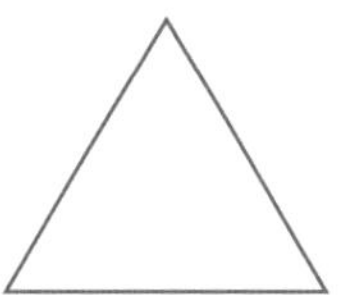

SOLUTION

a

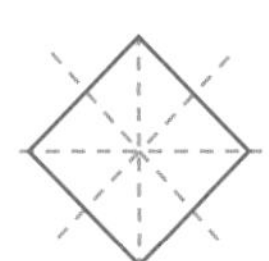

b

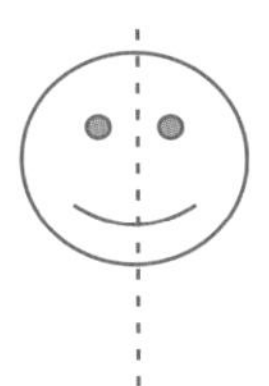

c

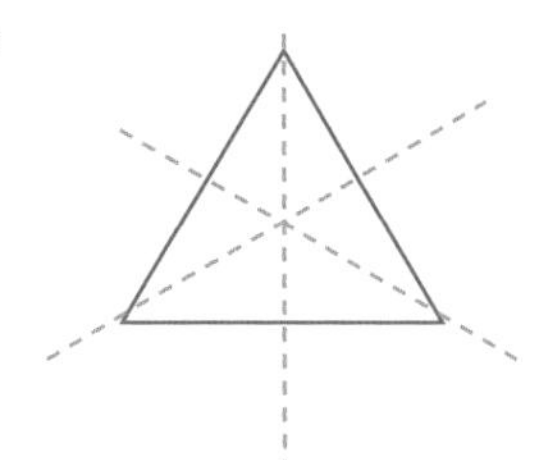

EXERCISE 10–05

1 A line of symmetry may be __________. Select the correct answer A, B, C or D.

A horizontal B diagonal C vertical D all of these

2 How many lines of symmetry does this star have? Select A, B, C or D.

A 1 B 3 C 5 D 4

3 Copy and complete each diagram using the line shown as the axis of symmetry.

a

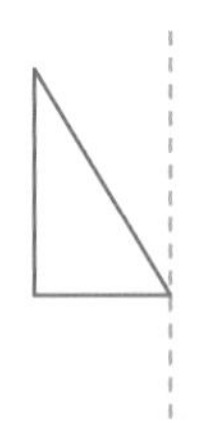

b

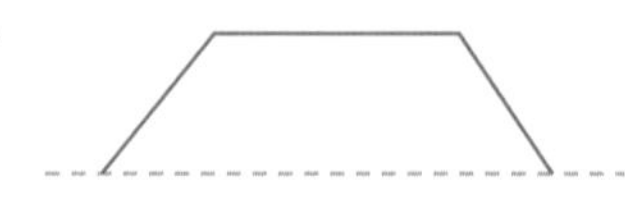

c

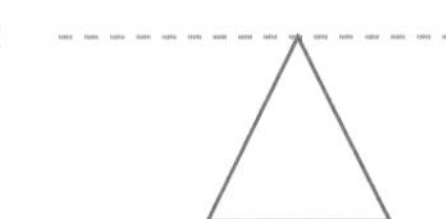

d

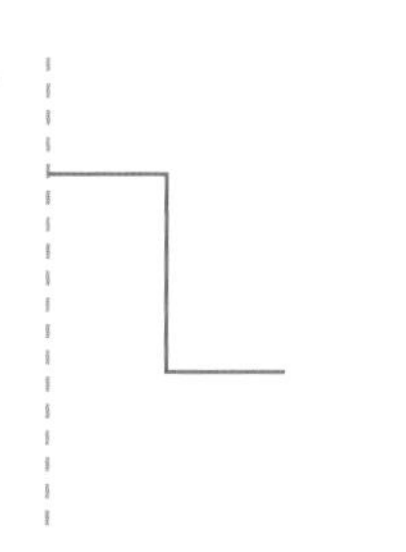

e

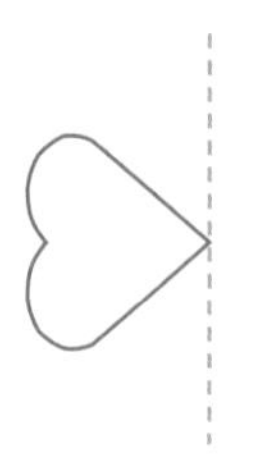

f

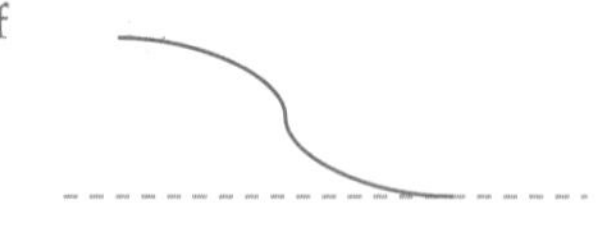

4 Copy each shape and draw the axes of symmetry.

a

b

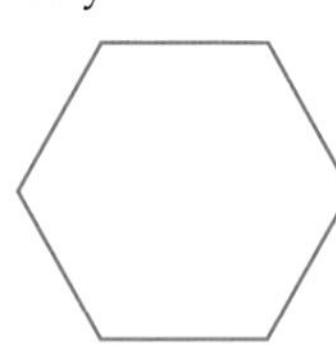

c

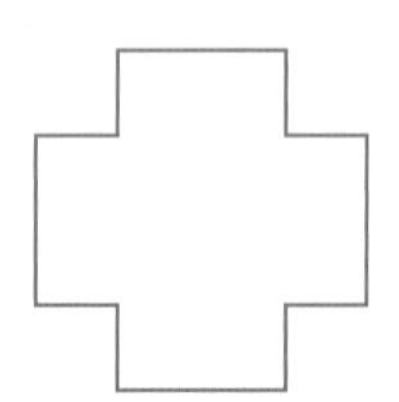

d

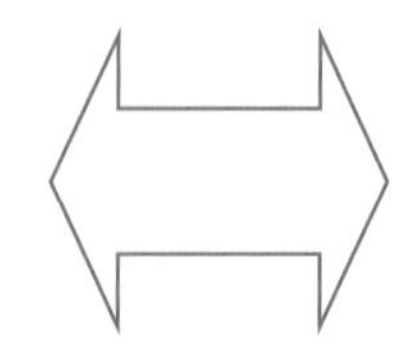

5 Write down the letters in the alphabet A to Z and draw in all lines of symmetry.

ISBN 9780170350969

10–06 Rotational symmetry

WORDBANK

rotational symmetry A shape has rotational symmetry if it can be spun around its centre so that it fits onto itself before a complete revolution.

This cross-shape has rotational symmetry because it fits onto itself when rotated through 90°, 180°, 270° and 360°. It has **rotational symmetry of order 4**.

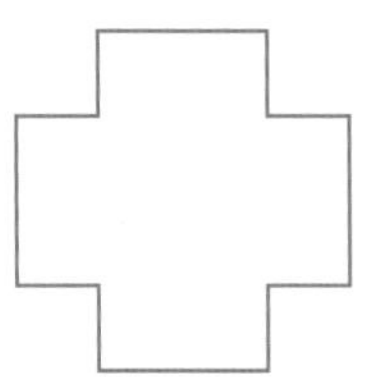

EXAMPLE 6

Does each figure have rotational symmetry? If so, state the order of rotational symmetry.

a 

b 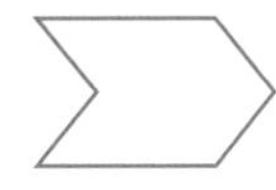

c 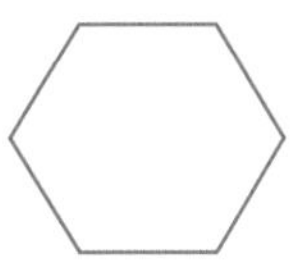

SOLUTION

a Yes, if turned through 180°. Order is 2.

b No

c Yes, if rotated through 60° turns. Order is 6.

EXAMPLE 7

State the order of rotational symmetry of each shape.

a 

b 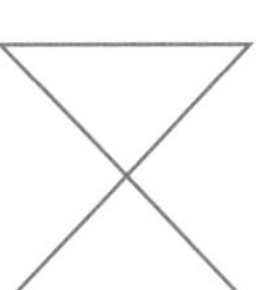

SOLUTION

a The star has rotational symmetry of order 5.

b The shape has rotational symmetry of order 2.

Shutterstock.com/alfocome

Shutterstock.com/florin oprea

Shutterstock.com/SnowWhiteimages

EXERCISE 10-06

1 A shape has rotational symmetry if it remains the same when rotated through what? Select the correct answer A, B, C or D.

A 360° B more than 360° C less than 180° D less than 360°

2 The number of times a shape can be rotated and remain the same is called its _______ of rotational symmetry. Select A, B, C or D.

A symmetry B rotation C order D orientation

3 For each shape, state if it has rotational symmetry.

a 

b 

c 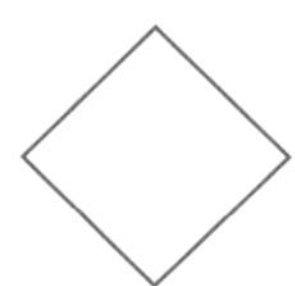

d 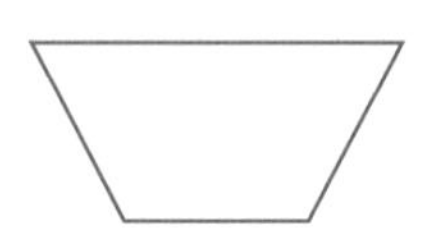

e 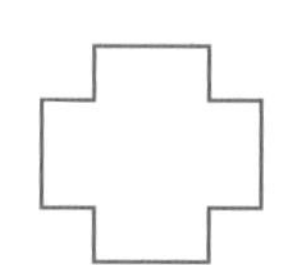

f 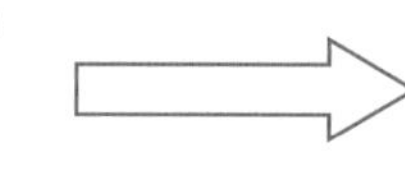

g 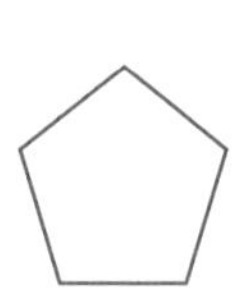

h 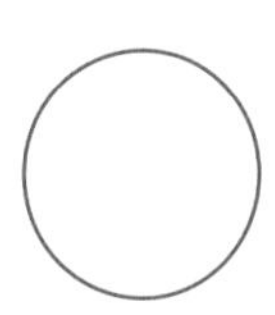

i 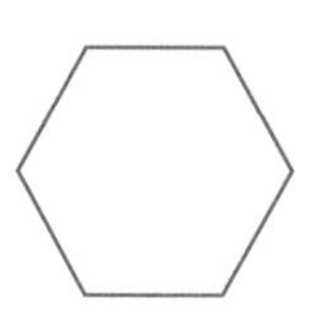

4 For each shape with rotational symmetry in Question 3, write down:

i how many degrees it must be turned first to fit onto itself

ii its order of rotational symmetry

5 How many degrees must each shape be turned first so that it fits onto itself?

a 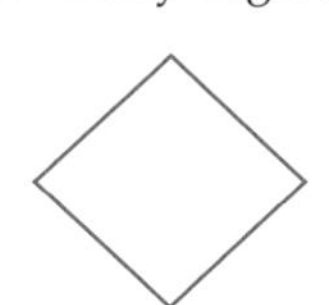

b 

c 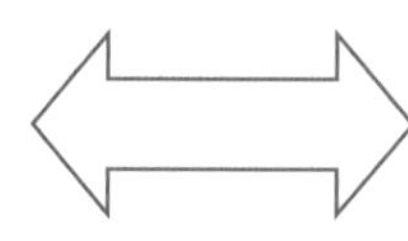

d 

e 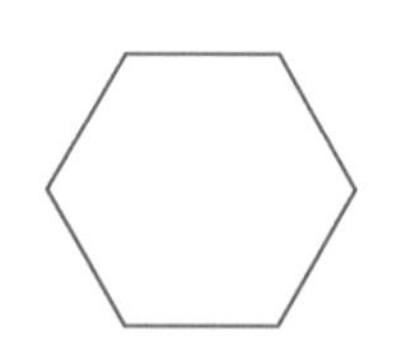

f 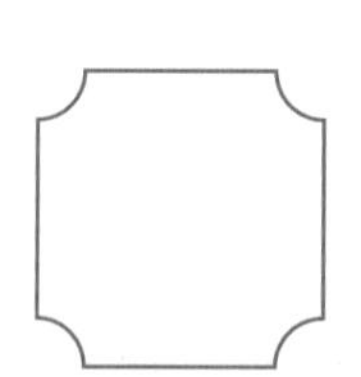

6 Write down the order of rotational symmetry for each shape in Question 5.

7 Which numbers from 0 to 9 have rotational symmetry?

ISBN 9780170350969

10-07 Prisms and pyramids

A **prism** is a solid shape with identical cross-sections that have straight sides.
That means if you cut across a prism, you will always get the same shape.
The name of the prism comes from the shape of its cross-sections.
A prism can be cut through as shown below and its cross-section will remain the same shape and size.

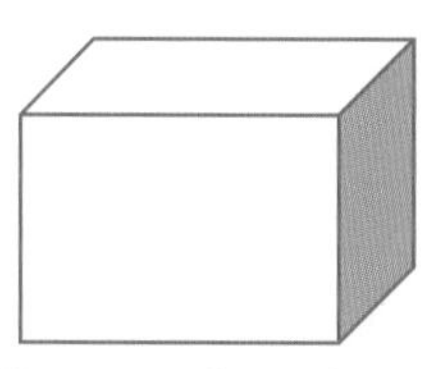

Rectangular prism

cross-section

identical faces

Hexagonal prism

A **pyramid** is a solid shape with a polygon as its base, and edges that start at the corners of the base and meet at a point above the base.
The side faces of a pyramid are triangles.
The name of the pyramid comes from the shape of its base.

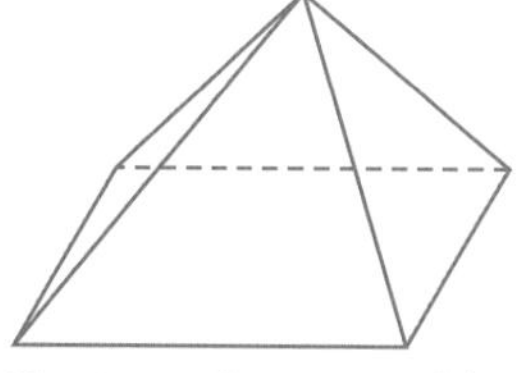

Rectangular pyramid

EXAMPLE 8

Name each prism.

a

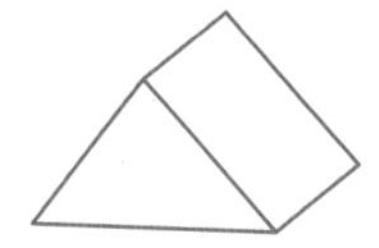

b

SOLUTION

a Triangular prism ← The cross-section is a triangle

b Pentagonal prism ← The cross-section is a pentagon

EXAMPLE 9

Name each pyramid.

a

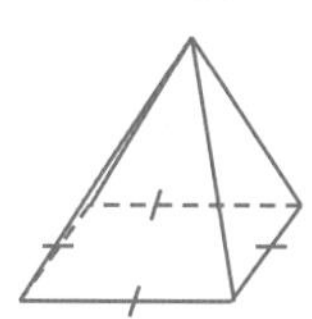

b

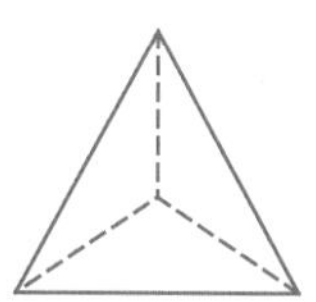

SOLUTION

a Square pyramid ← The base is a square

b Triangular pyramid ← The base is a triangle

EXERCISE 10–07

1 Name the solid shape with sides sloping and meeting at a point.
Select the correct answer A, B, C or D.

A polygon B pyramid C prism D pentagon

2 Name this solid. Select A, B, C or D.

A pyramid B triangular pyramid
C polygon D triangular prism

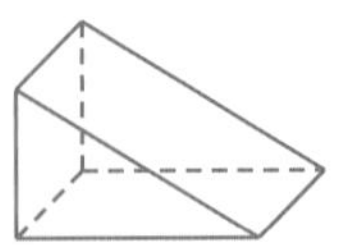

3 For each of the figures below, state if it is a prism, pyramid, or neither.

a 

b 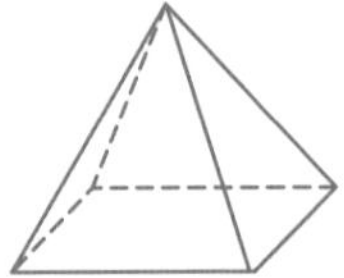

c 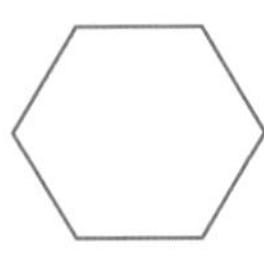

d 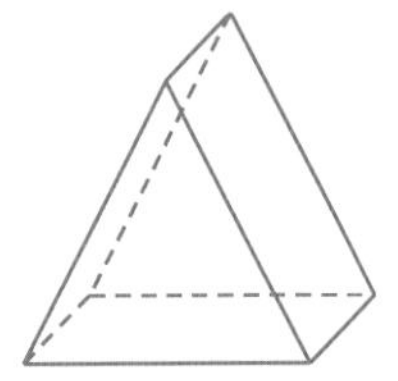

e 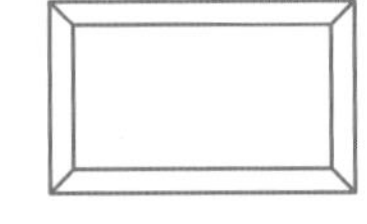

f 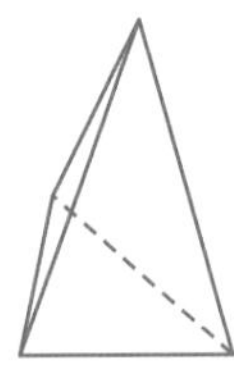

g 

h 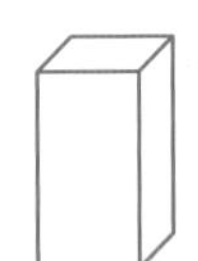

i 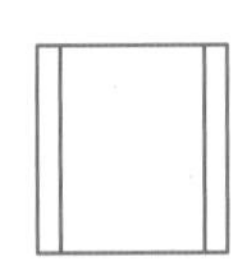

4 Name each of the prisms or pyramids in Question 3 with its correct name.

5 What shape is the cross-section for each of the prisms below?

a 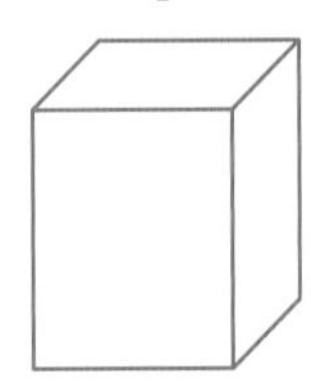

b 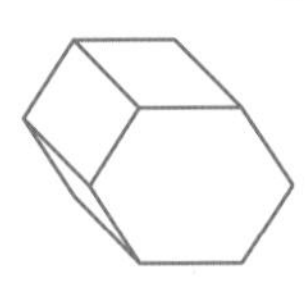

c 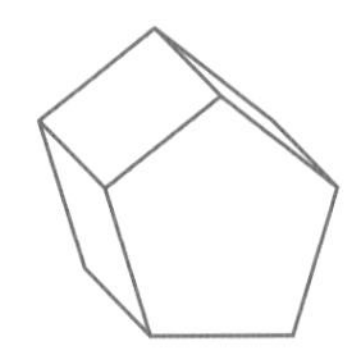

6 Name each of the prisms in Question 5.

7 What is the difference between a prism and a pyramid?

ISBN 9780170350969

LANGUAGE ACTIVITY

SPOT THE DIFFERENCE

There is 1 difference from one shape to another.

Spot the difference and then name each shape.

1

2

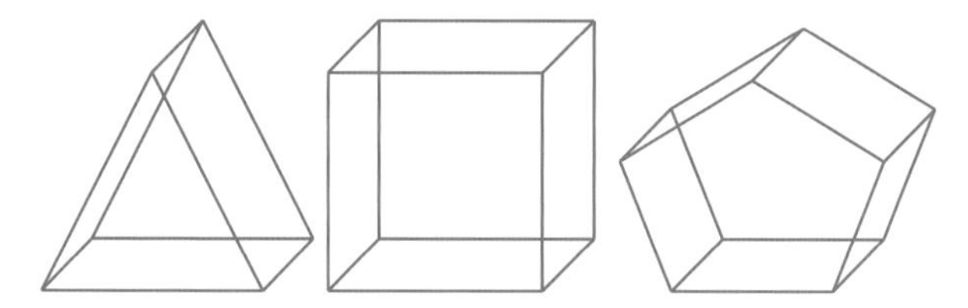

3

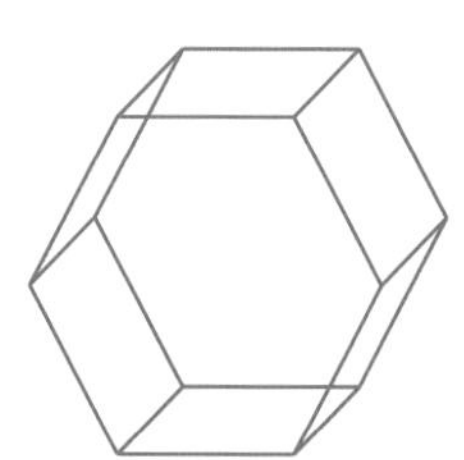

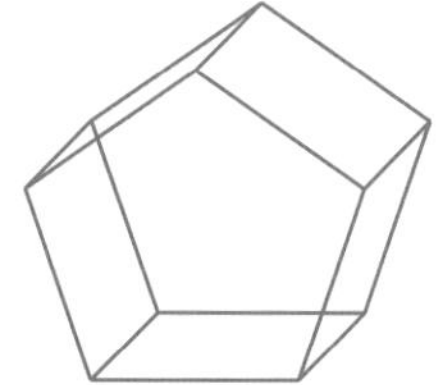

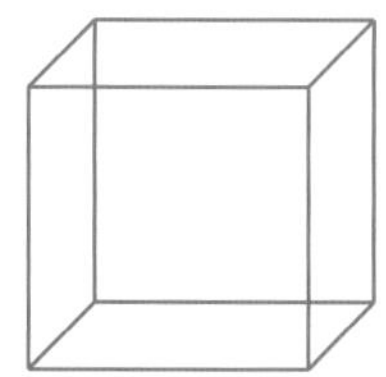

4

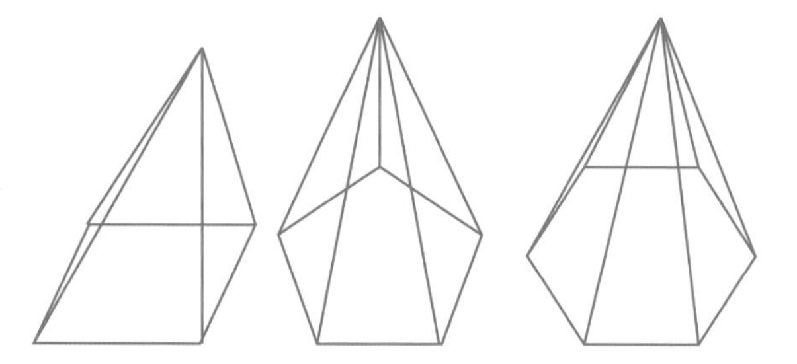

ISBN 9780170350969

PRACTICE TEST 10

Part A General topics

Calculators are not allowed.

1 If $x = 4$ and $y = 6$, evaluate $x^2 + 3y$.

2 What are the possible outcomes when a coin is tossed?

3 Copy and complete this pattern:
8, 4, 2, 1, ____

4 Evaluate 9×12.

5 Evaluate $-2 - (-3) + (-1)$.

6 What is the value of $a + b$ for this diagram?

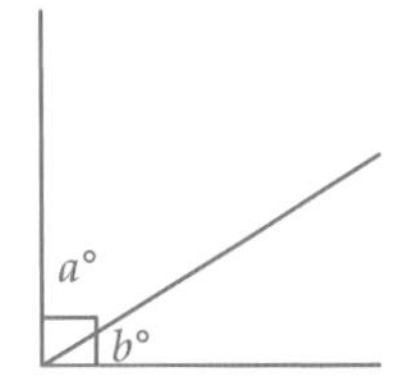

7 A plane flight started at 8:05 p.m. and lasted 2 hours 15 minutes. What time did it finish?

8 Copy and complete: 685 cm = ____ m.

9 Convert 0.8 to a simple fraction.

10 Use a factor tree to write 54 as a product of its prime factors.

Part B Shapes and symmetry

Calculators are allowed.

10–01 Polygons

11 What is a 5-sided polygon called? Select the correct answer A, B, C or D.

A pentagon B quadrilateral C hexagon D decagon

12 What is the name of this shape? Select A, B, C or D.

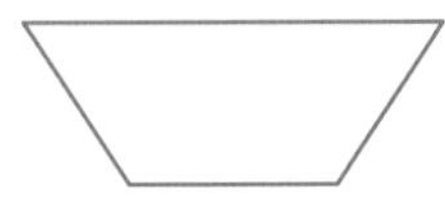

A pentagon B quadrilateral C hexagon D decagon

10–02 Translation and reflection

13 If a shape is slid up 2 units then the movement is called a ______. Select A, B, C or D.

A translation B reflection C transformation D enlargement

14 Describe a reflection in words.

10–03 Rotation

15 A shape rotated 180° will appear what? Select A, B, C or D.

A back-to-front B upside-down C enlarged D sideways

ISBN 9780170350969

10-04 Composite transformations

16 Copy the shape onto grid paper and translate it 4 units right and then reflect it in the line *AB*.

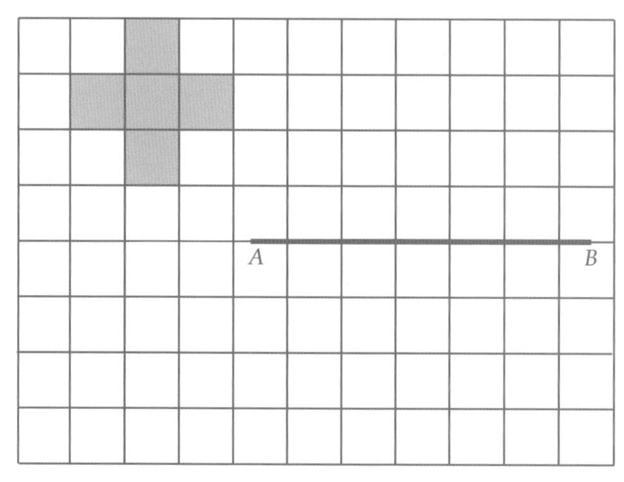

10-05 Line symmetry

17 Copy and complete each diagram diagram using the line shown as the axis of symmetry:

a

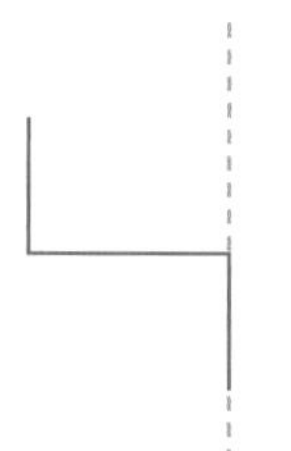

b

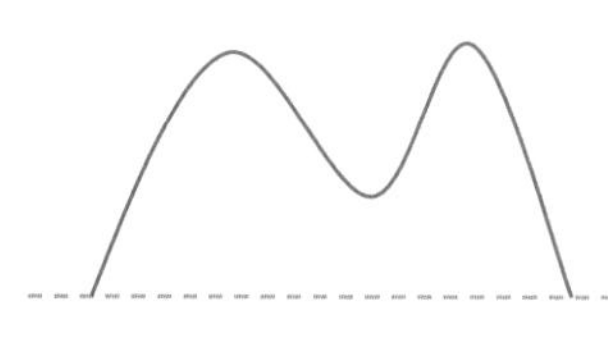

18 Copy the shapes below and draw in any lines of symmetry.

a

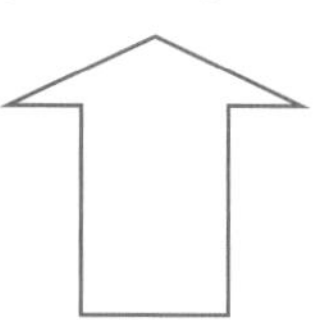

b

10-06 Rotational symmetry

19 Do the following figures have rotational symmetry? If so, state the degree of rotational symmetry.

a

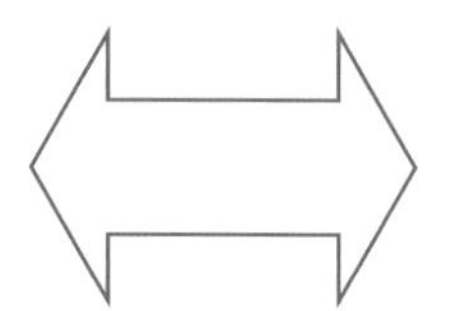

b

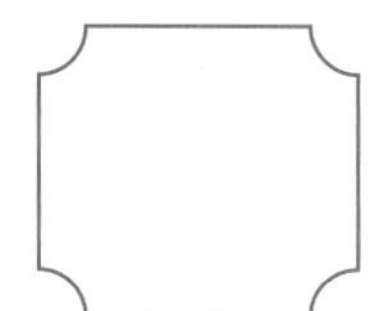

ISBN 9780170350969

10–07 Prisms and pyramids

20 Name each solid.

a 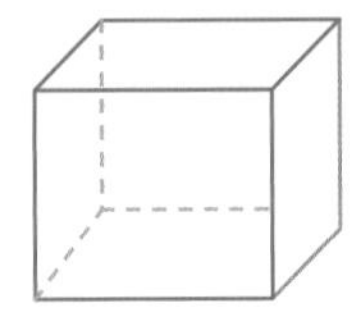

b 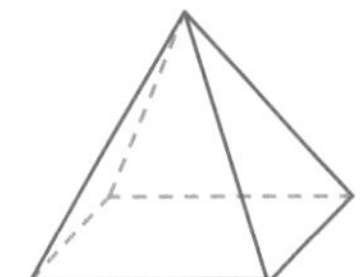

c 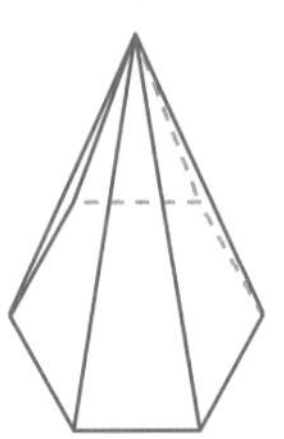

 ISBN 9780170350969

GEOMETRY

11

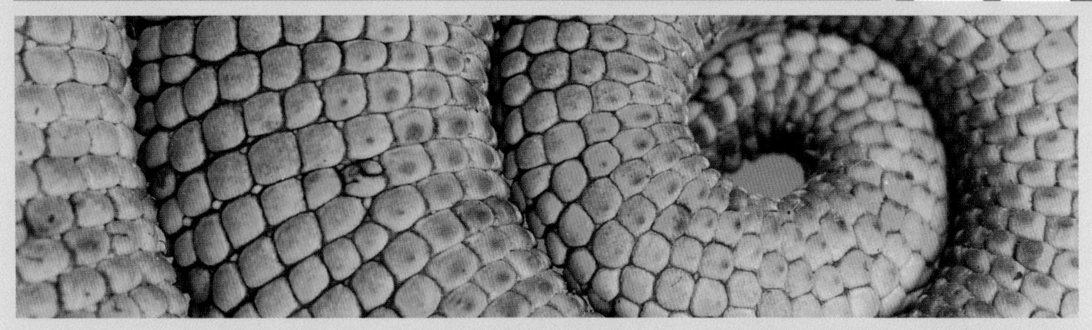

WHAT'S IN CHAPTER 11?

11-01 Types of triangles
11-02 Angle sum of a triangle
11-03 Exterior angle of a triangle
11-04 Types of quadrilaterals
11-05 Angle sum of a quadrilateral
11-06 Properties of quadrilaterals

IN THIS CHAPTER YOU WILL:

- name and classify triangles and their properties
- find the angle sum of a triangle
- find the exterior angle of a triangle
- name and classify quadrilaterals
- find the angle sum of a quadrilateral
- find properties of the sides, angles and diagonals of quadrilaterals
- solve geometry problems involving the properties of triangles and quadrilaterals

Shutterstock.com/Evgeny Tsapov

ISBN 9780170350969

11–01 Types of triangles

WORDBANK

triangle A shape with three straight sides.

There are three types of triangles, according to the lengths of their sides:

- **Scalene:** all sides different (no equal sides)
- **Isosceles:** two equal sides
- **Equilateral:** three equal sides

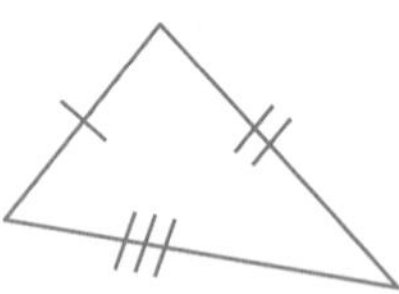

Scalene triangle

Isosceles triangle

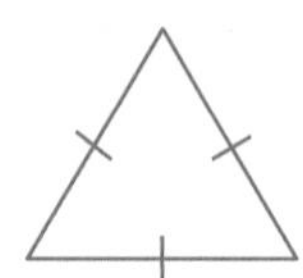

Equilateral triangle

The dashes are used on each side to indicate which sides are equal.

There are also three types of triangles according to size of their angles:

- **Acute-angled:** all angles are acute (less than 90°)
- **Obtuse-angled:** one of the angles is obtuse (more than 90° and less than 180°)
- **Right-angled:** one of the angles is a right angle (90°)

Acute-angled triangle

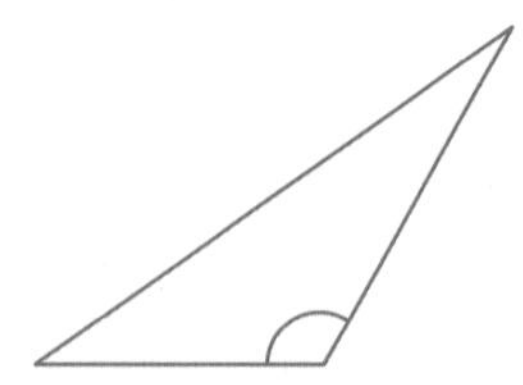

Obtuse-angled triangle

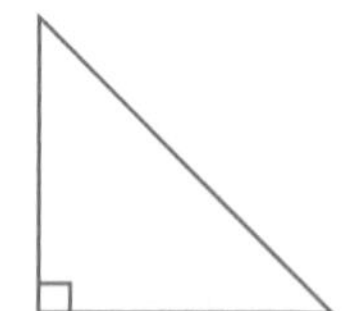

Right-angled triangle

EXAMPLE 1

Classify each triangle by side and by angle.

a

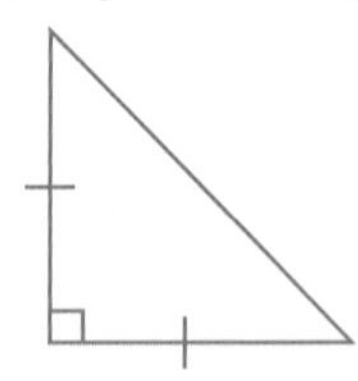

b

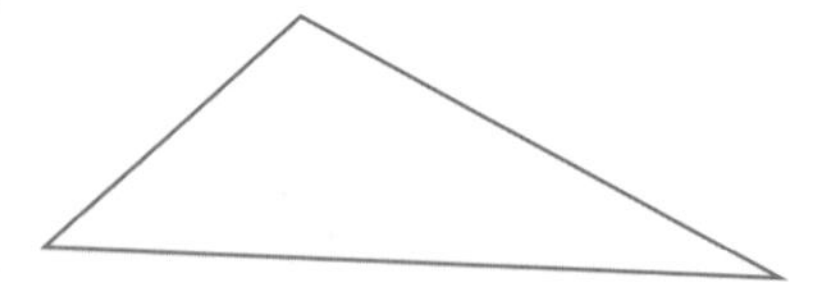

SOLUTION

a Triangle is isosceles and right-angled. ⟵ Two equal sides

b Triangle is scalene and obtuse-angled. ⟵ No equal sides

ISBN 9780170350969

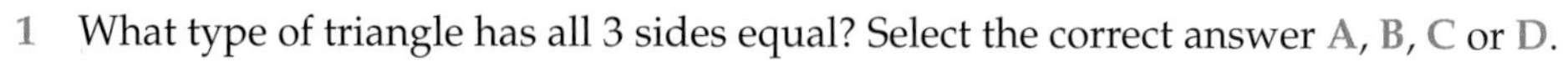

EXERCISE 11-01

1 What type of triangle has all 3 sides equal? Select the correct answer A, B, C or D.

A isosceles B acute-angled C obtuse-angled D equilateral

2 What type of triangle has all angles less than 90°? Select A, B, C or D.

A isosceles B acute-angled C obtuse-angled D equilateral

3 Classify the triangles below according to the length of their sides:

a 

b 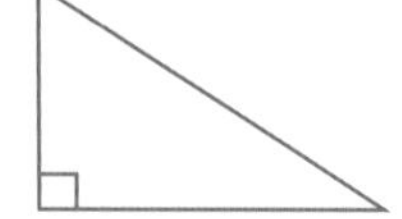

c 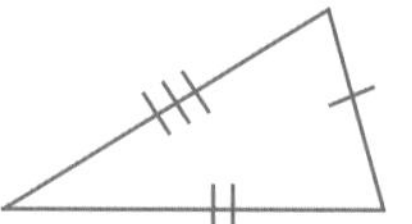

4 Draw an example of a triangle that is:

a equilateral b acute-angled c obtuse-angled

5 To name a triangle we use 3 capital letters that are the vertices of the triangle. We write △ before the 3 letters so that we know it is a triangle and not an angle, for example $\triangle ABC$. Name the triangles below.

a 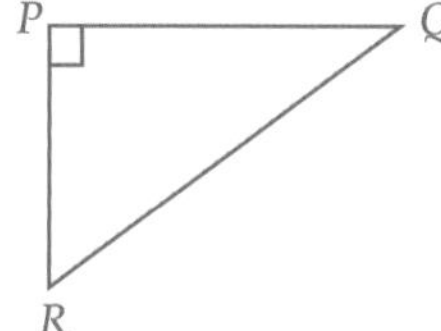

b 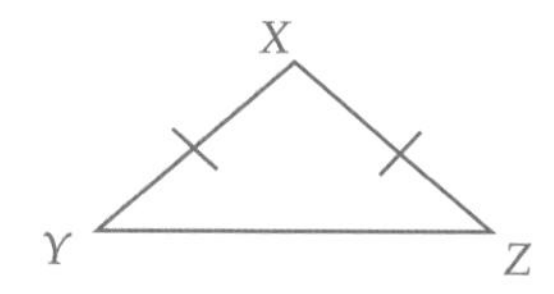

c 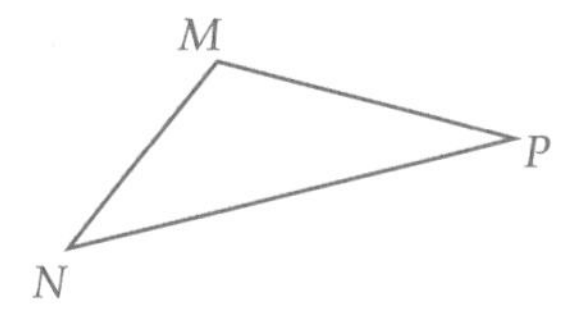

6 Classify each of the triangles in Question 5 by side and by angle.

7 Draw the triangles described below:

a Scalene $\triangle ABC$

b Acute-angled $\triangle PQR$

c Obtuse-angled $\triangle DEF$

d Right-angled isosceles $\triangle KLM$

e Acute-angled equilateral $\triangle RST$

8 Is each statement true or false?

a An isosceles triangle must be acute-angled.

b A right-angled triangle can also have an obtuse angle.

c An obtuse-angled triangle must also have acute angles.

d An equilateral triangle must have all acute angles.

e An acute-angled triangle cannot be isosceles.

f A right-angled triangle can also be equilateral.

9 In the diagram below count the number of triangles altogether and state what types of triangles you can see.

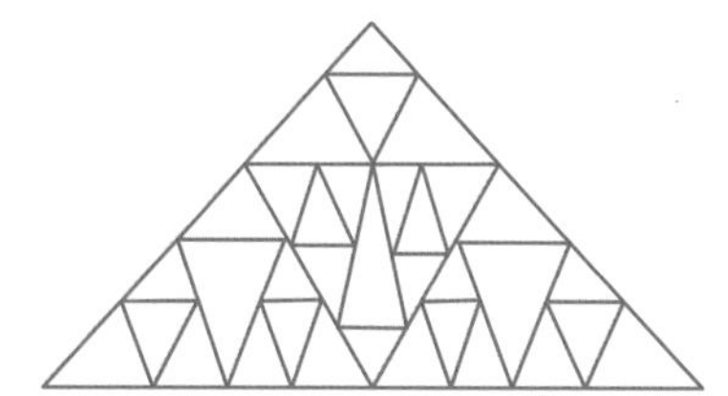

ISBN 9780170350969

11-02 Angle sum of a triangle

The **angle sum** of a triangle is the total when the three angles of a triangle are added together.

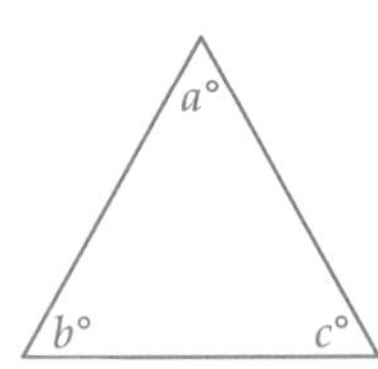

Angle sum $= a° + b° + c°$
$= ?$

The angle sum can easily be found by drawing any triangle on paper and then tearing off each angle and placing them together as shown:

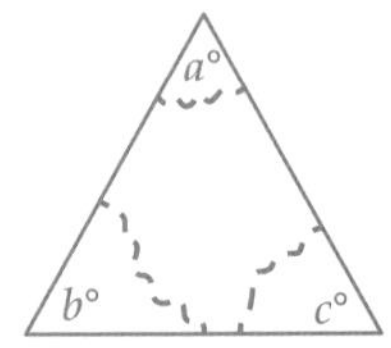

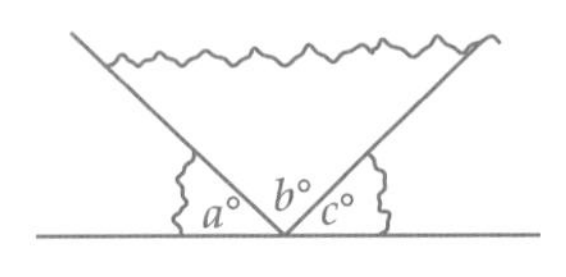

The angles when placed together form a straight line, which is 180°.

The angle sum of any triangle is 180°.

EXAMPLE 2

Find the values of the pronumerals in the diagrams below.

a

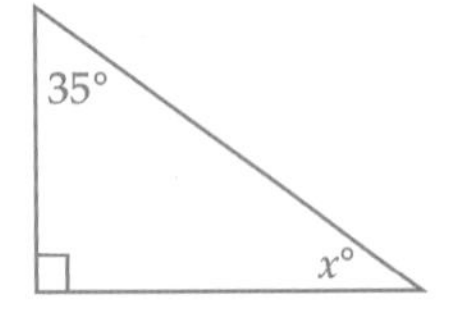

b

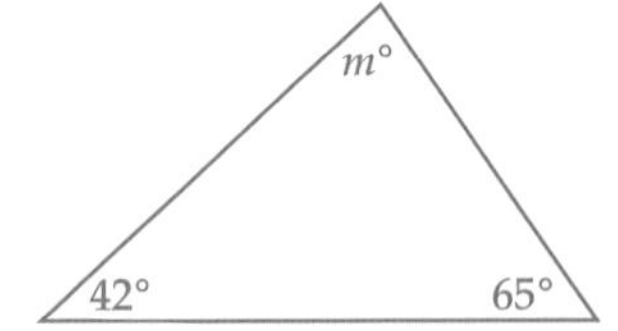

SOLUTION

a
$$x + 35 + 90 = 180$$
$$x + 125 = 180$$
$$x = 180 - 125$$
$$x = 55$$

b
$$m + 42 + 65 = 180$$
$$m + 107 = 180$$
$$m = 180 - 107$$
$$m = 73$$

 ISBN 9780170350969

EXERCISE 11–02

1 What is the size of n in the triangle? Select the correct answer A, B, C or D.

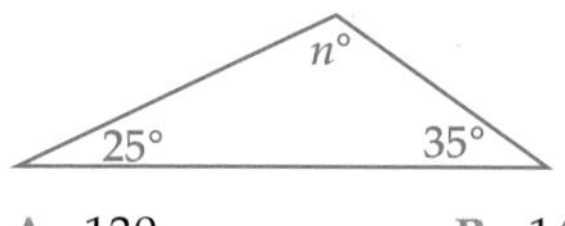

A 130 B 145 C 120 D 110

2 What is the angle sum of a right-angled triangle? Select A, B, C or D.

A 90° B 150° C 360° D 180°

3 Find the value of the pronumerals in the diagrams below.

a
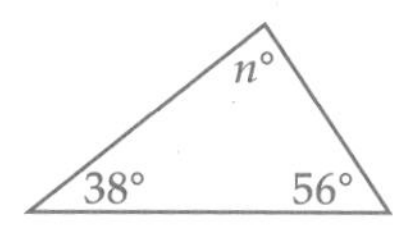

b
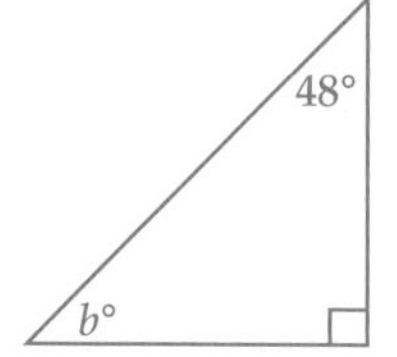

c
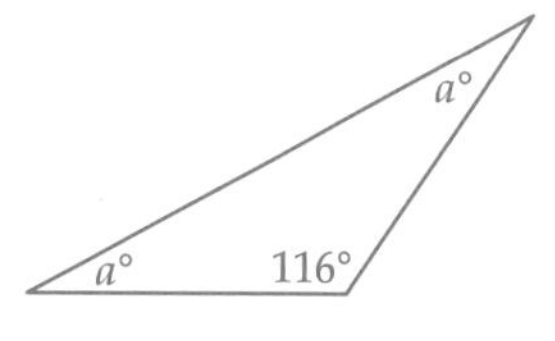

d
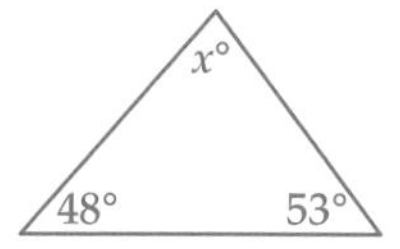

e
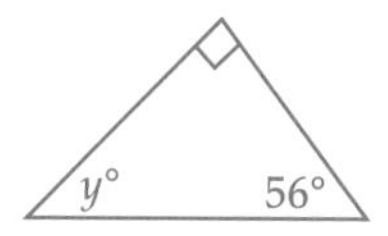

f
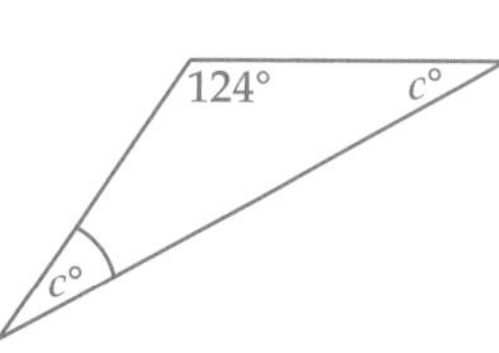

4 An equilateral triangle has all 3 sides and all 3 angles equal in size.

a What is the size of each angle in an equilateral triangle?

b Illustrate this on a diagram.

5 An isosceles triangle has two equal angles, opposite the two equal sides.

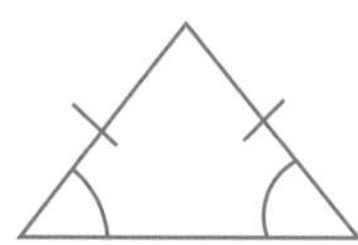

Use this fact to find the size of the pronumerals below.

a
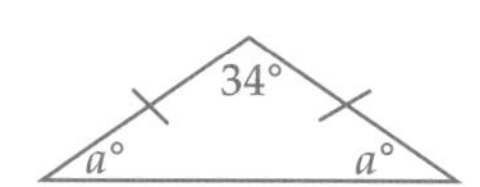

b
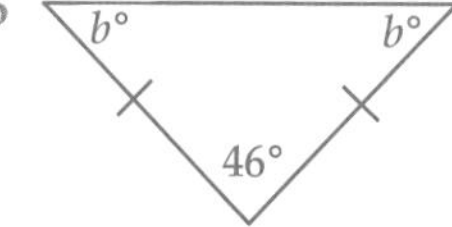

c
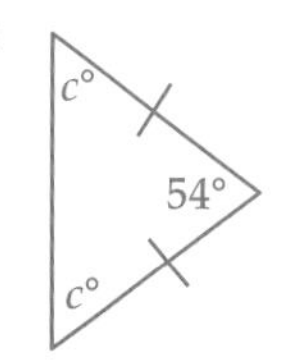

6 Draw two possible obtuse-angled scalene triangles, showing the size of each angle.

7 Draw two possible right-angled triangles, showing the size of each angle.

8 Draw an example, showing angle sizes, of:

a an equilateral triangle

b an isosceles triangle.

11-03 Exterior angle of a triangle

WORDBANK

exterior angle An angle outside a shape created by extending one side of the shape. The diagram below shows the exterior angle of a triangle.

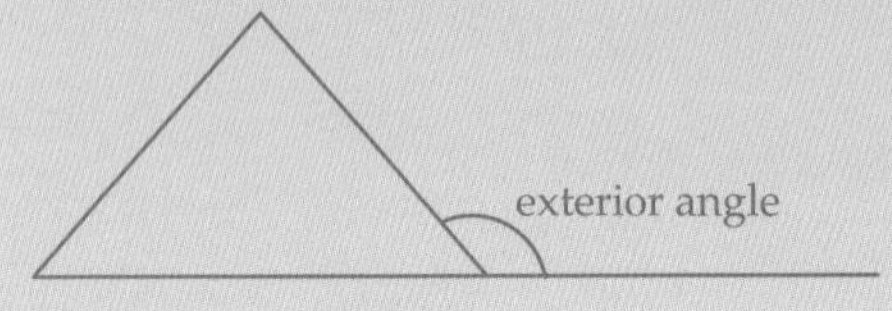

In the diagram below, the exterior angle is $d°$.

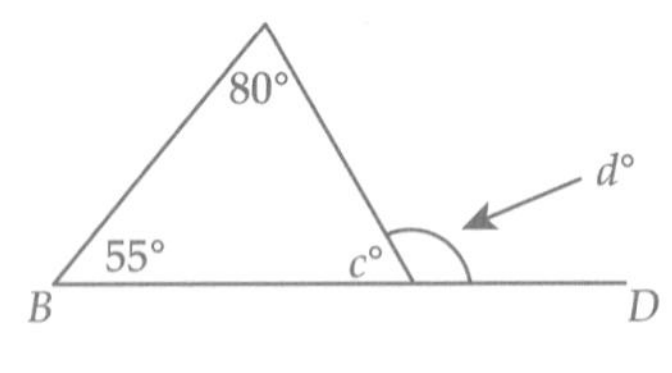

$55 + 80 + c = 180$ ← angle sum of a triangle
But also $d + c = 180$ ← angles on a straight line

So $d = 55 + 80$
So $d = 135$

The exterior angle of a triangle is equal to the sum of the two interior opposite angles (55° and 80° in the diagram above).

EXAMPLE 3

Find the value of each pronumeral.

a
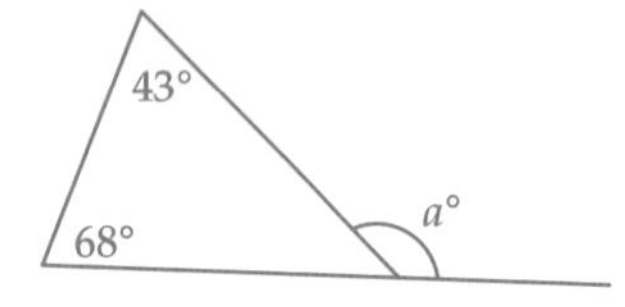

b
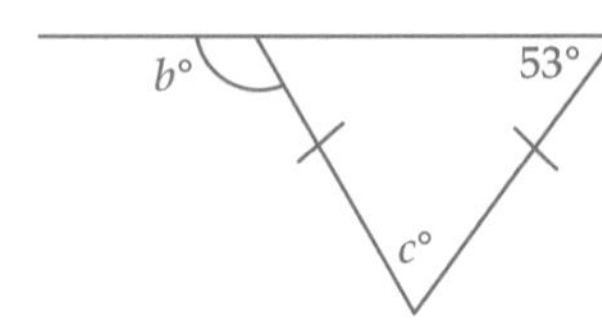

SOLUTION

a $a = 68 + 43$ ← exterior angle rule
$a = 111$

b $c = 180 - 2 \times 53$ ← angle sum of an isosceles triangle
$c = 74$
$b = 53 + 74$ ← exterior angle rule
$b = 127$

iStockphoto/chrisho

 ISBN 9780170350969

EXERCISE 11–03

1 The exterior angle of a triangle equals ____. Select the correct answer A, B, C or D.

A the opposite angle
B two of the interior angles
C the sum of all the angles
D the sum of the two interior opposite angles

2 What is the size of an interior angle of an equilateral triangle? Select A, B, C or D.

A 60°
B 120°
C 180°
D 90°

3 For the diagram below:

a name the interior angles

b name the exterior angle

c write down an equation to find the size of the exterior angle.

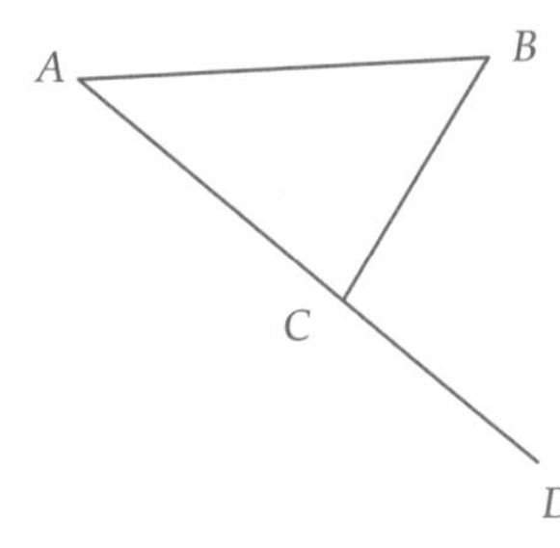

4 Find the value of the pronumerals in the diagrams below:

a

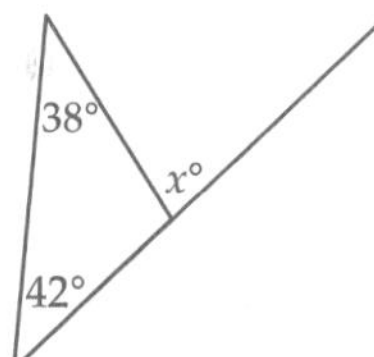

b

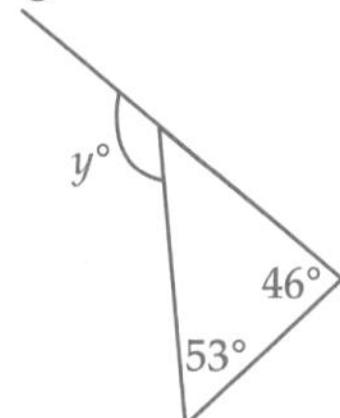

c

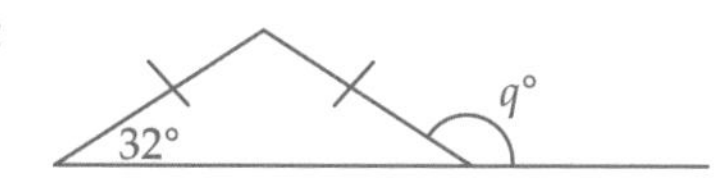

d

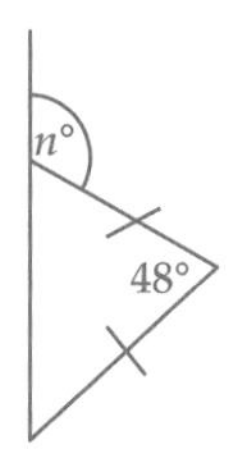

5 Draw a diagram showing an equilateral triangle with an exterior angle. Find the size of the exterior angle. Will it always be the same size? Why, or why not?

6 An isosceles triangle has base angles 55° each as shown below.

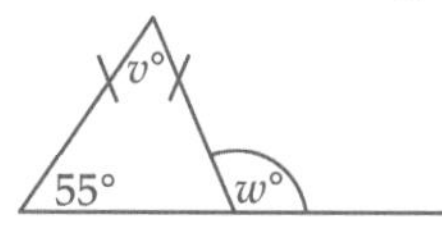

a Find the size of v.

b Find the size of the exterior angle w.

11–04 Types of quadrilaterals

WORDBANK

quadrilateral A shape with four straight sides.

There are many types of quadrilaterals. They are displayed in the table below:

NAME OF QUADRILATERAL	DIAGRAM	FEATURES
Parallelogram		Opposite sides parallel Opposite sides equal Opposite angles equal
Rhombus		All sides equal in length A special type of parallelogram
Trapezium		One pair of sides parallel
Kite		Adjacent sides equal One pair of opposite angles equal
Rectangle		All angles are 90° A special type of parallelogram
Square		All sides are equal All angles are 90° A special type of parallelogram and rhombus
Convex quadrilateral		Any quadrilateral where all vertices (corners) point outwards All diagonals lies inside the shape
Non-convex quadrilateral		Any quadrilateral where one vertex points inwards. One diagonal lies outside the shape. One angle is more than 180°.

ISBN 9780170350969

EXERCISE 11–04

1 If the opposite sides of a quadrilateral are equal, what must the quadrilateral be? Select the correct answer A, B, C or D.

A a square B a trapezium C a parallelogram D a rhombus

2 If all 4 angles of a quadrilateral are 90°, what must the quadrilateral be? Select A, B, C or D.

A a square B a trapezium C a parallelogram D a rhombus

3 Draw an example of each shape.

a a rhombus b a non-convex quadrilateral c a square

4 A closed figure starts and ends at the same point. Which figures below are closed?

a 

b 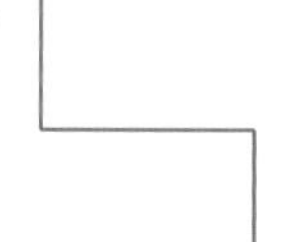

c 

5 Name each quadrilateral.

a 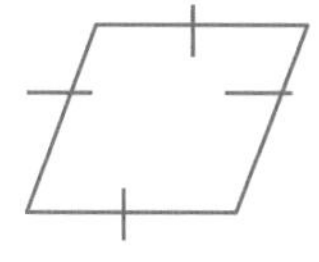

b 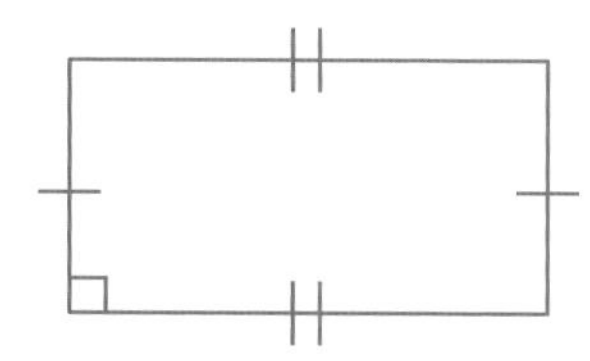

c 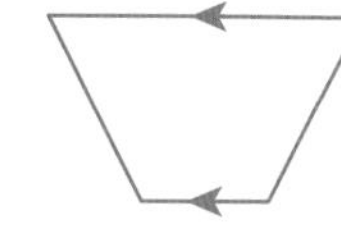

d 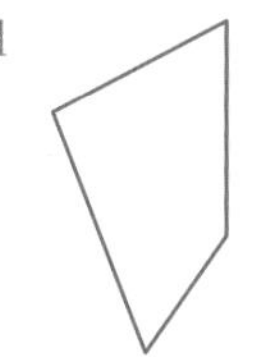

e 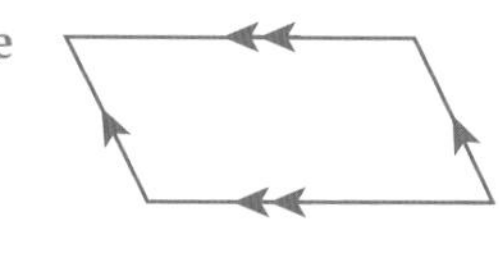

f 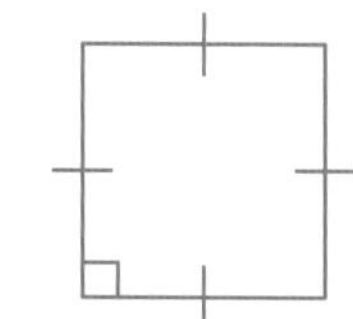

6 Describe each quadrilateral in words.

a a trapezium b a rhombus c a rectangle

7 Is each statement true or false?

a A square is a rectangle.

b A rectangle is a square.

c A rhombus is a parallelogram.

d A parallelogram is a rhombus.

e A square is a regular quadrilateral.

f A trapezium has two equal sides.

8 Make a copy of the shapes below, cut them out and join them to form a rectangle and a parallelogram:

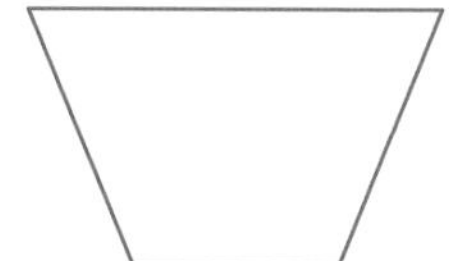

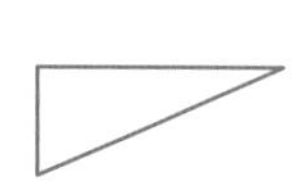

11-05 Angle sum of a quadrilateral

The **angle sum** of a quadrilateral is the total when its four angles are added together.

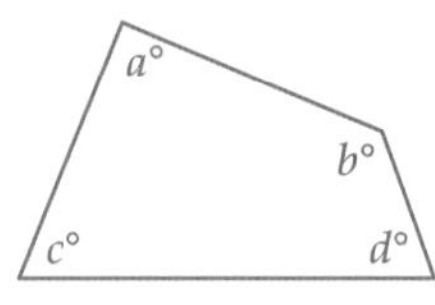

$$\text{Angle sum} = a° + b° + c° + d°$$
$$= ?$$

The angle sum can easily be found by drawing any quadrilateral on paper and then tearing off each angle and placing them together as shown:

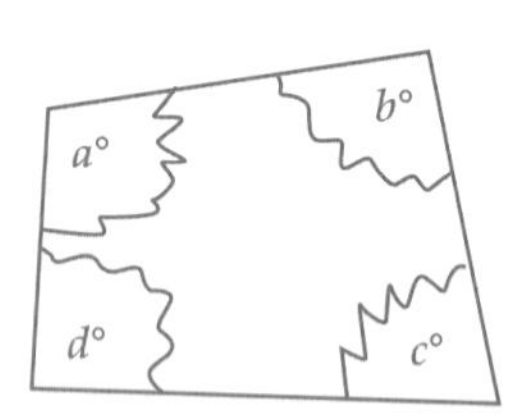

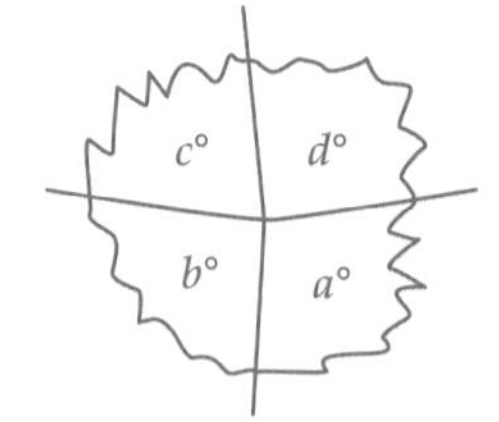

The four angles when placed together form a revolution, which is 360°.

> The angle sum of any quadrilateral is 360°.

EXAMPLE 4

Find the values of the pronumerals in the diagrams below:

a

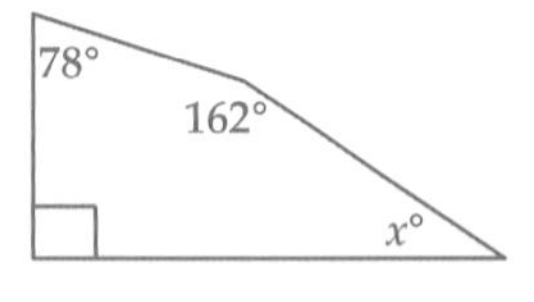

b

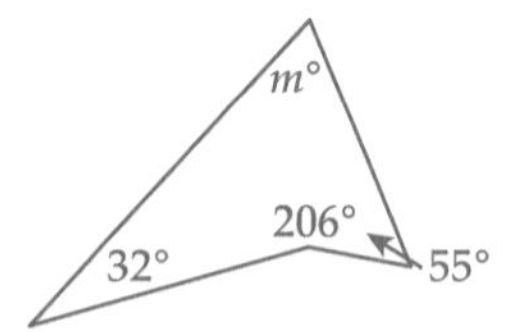

SOLUTION

a $78 + 90 + x + 162 = 360$

$x + 330 = 360$

$x = 360 - 330$

$x = 30$

b $m + 32 + 206 + 55 = 360$

$m + 293 = 360$

$m = 360 - 293$

$m = 67$

ISBN 9780170350969

EXERCISE 11–05

1 What is the angle sum of a rectangle? Select the correct answer A, B, C or D.

A 180° B 360° C 120° D 90°

2 A quadrilateral has 2 equal angles of 85° and a right angle. What is the size of its fourth angle? Select A, B, C or D.

A 100° B 80° C 90° D 110°

3 Find the value of the pronumerals in the diagrams below.

a
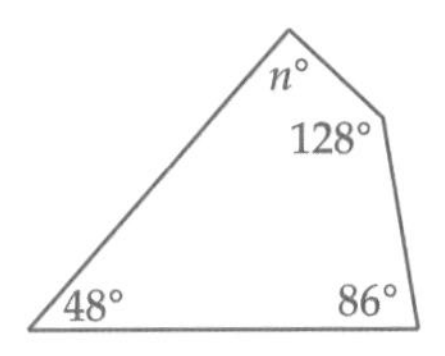

b
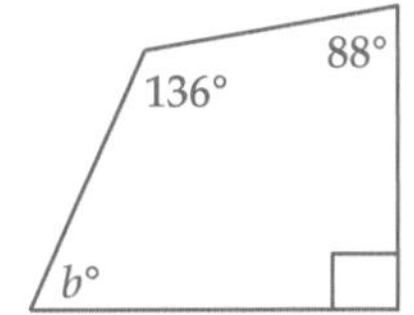

c
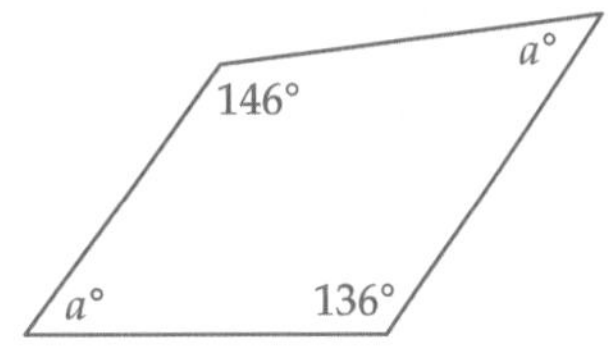

d
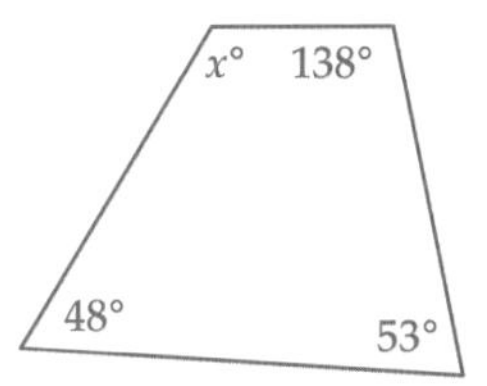

e
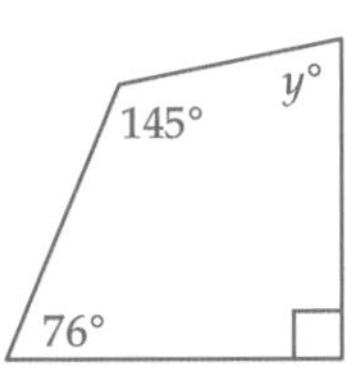

f
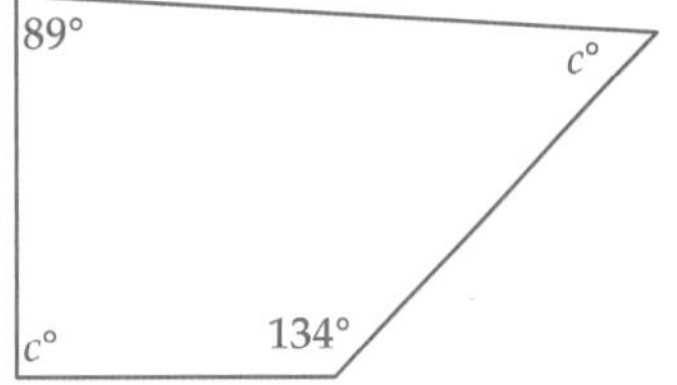

4 Find the value of each of the pronumerals in the quadrilaterals below.

a
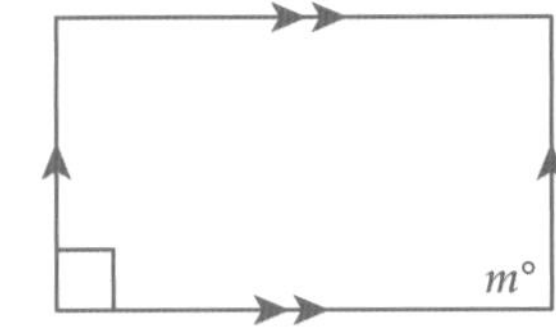

b
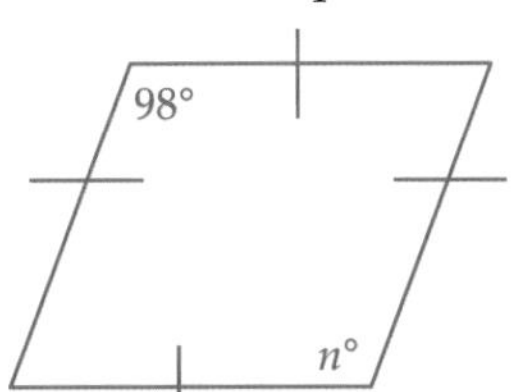

c
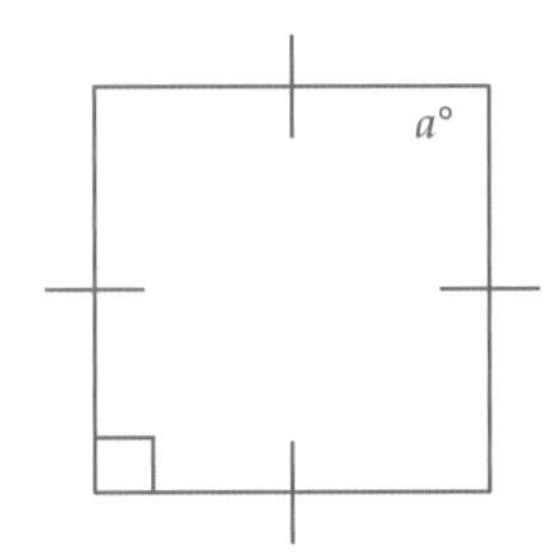

5 Find the size of the other two angles in Question 4b.

6 Draw a parallelogram and mark in the equal angles.

7 Find the value of all unknown angles in each quadrilateral.

a
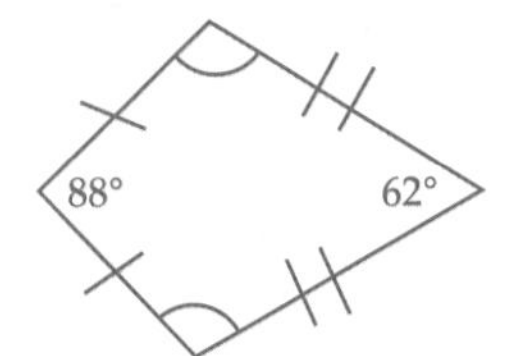

b
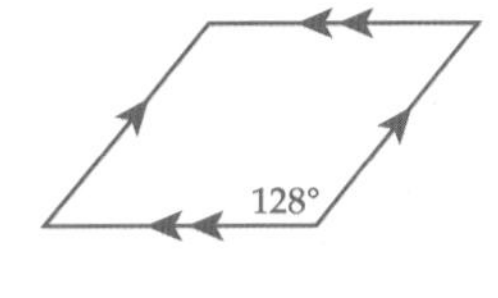

11-06 Properties of quadrilaterals

The diagram below shows how the special quadrilaterals are related to each other.

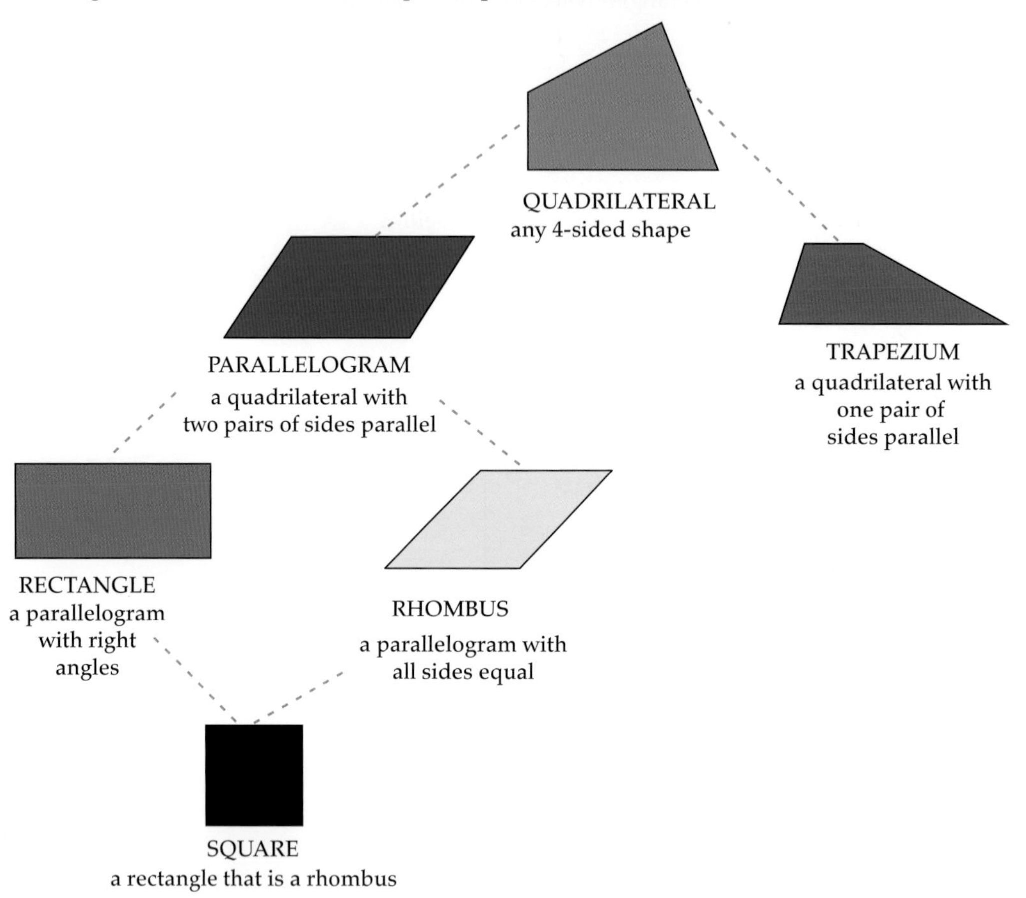

From this you can see that a rhombus is a special parallelogram. A square is a special rhombus and a special rectangle as well.

EXAMPLE 5

Write down the difference between:

a a rectangle and a parallelogram

b a square and a rectangle

SOLUTION

a A rectangle has all angles 90° and a parallelogram does not.

b A square has all 4 sides equal and a rectangle does not.

iStockphoto/Fotmen

ISBN 9780170350969

11–06 Properties of quadrilaterals

Properties of the diagonals of the special quadrilaterals

Parallelogram		The diagonals bisect each other
Rectangle		The diagonals are equal in length The diagonals bisect each other
Kite		One diagonal bisects the other at right angles
Rhombus		The diagonals bisect each other at right angles The diagonals bisect the angles of the rhombus
Square		The diagonals bisect each other at right angles The diagonals bisect the angles of the square The diagonals are equal in length

Shutterstock.com/attem

ISBN 9780170350969

EXERCISE 11–06

1 Which of these quadrilaterals has equal diagonals? Select the correct answer A, B, C or D.

A rhombus B kite C rectangle D trapezium

2 Which of these quadrilaterals is also a rectangle? Select A, B, C or D.

A rhombus B parallelogram C square D kite

3 Is each statement true or false?

a A square is a special rhombus.

b A rectangle is a special trapezium.

c A rhombus is a special parallelogram.

d A kite is a special rhombus.

e A rhombus is a special rectangle.

4 Write down the difference between:

a a square and a rhombus

b a parallelogram and a quadrilateral

c a rhombus and a parallelogram

d a parallelogram and a trapezium

5 Copy and complete the table.

Quadrilateral	Angles 90°	Equal sides	Equal diagonals
Parallelogram	No		
Rhombus		Yes	
Rectangle			Yes
Kite	No		
Square		Yes	
Trapezium			No

6 a Write down which quadrilaterals have diagonals bisecting each other at 90°.

b Draw a diagram of each of these quadrilaterals.

7 Copy and complete the table.

Quadrilateral	Sides	Angles	Diagonals
Parallelogram	Opposite sides equal and parallel	Opposite angles are equal	Diagonals bisect each other
Trapezium		All angles different	
Rhombus	All sides equal		
Rectangle		All angles are 90°	
Square			

ISBN 9780170350969

CODE PUZZLE

Find the unknown angle in each shape.

Match the letter of each shape to each angle size shown in the message next page to decode the message.

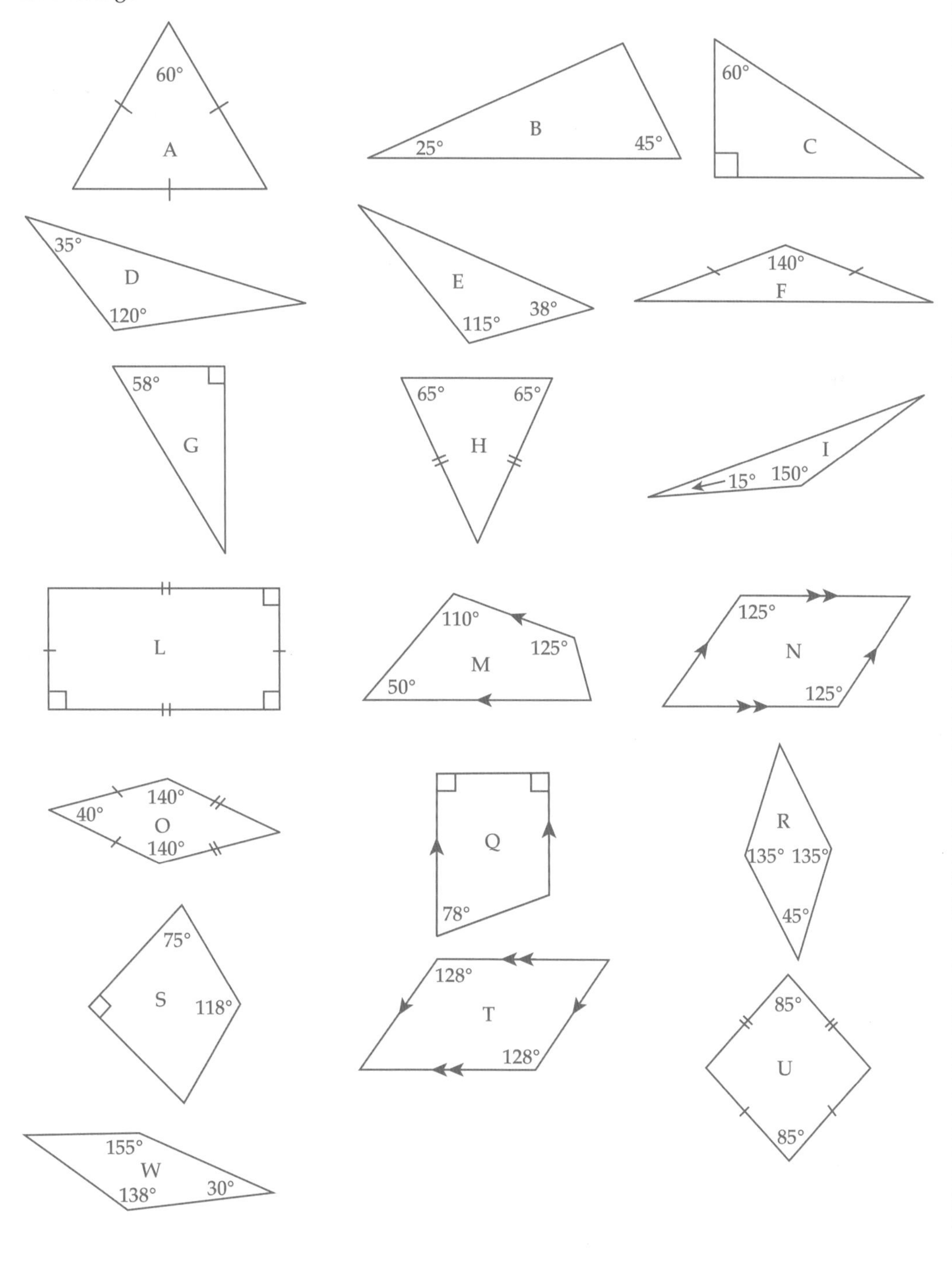

37°-50°-27°-55° 20°-15°-55°-25°-15°-55°-32° 60°-55°-32°-90°-27°-77°

15°-55° 102°-95°-60°-25°-45°-15°-90°-60°-52°-27°-45°-60°-90°-77°

60°-55°-25° 52°-45°-15°-90°-27°-77°-60°-55°-32°

45°-27°-75°-25°-75°-110°-27°-45° 60°-27°-55°-32°-90° 77°-95°-75°

60°-55°-25° 52°-50°-27° 60°-55°-77°-37°-27°-45° 37°-15°-90°-90° 30°-40°-75°-27°.

ISBN 9780170350969

PRACTICE TEST 11

Part A General topics

Calculators are not allowed.

1 Evaluate $-4 \times 3 + (-2)$.

2 Convert $\frac{1}{5}$ to a decimal.

3 A square has an area of 36 cm². What is the length of one side?

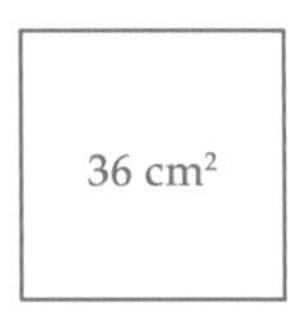

4 Find the lowest common multiple of 9 and 3.

5 In which quadrant of the number plane would the point (2, –4) be: 1st, 2nd, 3rd or 4th?

6 Simplify $3 \times a \times 6 \times y$.

7 What angle is supplementary to 55°?

8 Evaluate 4.85×6.

9 What is the time 1 hour 15 minutes before 12:30 p.m.?

10 Evaluate $\frac{2}{3} - \frac{4}{7}$.

Part B Geometry

Calculators are allowed.

11-01 Types of triangles

11 Which diagram below shows an isosceles triangle? Select the correct answer A, B, C or D.

A 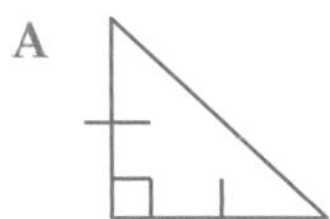 B C 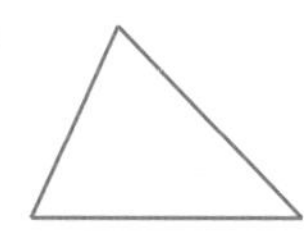 D

12 Which triangle in Question 11 above is obtuse-angled? Select A, B, C or D.

11-02 Angle sum of a triangle

13 Find the value of x. Select A, B, C or D.

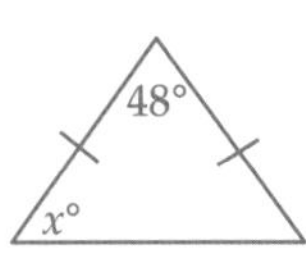

A 56 B 132 C 66 D 48

14 What is the angle sum of an equilateral triangle? Select A, B, C or D.

A 60° B 120° C 180° D 360°

11-03 Exterior angle of a triangle

15 Find the value of n.

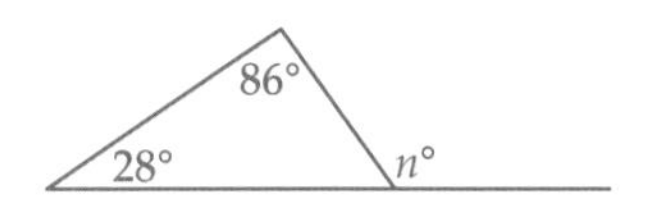

ISBN 9780170350969

PRACTICE TEST 11

11-04 Types of quadrilaterals

16 Name each quadrilateral.

a 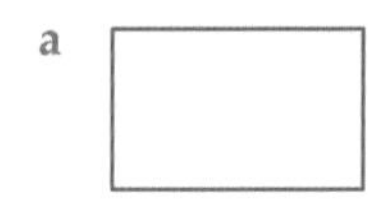 b 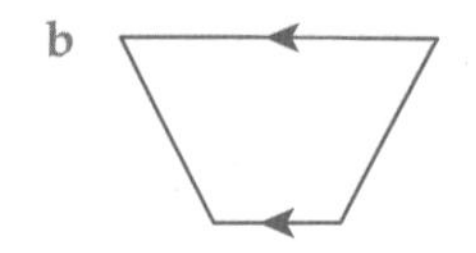 c 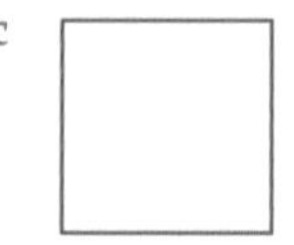 d 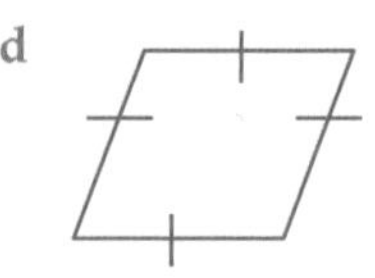

17 Which two quadrilaterals have all angles equal to 90°?

11-05 Angle sum of a quadrilateral

18 Find the value of c.

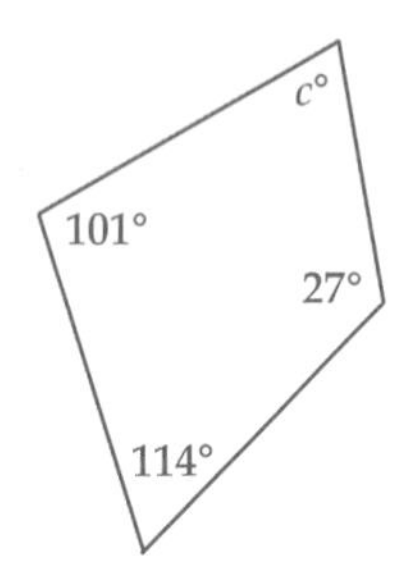

11-06 Properties of quadrilaterals

19 Is each statement true or false?

a The diagonals of a rectangle are equal.

b The diagonals of a parallelogram bisect each other at right angles.

c The diagonals of a rhombus bisect each other.

 ISBN 9780170350969

LENGTH AND TIME

12

WHAT'S IN CHAPTER 12?

12-01 The metric system

12-02 Measuring length

12-03 Perimeter

12-04 Perimeter of composite shapes

12-05 Time

12-06 24-hour time

12-07 Time calculations

12-08 Timetables

IN THIS CHAPTER YOU WILL:

- learn the metric units for length, capacity and mass
- measure and calculate lengths
- find perimeters of shapes, including composite shapes
- learn the metric units for time
- convert between 24-hour time and 12-hour (a.m./p.m.) time
- round times to the nearest minute or hour
- calculate time differences
- read and interpret timetables

Shutterstock.com/wongwean

ISBN 9780170350969

12-01 The metric system

The metric system is based on powers of ten. Milli- means $\frac{1}{1000}$, centi- means $\frac{1}{100}$, kilo- means 1000 and mega- means 1 000 000.

Unit	Abbreviation	Size	Example
Length			
metre	m	base unit	Height of a kitchen bench
millimetre	mm	1000 mm = 1 m	Smallest gap on your ruler
centimetre	cm	100 cm = 1 m 10 mm = 1 cm	Width of a pen
kilometre	km	1000 m = 1 km	Distance between 2 bus stops
Capacity			
litre	L	base unit	A carton of milk
millilitre	mL	1000 mL = 1 L	A large drop of water
kilolitre	kL	1000 L = 1 kL	Amount of water in a spa
megalitre	ML	1 000 000 L = 1 ML	Amount of water in 2 Olympic-sized swimming pools
Mass			
kilogram	kg	base unit	A packet of sugar
gram	g	1000 g = 1 kg	A tablet
milligram	mg	1000 mg = 1 g	A grain of salt
tonne	t	1000 kg = 1 t	A small car

To convert units use the initials below to remember:
SOLD: Small Over to Large Divide, that is, small unit to large unit divide.
LOSM Large Over to Small Multiply, that is large unit to small unit multiply.

EXAMPLE 1

Convert:

a 300 cm = ____ m b 5 L = ____ kL c 480 kg = ____ g

SOLUTION

Use the table of units to convert.

a 300 cm = 300 ÷ 100 m ← SOLD: Small Over to Large Divide, 100 cm = 1 m
= 3 m

b 5 L = 5 ÷ 1000 kL ← SOLD: Small Over to Large Divide, 1000 L = 1kL
= 0.005 kL

c 480 kg = 480 × 1000 g ← LOSM: Large Over to Small Multiply, 1000 g = 1 kg
= 480 000 g

ISBN 9780170350969

EXERCISE 12-01

1 What is the height of a door? Select the closest answer A, B, C or D.

A 2.5 mm B 2.5 cm C 2.5 m D 2.5 km

2 What is the capacity of a bath? Select A, B, C or D.

A 150 mL B 150 L C 150 kL D 150 ML

3 Which unit of measure would be the most suitable to measure the following?

a the length of a desk
b the thickness of a calculator
c the height of a door
d the amount of water in a sink
e the distance between 2 towns
f the amount a cup can hold
g the mass of a truck
h the mass of a grain of salt
i the length of the basketball court
j the mass of a bag of potatoes

4 Copy and complete the table.

Millimetres	Centimetres	Metres	Kilometres
3000			
	5500		
		250	
			80
		76.8	
	820		
750 000			

5 Copy and complete:

a 756 cm = ____ m
b 125 mm = ____ cm
c 26.8 m = ____ km
d 8.5 m = ____ cm
e 59.4 km = ____ m
f 42 cm = ____ mm
g 5678 mm = ____ m
h 7.6 km = ____ cm

6 Convert:

a 6000 mL = ____ L
b 150 mg = ____ g
c 2800 g = ____ kg
d 4.5 kL = ____ L
e 3.5 L = ____ kL
f 7.25 g = ____ mg
g 8 ML = ____ kL
h 4.75 kg = ____ g
i 9.6 t = ____ kg
j 65 mg = ____ g
k 84 t = ____ kg
l 2.8 g = ____ mg

7 Change each length to metres and add:

4.8 km, 356 cm, 78 m and 1120 mm

8 Is each statement true or false?

a 550 mL = 0.55 L
b 240 mg = 2.4 g
c 350 g = 35 kg
d 6.8 kL = 6800 L
e 3.5 km = 350 m
f 6.25 kg = 6250 t

12-02 Measuring length

When we measure **length** we are interested to find out how **long** an object is.

To measure length we use the most suitable unit of length from the metric table.
For instance:
- to measure a pin use millimetres
- to measure a photograph use centimetres
- to measure the length of a playground use metres
- to measure the distance travelled in a car use kilometres.

EXAMPLE 2

What unit of length would you use to measure:

a your height? b the length of a swimming pool?

SOLUTION

a centimetres b metres

EXAMPLE 3

Measure the intervals below correct to the nearest millimetre:

a ______________________ b _____________

SOLUTION

a 35 mm b 20 mm

Note that all measurements are approximate because they depend upon the accuracy of the measuring instrument as well as the skill of the person measuring.

EXAMPLE 4

The diagram shows a road that goes from Green Valley to Mt Hutt:

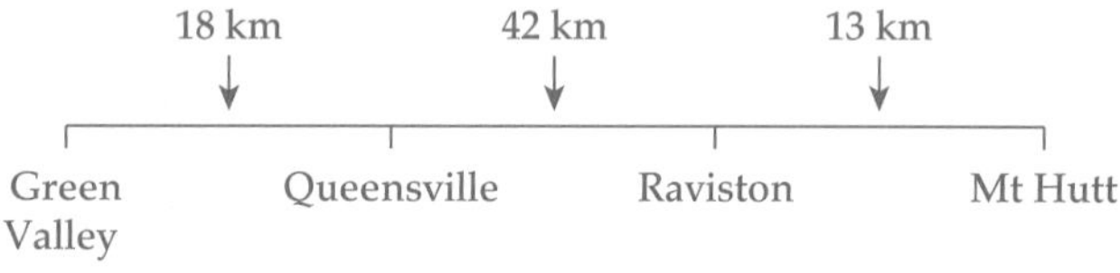

How far is it from Queensville to Mt Hutt?

SOLUTION

From Queensville to Mt Hutt it is: 42 km + 13 km = 55 km.

Shutterstock.com/Steve Lovegrove

ISBN 9780170350969

EXERCISE 12-02

1 What unit would you use to measure the distance between two railway stations? Select the correct answer A, B, C or D.

A millimetres B centimetres C metres D kilometres

2 What unit would you use to measure the length of a toothpick? Select A, B, C or D.

A millimetres B centimetres C metres D kilometres

3 What unit of length would you use to measure:

a the length of your fingernail?

b the height of your desk?

c the distance between your home and school?

d the length of your hand?

e the width of your textbook?

f the distance from your home to the beach?

g your height?

h the length of a railway carriage?

i the thickness of a coin?

j the distance from Sydney to Newcastle?

4 Measure the lines below correct to the nearest millimetre.

a ______________________ b ___________

c ______________________________ d ____________________

e ______________ f ____________________________________

5 Round your answers in Question 4 correct to the nearest centimetre.

6 This line represents a road from Red Cliffs to Forestville, a distance of 120 km.

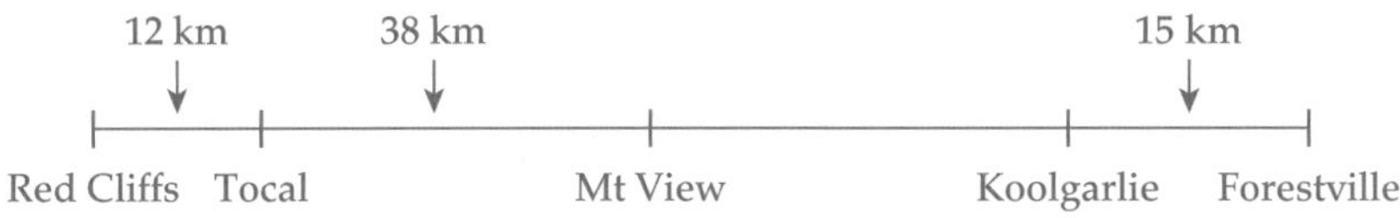

a How far is it from Mt View to Red Cliffs?

b What is the distance from Mt View to Koolgarlie?

c Mel walks from Koolgarlie to Forestville in $2\frac{1}{2}$ hours. What speed did she walk in km/h?

d Which is further, Tocal to Mt View or Mt View to Koolgarlie?

e Greg rides from Tocal to Forestville in 7 hours. What speed was he cycling in km/h?

12-03 Perimeter

WORDBANK

perimeter The distance around the edges of a shape. Perimeter is measured using a length unit such as mm, cm, m or km.

To calculate the perimeter of a shape, add the length of each side of the shape.

EXAMPLE 5

Find the perimeter of each shape.

a

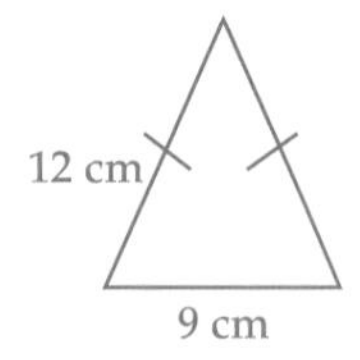

b

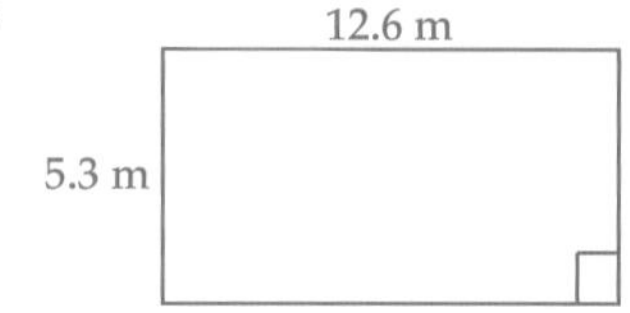

SOLUTION

a Perimeter $= 12 + 12 + 9$ ← Isosceles triangle so 2 sides equal in length

$= 33$ cm

b Perimeter $= 12.6 + 5.3 + 12.6 + 5.3$ ← Rectangle has opposite sides equal in length

$= 35.8$ m

There is a formula we can use for the perimeter of a rectangle:

Formula for the perimeter of a rectangle

Perimeter $= 2 \times$ length $+ 2 \times$ width

$P = 2l + 2w$

Using the formula for the rectangle in Example 5b:

Perimeter $= 2l + 2w$

$= 2 \times 12.6 + 2 \times 5.3$

$= 35.8$ m

ISBN 9780170350969

EXERCISE 12-03

1 What is the perimeter of a square of side length 5 cm? Select the correct answer A, B, C or D.

A 15 cm B 20 cm C 25 cm D 30 cm

2 What is the perimeter of a rectangle with length 6 mm and breadth 3 mm? Select A, B, C or D.

A 9 mm B 12 mm C 18 mm D 15 mm

3 Find the perimeter for each shape described in the table.

Shape	Side lengths
a Rectangle	$l = 6$ cm, $w = 4$ cm
b Square	$s = 9$ mm
c Equilateral triangle	$s = 5$ cm
d Rhombus	$s = 7$ m
e Rectangle	$l = 9$ cm and $w = 4$ cm
f Triangle	2 sides = 12 cm and the other side = 7 cm

4 Use the formula to find the perimeter of each rectangle.

a
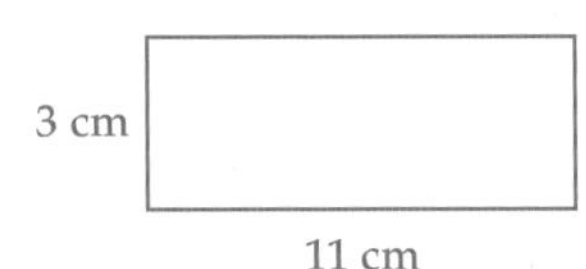

b
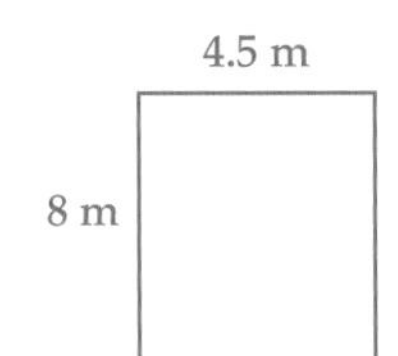

c
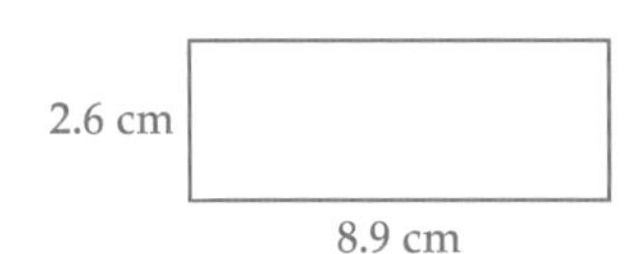

5 Find the perimeter of each shape.

a
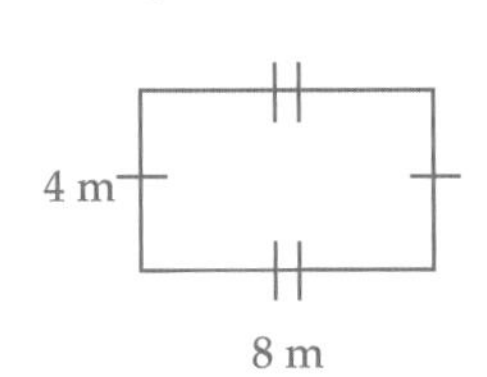

b
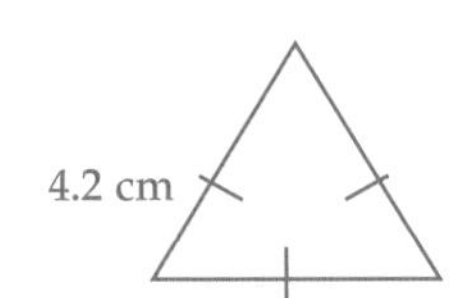

c
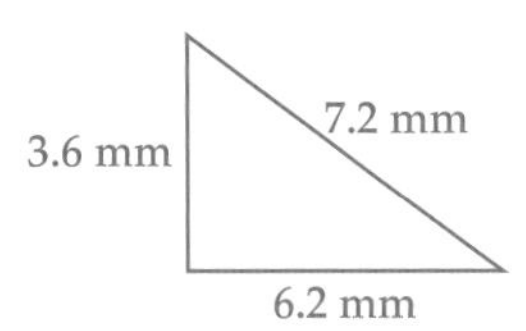

d
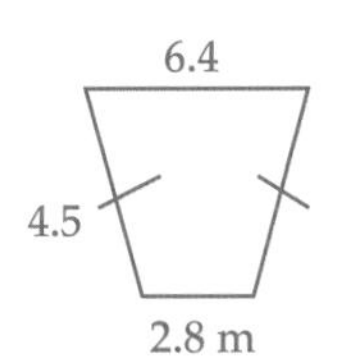

e
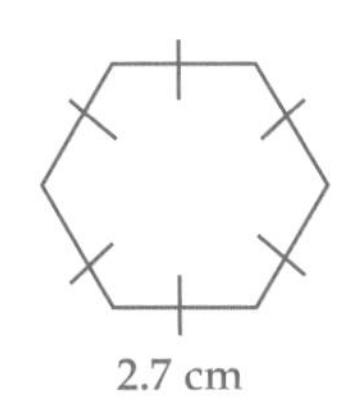

f
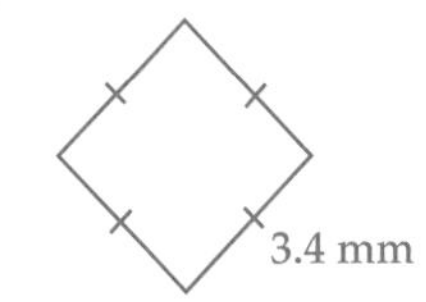

12–04 Perimeter of composite shapes

WORDBANK

composite shape A shape made up of 2 or more shapes. For example, a square and a rectangle.

EXAMPLE 6

Find the perimeter of these shapes.

a

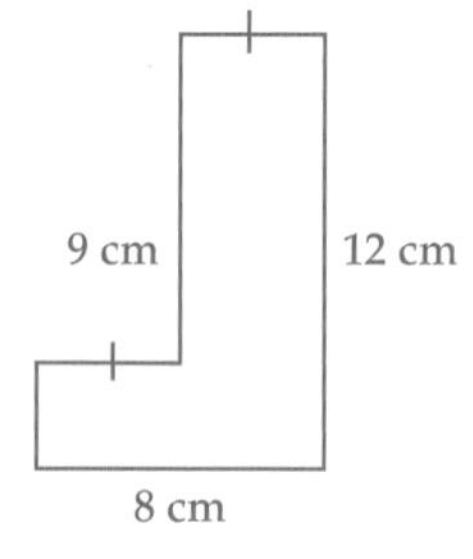

b

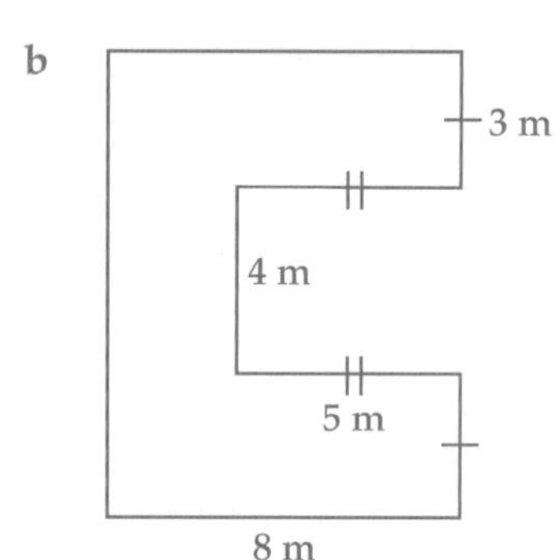

SOLUTION

First, find the unknown sides.

a

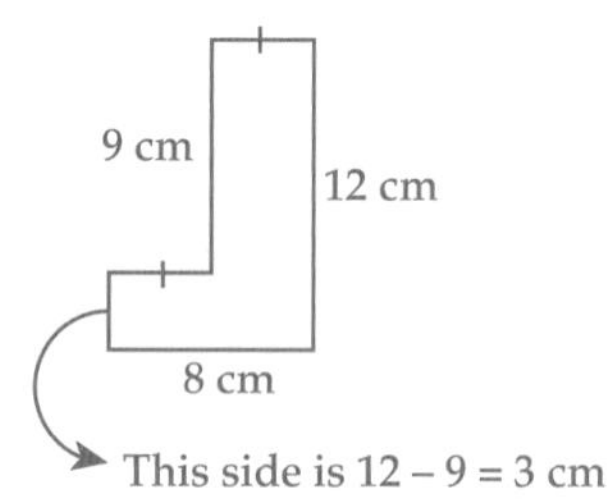

The equal sides are 4 cm each as $8 \div 2 = 4$ cm.

$$P = 3 + 4 + 9 + 4 + 12 + 8$$
$$= 40 \text{ cm}$$

b

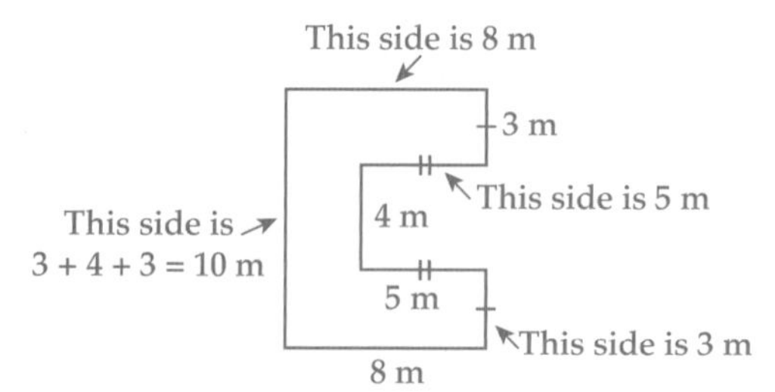

$$P = 3 + 5 + 4 + 5 + 3 + 8 + 10 + 8$$
$$= 46 \text{ m}$$

ISBN 9780170350969

EXERCISE 12–04

1 What is the length of the missing side? Select the correct answer A, B, C or D.

A 12 m B 11 m C 9 m D 10 m

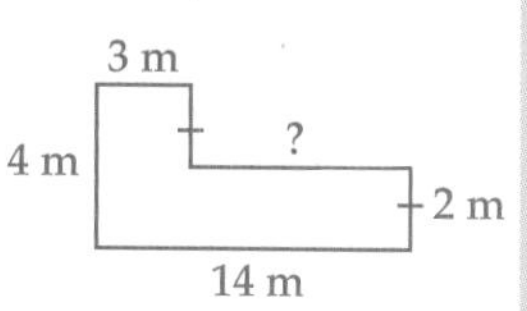

2 What is the perimeter of the shape in Question 1? Select A, B, C or D.

A 23 m B 37 m C 38 m D 36 m

3 Find the perimeter of each shape.

a
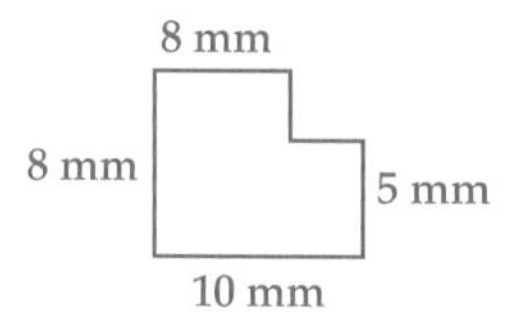

b
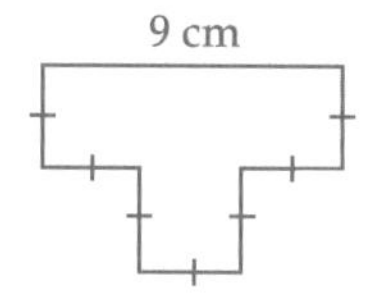

c
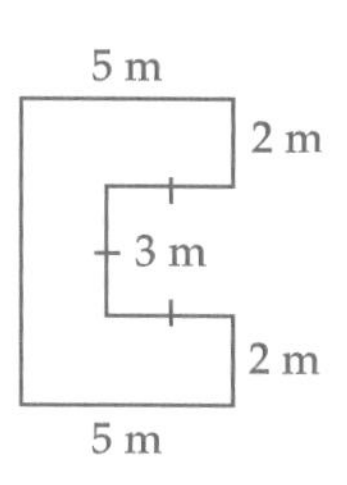

d
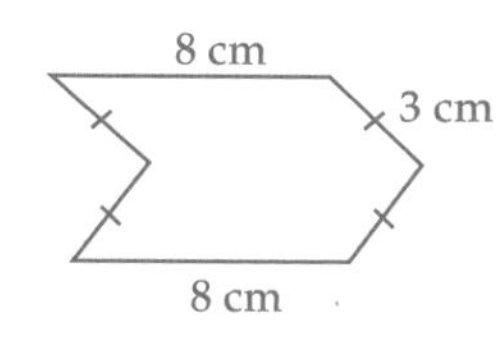

e
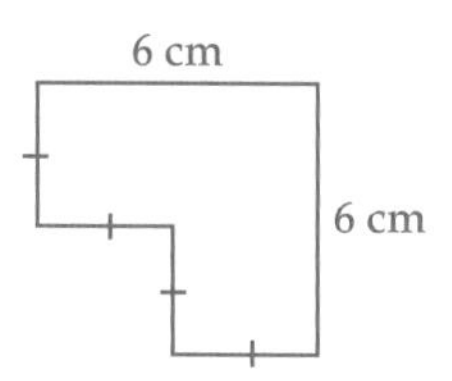

f
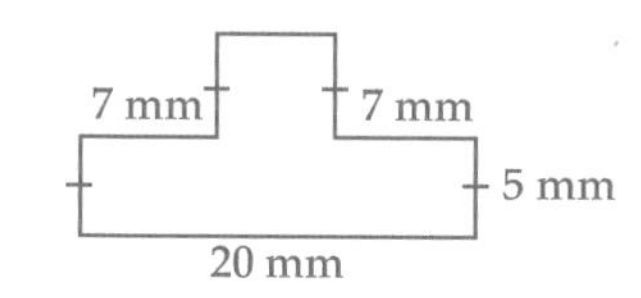

4 Find the perimeter of each shape.

a
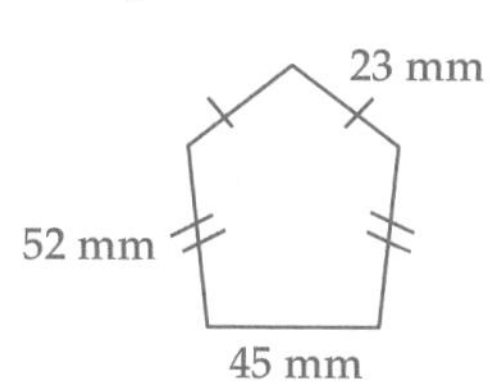

b
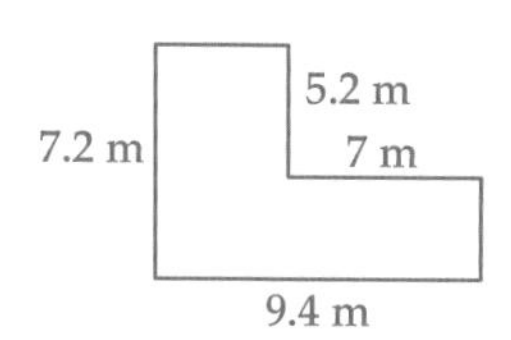

c
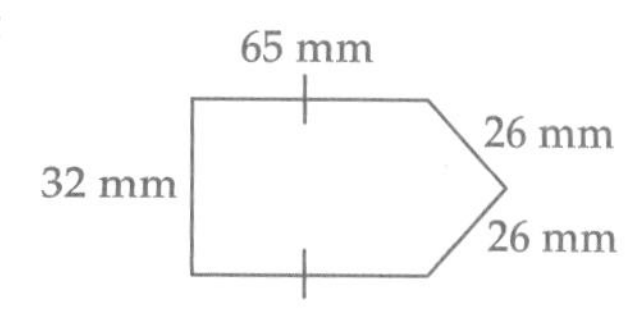

d
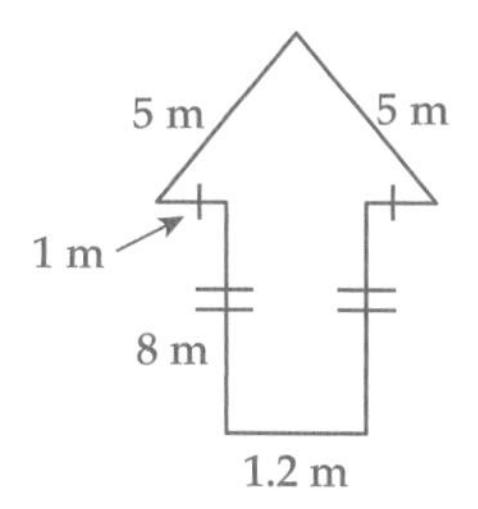

e
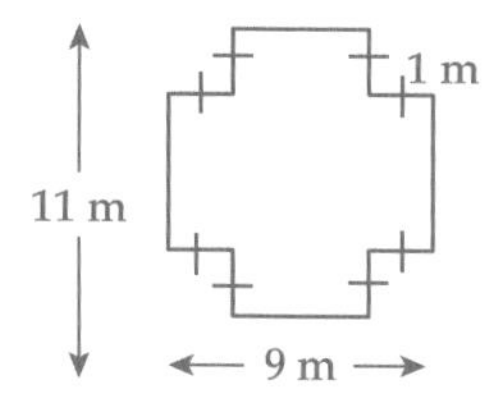

f
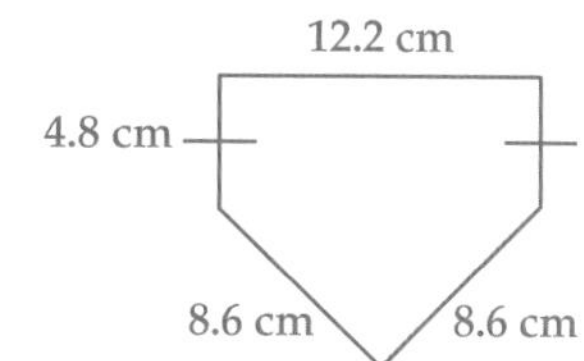

ISBN 9780170350969

12-05 Time

60 seconds = 1 minute	366 days = 1 leap year
60 minutes = 1 hour	52 weeks = 1 year
24 hours = 1 day	12 months = 1 year
7 days = 1 week	10 years = 1 decade
2 weeks = 1 fortnight	100 years = 1 century
365 days = 1 year	1000 years = 1 millennium

To convert units use the initials below to remember:
SOLD: Small Over to Large Divide; that is, small unit to large unit divide.
LOSM: Large Over to Small Multiply; that is, large unit to small unit multiply.

EXAMPLE 7

Complete the following time conversions.

a 2 minutes = ____ seconds
b 90 seconds = ____ minutes
c 3 hours = ____ minutes
d 2 weeks = ____ days
e 48 hours = ____ days

SOLUTION

a $2 \text{ minutes} = 2 \times 60 \text{ seconds}$
$= 120 \text{ seconds}$ ← LOSM

b $90 \text{ seconds} = 60 \text{ seconds} + 30 \text{ seconds}$
$= 1 \text{ minute} + \frac{1}{2} \text{ minute}$
$= 1\frac{1}{2} \text{ minutes}$

or $90 \text{ seconds} = \frac{9\not{0}}{6\not{0}} \text{ minutes}$ ← SOLD
$= \frac{3}{2} \text{ minutes}$
$= 1\frac{1}{2} \text{ minutes}$

c $3 \text{ hours} = 3 \times 60 \text{ minutes}$ ← LOSM
$= 180 \text{ minutes}$

d $2 \text{ weeks} = 2 \times 7 \text{ days}$
$= 14 \text{ days}$

e $48 \text{ hours} = \frac{48}{24} \text{ days}$
$= 2 \text{ days}$

ISBN 9780170350969

EXERCISE 12-05

1 How many minutes in 4 hours? Select the correct answer A, B, C or D.

A 60 B 120 C 180 D 240

2 How many days in 72 hours? Select A, B, C or D.

A 2 B 3 C 4 D 5

3 How many seconds are there in:

a 3 minutes? b 5 minutes? c 25 minutes?

4 Copy and complete:

a 2 days = ____ hours
b $1\frac{1}{2}$ hours = ____ minutes
c 120 minutes = ____ hours
d 3 weeks = ____ days
e 28 days = ____ weeks
f 1 fortnight = ____ days
g 48 months = ____ years
h 104 weeks = ____ years
i 1 century = ____ years
j 1 leap year = ____ days

5 How many minutes are there in:

a 2 hours?
b $3\frac{1}{2}$ hours?
c 1 hour 17 minutes?
d three-quarters of an hour?

6 Name the seven days of the week, starting with Sunday.

7 Name the 12 months of the year, starting with January.

8 How many years are there in:

a 3 decades? b 5 millennia? c 4.5 centuries? d 5.6 decades?

9 How many decades make up a century?

10 How many centuries make up a millennium?

Alamy/Paul Kingsley

12–06 24-hour time

Digital time describes the time with the hour first, followed by the minutes, such as 2:45 a.m. or 8:20 p.m.

For **12-hour time**, it is necessary to write a.m. or p.m. so we know if it is before 12 noon (a.m.) or after 12 noon (p.m.).

With **24-hour time**, the hours of a day are numbered from 0 to 23 so we do not need to write a.m. or p.m.

To write 24-hour time:
- use 4 digits
- count the hours from midnight (0000 hours) to 12 noon (1200)
- continue 1300 hours for 1 p.m. up to 2300 hours for 11 p.m.

Here is a table to show how to write in 24-hour time:

1 a.m.	0100	1 p.m.	1300
2 a.m.	0200	2 p.m.	1400
3 a.m.	0300	3 p.m.	1500
4 a.m.	0400	4 p.m.	1600
5 a.m.	0500	5 p.m.	1700
6 a.m.	0600	6 p.m.	1800
7 a.m.	0700	7 p.m.	1900
8 a.m.	0800	8 p.m.	2000
9 a.m.	0900	9 p.m.	2100
10 a.m.	1000	10 p.m.	2200
11 a.m.	1100	11 p.m.	2300
12 noon	1200	12 midnight	0000

EXAMPLE 8

Write in 24-hour time:

a 11:30 a.m. b 2:25 p.m. c 8:05 p.m. d 7:54 p.m.

SOLUTION

a 1130 b 1425 c 2005 d 1954

ISBN 9780170350969

EXERCISE 12–06

1 What is 3:05 p.m. in 24-hour time? Select the correct answer A, B, C or D.

A 0305 B 3050 C 1505 D 1550

2 Write 2345 in 12-hour time. Select A, B, C or D.

A 23:45 B 2:34 p.m. C 3:45 p.m. D 11:45 p.m.

3 Write the following times in 12-hour time and 24-hour time.

a 2 o'clock in the morning
b quarter past three in the afternoon
c half past 5 in the afternoon
d 5:50 in the morning
e 7:20 in the morning
f 9:30 in the evening
g 7:55 in the morning
h 11:15 in the evening

4 Convert to 24-hour times.

a 5:20 a.m. b 6:15 p.m. c 3:30 a.m. d 7:25 p.m.
e 11:15 a.m. f 9:08 p.m. g 1:22 a.m. h 12:20 p.m.
i 10:36 p.m. j 5:45 a.m. k 5:56 p.m. l 12:28 a.m.

5 Convert to 12-hour times.

a 0320 b 1450 c 2215 d 0704
e 2152 f 0432 g 1950 h 1126
i 2345 j 0036 k 0652 l 2008

6 a The bus leaves Alice Springs at 0900. Is it leaving in the morning or afternoon?

iStockphoto/bloodstone

b The bus arrives at Uluru at 1350. What time is this?

c Australian troops have to meet at 0500 hours. Write this in 12-hour time.

d A train travels from Sydney to Canberra in 4 hours 15 minutes. If it leaves Sydney at 2010, what time will it arrive in Canberra? Write the answer in 12-hour time and 24-hour time.

7 Nelly needs to catch a plane that will leave Melbourne airport at 2034. She has to be at the airport half an hour before the flight leaves and it takes her two-and-a-half-hours to get to the airport from home. What time does she need to leave home? Answer in 12-hour time.

12–07 Time calculations

To round time to the nearest minute:
- look at the number of seconds
- if it is 30 seconds or more, round up
- if it is less than 30 seconds, round down.

To round time to the nearest hour:
- look at the number of minutes
- if it is 30 minutes or more, round up
- if it is less than 30 minutes, round down.

EXAMPLE 9

Round:

a 4 h 36 min to the nearest hour b 18 min 24 s to the nearest minute

SOLUTION

a 36 min is more than 30 min (half an hour) so round up.

4 h 36 min ≈ 5 h

b 24 s is less than 30 s (half a minute) so round down.

18 min 24 s ≈ 18 min

EXAMPLE 10

Calculate the time difference from 8:35 a.m. to 3:10 p.m.

SOLUTION

Use a number line and 'build bridges'.

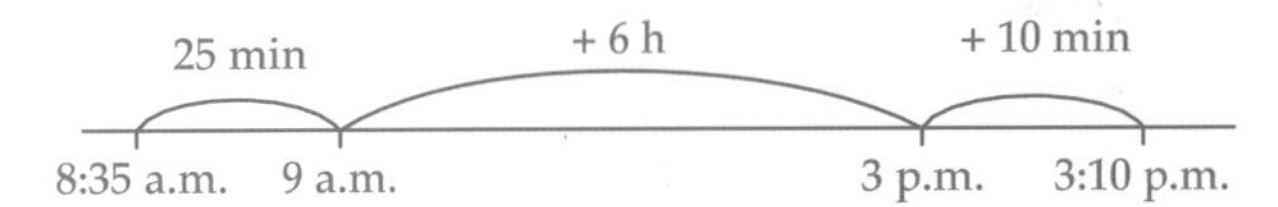

Time difference = 25 min + 6 h + 10 min
= 6 h + 35 min

OR: convert to 24-hour time and use the calculator's [° ' "] or [DMS] keys:

8:35 a.m. = 0835, 3:10 p.m. = 1510, so enter 15 [° ' "] 10 [° ' "] [−] 8 [° ' "] 35 [° ' "] [=]

So 6 hours 35 minutes is the time difference.

ISBN 9780170350969

EXERCISE 12–07

1 What is 12 minutes 31 seconds rounded to the nearest minute? Select the correct answer A, B, C or D.

A 12 B 31 C 13 D 11

2 What is the time difference from 4:25 p.m. to 9:18 p.m.? Select A, B, C or D.

A 4 h 53 min B 5 h 53 min C 5 h 7 min D 4 h 7 min

3 Round:

a 2 h 28 min to the nearest hour
b 6 h 43 min to the nearest hour
c 3 h 21 min to the nearest hour
d 6 h 13 min to the nearest hour
e 11 min 10 s to the nearest minute
f 25 min 35 s to the nearest minute
g 8 min 12 s to the nearest minute
h 22 min 42 s to the nearest minute

4 Copy and complete this table.

Time	Hours and minutes	Time	Minutes and seconds
129 minutes		86 seconds	
183 minutes		290 seconds	
958 minutes		421 seconds	

5 Find the time difference between each pair of times.

a 2:20 a.m. and 6:45 p.m.
b 1:50 a.m. and 10:59 a.m.
c 8:21 p.m. and 8:56 a.m. the next day
d 0220 to 1654
e 1227 to 2048
f 1156 to 0317 the next day

6 Senait boarded a train at 0540 and arrived at her destination at 1335. How long was her train journey?

7 Zain was baking cookies and needed to time them in the oven to the nearest minute to get the best taste. If he placed them in the oven at 5:23 p.m. and they needed 26 minutes to bake, what time should he take them out?

Shutterstock.com/Michael Shake

8 Find the sum of 2 h 25 min, 9 h 42 min and 5.5 hours.

9 What is the total time from 0445 to 1832?

12–08 Timetables

EXAMPLE 11

Part of a train timetable is shown below.

Station	a.m.	a.m.	a.m.
Cardiff	9:40	10:25	11:15
Hamilton	9:52	10:37	
Wickham	10:06	10:51	
Civic	10:10	10:55	11:30
Newcastle	10:22	11:07	11:34

a How long does it take the 9:40 train from Cardiff to get to Newcastle Station?

b Do the other two trains take the same time? Why, or why not?

c Why are there no times on some of the stations?

d If I wanted to be at Civic Station by 10:30, which train should I catch from Hamilton?

e Which train journey is the quickest?

SOLUTION

a 42 min — The time difference between 9:40 and 10:22

b No, the 10:25 takes 42 min but the 11:15 only takes 19 min. The 11:15 is faster as it doesn't stop at all the stations.

c The train doesn't stop there.

d The 9:52 train

e The 11:15 train

EXAMPLE 12

How would I travel from Eden to Shores on the train lines below?

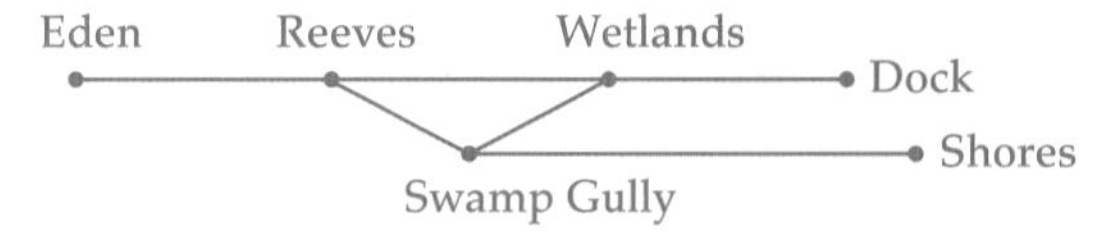

SOLUTION

Start at Eden and change at Reeves to get to Shores.

Alamy/Suzanne Long

 ISBN 9780170350969

EXERCISE 12–08

Use the train timetable in Example 11 to answer Questions 1 and 2.

1 Which train would I catch if I wanted to be at Wickham by 10:10 am? Select the correct answer A, B, C or D.

A Cardiff 10:25 B Hamilton 9:52

C Cardiff 11:15 D Hamilton 10:37

2 How long does the train take to travel from Wickham to Newcastle? Select A, B, C or D.

A 7 min B 8 min C 16 min D 22 min

3 Answer the questions for the bus timetable below.

Bus stop	Bus 1	Bus 2	Bus 3
King St	8:45	9:05	9:48
Wall St	8:58		10:01
Brown St	9:10	9:20	10:13
Rose Ave	9:22		10:25
Downie La	9:35	9:35	10:38

a How long does Bus 1 take to travel from King St to Downie Lane?

b Do the other two buses take the same time? How long do they take?

c If I wanted to be at Brown St by 9:20, which bus could I catch from Wall St?

d If I had to be at Downie Lane by 9:40, which bus would I catch from King St?

e If I arrived at King St by 8:40 and wanted to go to Rose Ave, which bus would I catch?

4 Answer the questions for the train lines below.

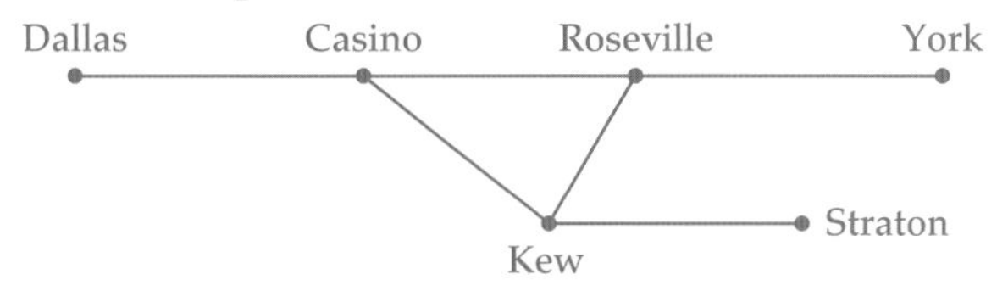

a How could you travel from Dallas to Kew?

b How could you travel from Straton to Roseville?

c Where would you change trains to travel from York to Straton?

d Where would you change trains to travel from Dallas to Straton?

5 Find a train or bus timetable on the internet and write down 5 questions about it.

CROSSWORD PUZZLE

Make a copy of this page and complete the crossword below using these 'time' words:

AM	ANALOGUE	CENTURY	DAY	DECADE
DIGITAL	FORTNIGHT	HOUR	LEAPYEAR	MILLENNIUM
MINUTE	MONTH	PM	SECOND	TIMELINE
TIMEZONES	WEEK	YEAR		

ISBN 9780170350969

PRACTICE TEST 12

Part A General topics

Calculators are not allowed.

1 Evaluate $-3 \times 8 \div 2$.

2 Evaluate 15×40.

3 What is an isosceles triangle?

4 Find $\frac{1}{5}$ of \$250.

5 Evaluate 0.09×0.6.

6 Evaluate $18 - 3 \times 5$.

7 Simplify $3 \times a \times 9 \times a \times b$.

8 Find w.

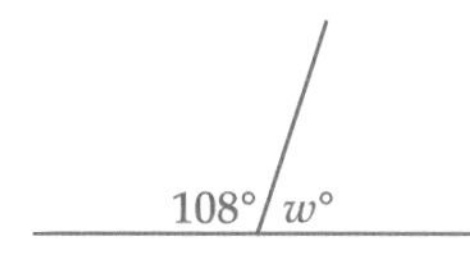

9 Evaluate $\frac{1}{4} - \frac{1}{5}$.

10 Write in ascending order: 0.4, 0.04, 0.402, 0.45.

Part B Length and time

Calculators are allowed.

12-01 The metric system

11 Which of the following is a unit of length? Select the correct answer A, B, C or D.

A litre B millilitre C kilometre D hectare

12 Which unit is used to measure a dose of medicine? Select A, B, C or D.

A litres B millimetres C centimetres D millilitres

12-02 Measuring length

13 1380 mm = ________ cm. Select A, B, C or D.

A 0.138 B 1.38 C 13.8 D 138

12-03 Perimeter

14 What is the perimeter of an equilateral triangle of side 4.6 cm?

15 Find the perimeter of a rectangle with length 12 m and width 5.2 m.

12-04 Perimeter of composite shapes

16 Find the perimeter of each shape.

a

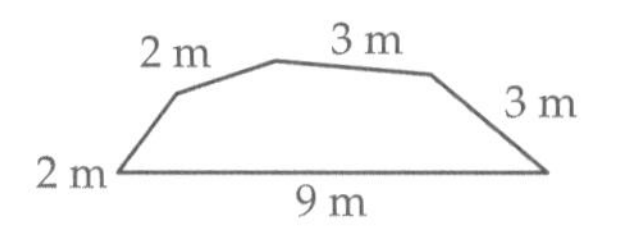

b

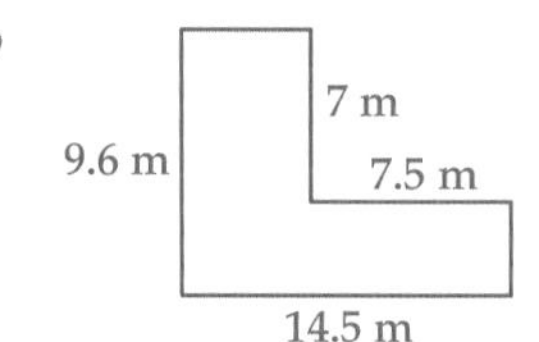

ISBN 9780170350969

12-05 Time

17 How many minutes in 5 hours 12 minutes?

18 Convert 223 minutes to hours and minutes.

12-06 24-hour time

19 Write 5:48 p.m. in 24-hour time.

20 Write 1953 in 12-hour time.

12-07 Time calculations

21 Find the sum of 12 h 45 min and 5 h 22 min.

12-08 Timetables

22 Use the train timetable below to calculate how long it takes the 11:15 train from Cardiff to arrive at Newcastle.

Station	a.m.	a.m.	a.m.
Cardiff	9:40	10:25	11:15
Hamilton	9:52	10:37	
Wickham	10:06	10:51	
Civic	10:10	10:55	11:30
Newcastle	10:22	11:07	11:34

ISBN 9780170350969

AREA AND VOLUME

13

WHAT'S IN CHAPTER 13?

13–01 Area
13–02 Area of a rectangle
13–03 Area of a triangle
13–04 Area of a parallelogram
13–05 Area of composite shapes
13–06 Volume
13–07 Volume of a rectangular prism
13–08 Volume and capacity

IN THIS CHAPTER YOU WILL:

- understand area and its metric units
- find the area of a square, rectangle and triangle
- find the area of a parallelogram
- find the area of composite shapes
- understand volume and its metric units
- find the volume of a rectangular prism
- understand the relationship between volume and capacity

Shutterstock.com/dotshock

ISBN 9780170350969

13-01 Area

WORDBANK

area Area is the amount of surface space inside a flat shape. Area is measured in square units, usually mm^2, cm^2, m^2 or km^2.

A **square centimetre (cm^2)** is the area of a square of length 1 cm.

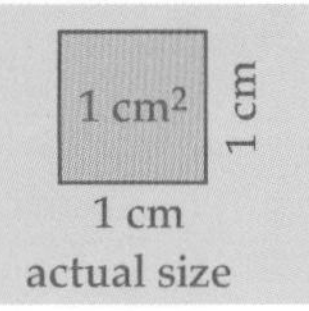

EXAMPLE 1

Find the area of each figure by counting the number of square centimetres.

a

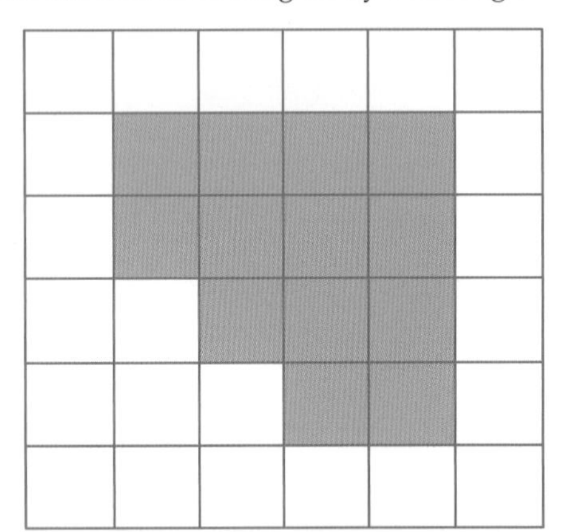

b

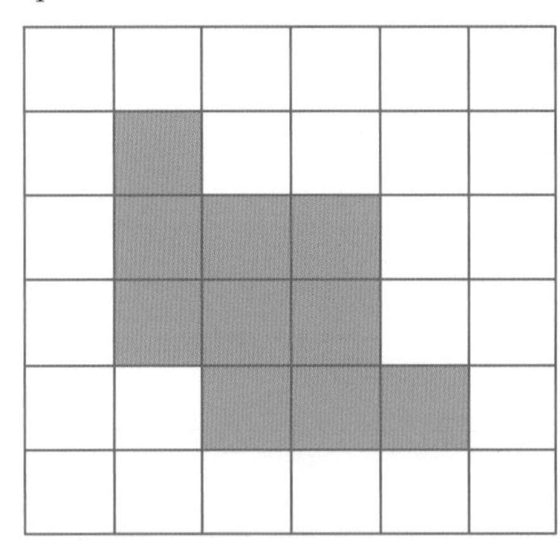

SOLUTION

Count the number of shaded squares.

a Area = 13 cm^2

b Area = 10 cm^2

EXAMPLE 2

Find the approximate area of the shapes below.

a

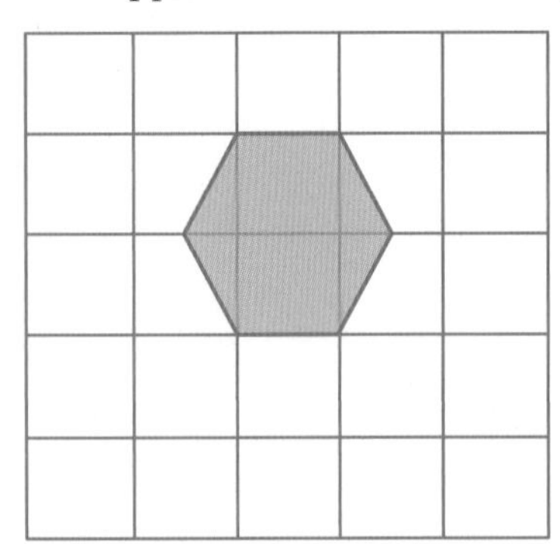

b

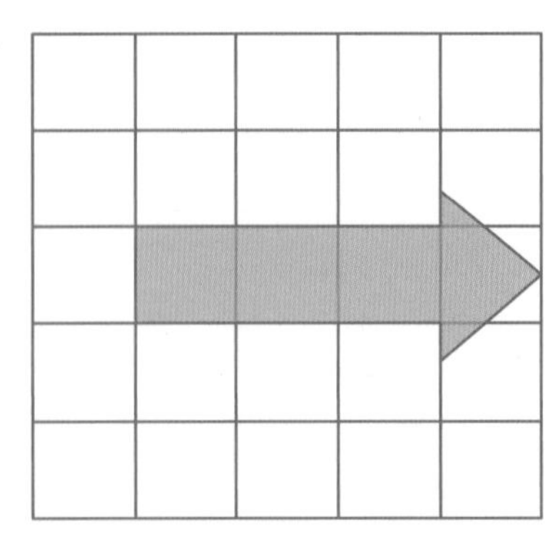

SOLUTION

Count whole squares and approximate the rest.

a Approximate area = 3 cm^2

b Approximate area = 4 cm^2

ISBN 9780170350969

EXERCISE 13–01

1 Which unit of measure could be used for area? Select the correct answer A, B, C or D.

A mL B mm C mm^2 D mL^2

2 Is each statement true or false?

a Area is the distance around a shape.
b Area is measured in square units.
c Perimeter is the size of a region.
d Perimeter and area are two different measurements.

3 Find the area of each shaded figure if each square is a square centimetre.

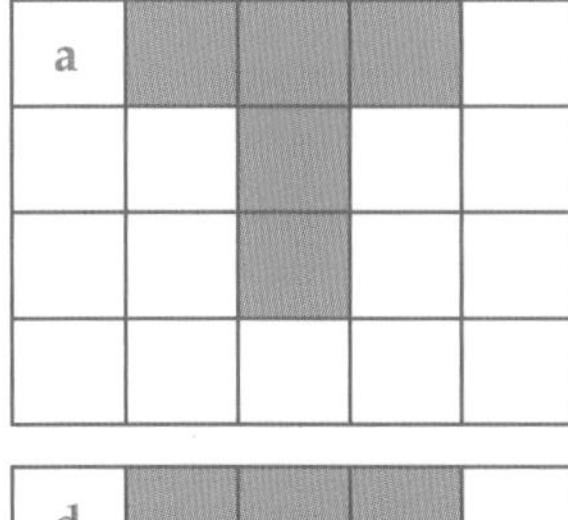

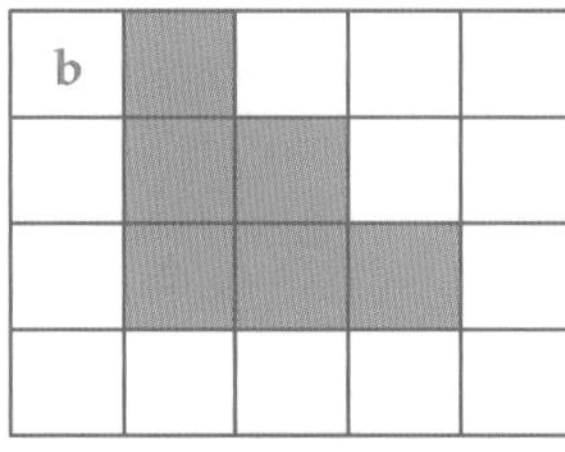

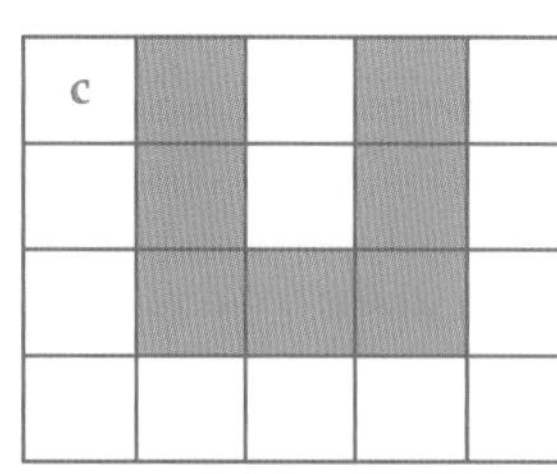

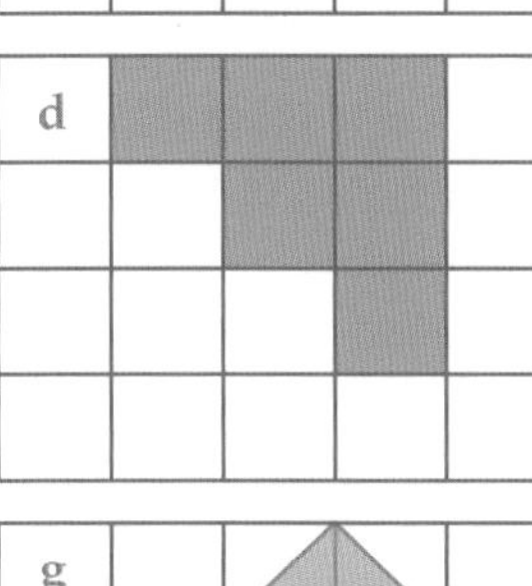

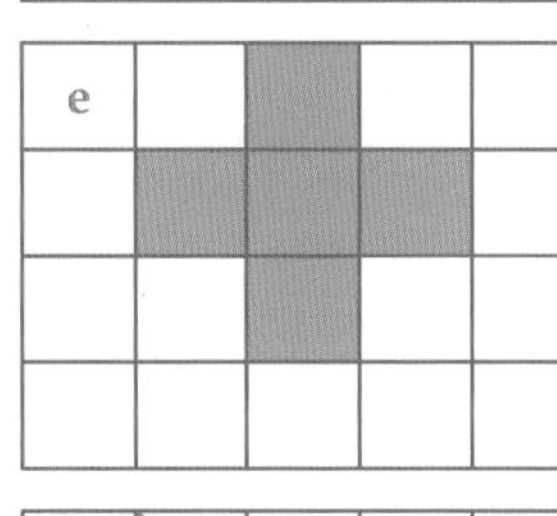

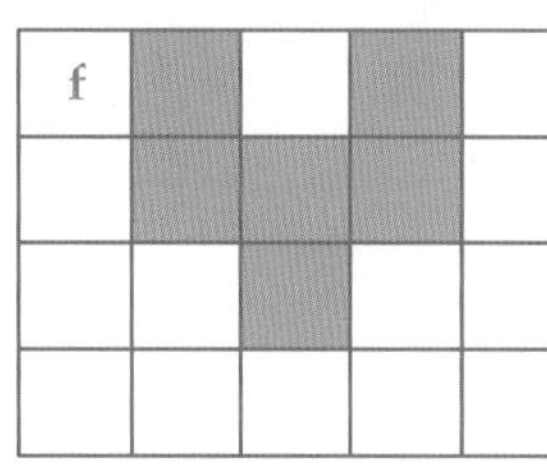

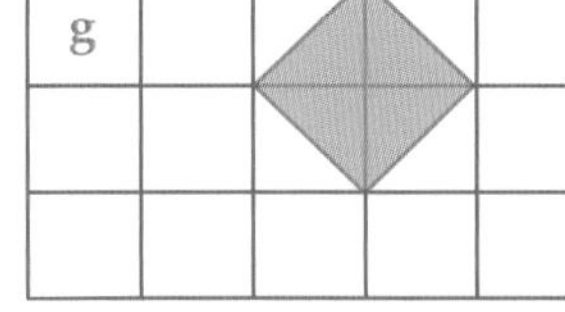

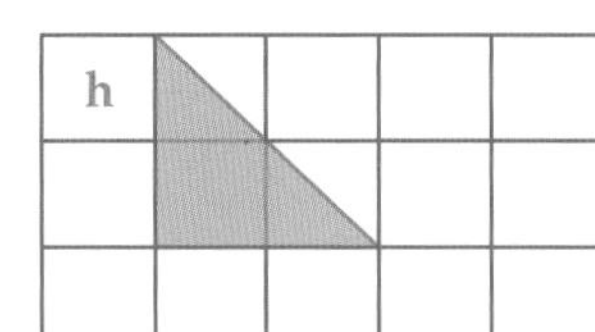

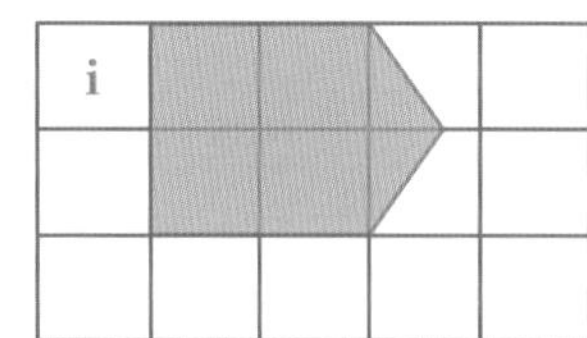

4 Find the approximate area of the shaded shapes below:

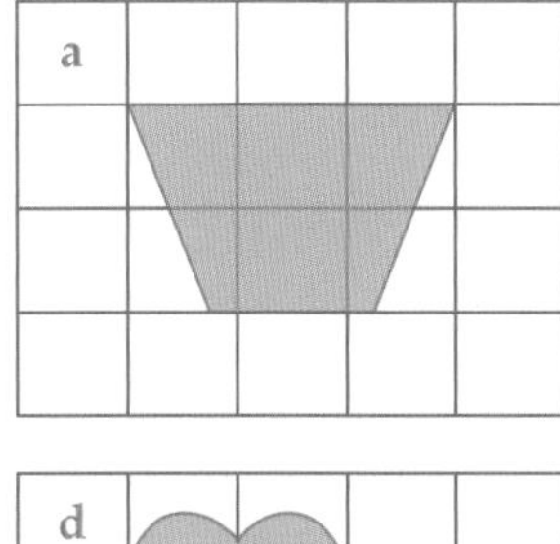

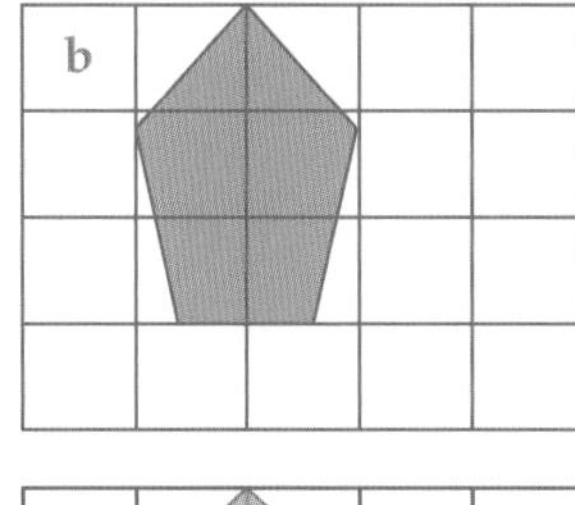

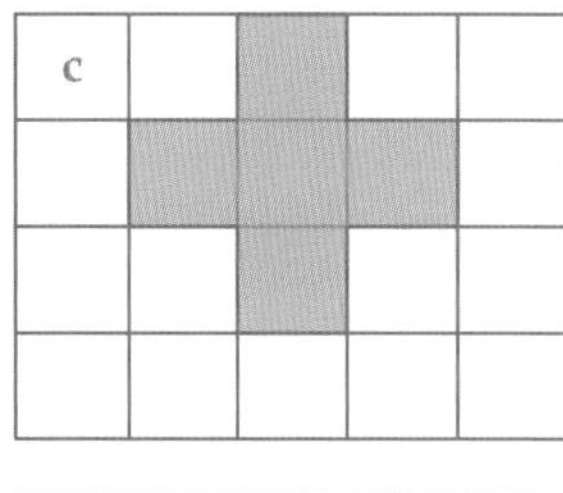

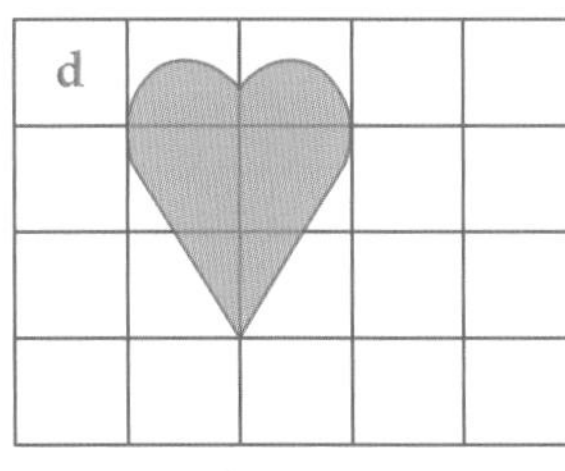

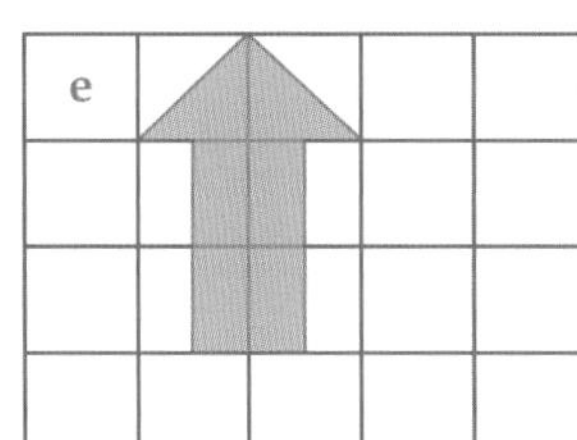

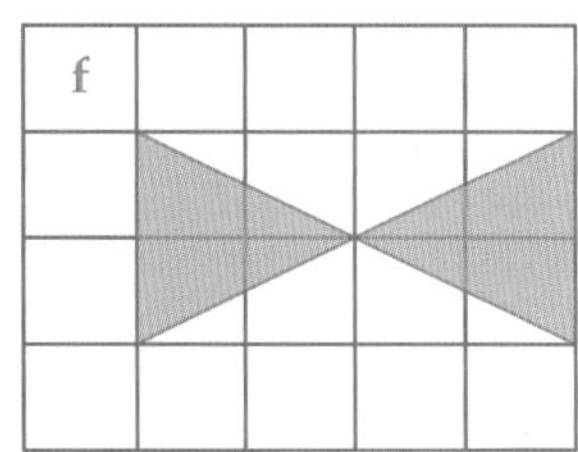

13-02 Area of a rectangle

This blue rectangle is 4 cm long and 2 cm wide.

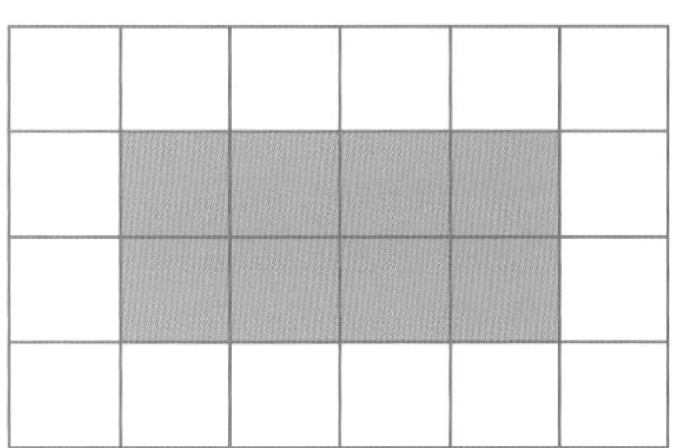

The area of the rectangle is 8 square centimetres if we count the squares.
The area can also be calculated by observing there are 2 rows of 4 squares, which is 2 lots of 4, or 4×2.
So the area of the rectangle is 4 (length of the rectangle) $\times$ 2 (width of the rectangle).

Formula for the area of a rectangle:
Area = length $\times$ width
$A = l \times w$

EXAMPLE 3

Find the area of each of the rectangles below:

a
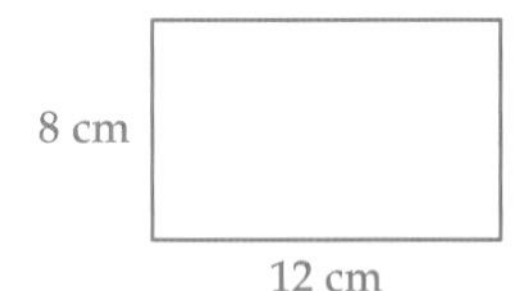

b

c
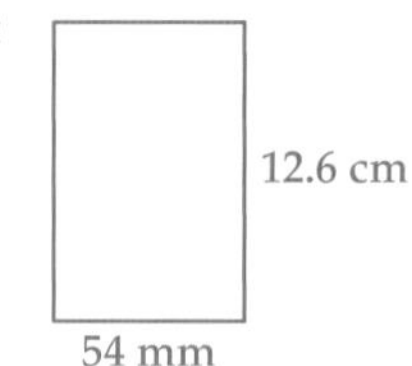

SOLUTION

a $A = l \times w$
$= 12 \times 8$
$= 96 \text{ cm}^2$

All units for area are square units.

b $A = l \times w$
$= 15.7 \times 3.2$
$= 50.24 \text{ m}^2$

c $A = l \times w$ ← 54 mm = 5.4 cm
$= 12.6 \times 5.4$
$= 68.04 \text{ cm}^2$

In **c** the units were not the same. When this happens, all units must be converted to the same unit before we can find the area.

 ISBN 9780170350969

EXERCISE 13–02

1 What is the area of a rectangle of length 8 cm and width 7 cm? Select the correct answer A, B, C or D.

A 48 cm^2 B 56 cm^2 C 64 cm^2 D 30 cm^2

2 What is the area of a rectangle if $l = 4.7$ m and $w = 3.6$ m? Select A, B, C or D.

A 17.02 m^2 B 29.61 m^2 C 26.64 m^2 D 16.92 m^2

3 Find the area of each rectangle.

a 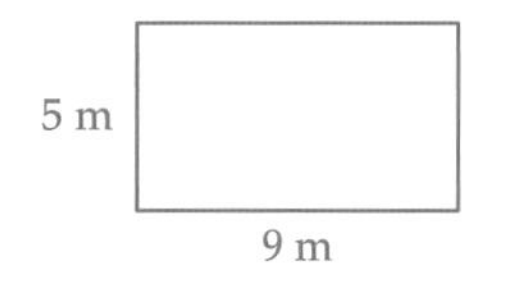
5 m
9 m

b 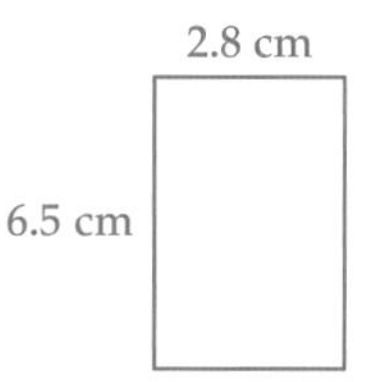
2.8 cm
6.5 cm

c 
2.1 m
11.4 m

d 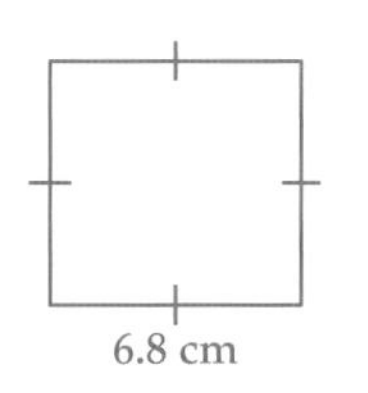
6.8 cm

e 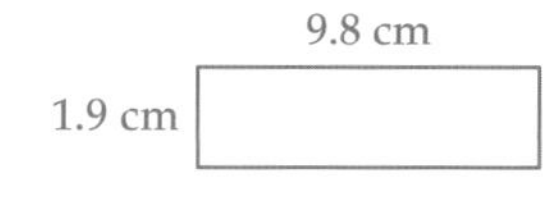
9.8 cm
1.9 cm

f 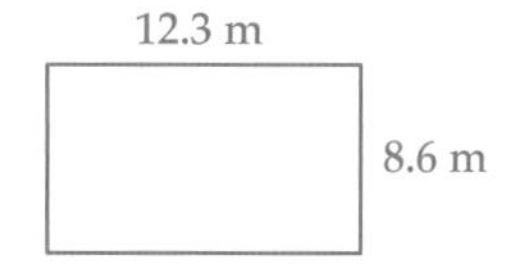
12.3 m
8.6 m

4 Find the area of each rectangle.

a 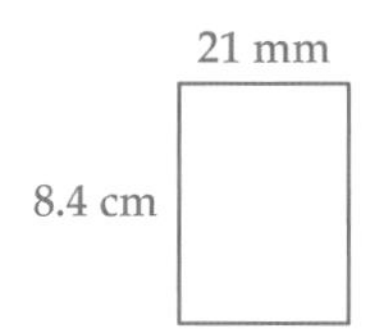
21 mm
8.4 cm

b 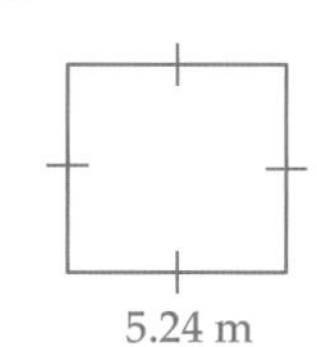
5.24 m

c 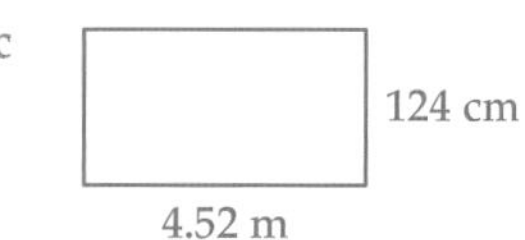
124 cm
4.52 m

5 Find the area of each figure described below.

a A rectangle with length 6.9 m and width 75 cm.

b A square with sides of length 7.3 mm.

c A rectangle with width 2.6 mm and length 0.98 m.

d A rectangle with length 826 cm and width 45.2 mm.

6 Georgina was planting a vegetable garden. She dug up a patch of land 5 m long and 3.5 m wide.

a What was the area of Georgina's garden?

b If she could plant 5 cabbages to each m^2, how many cabbages can Georgina plant?

Alamy/Arco Images / Farkaschovsky, H.

13-03 Area of a triangle

This triangle has a base of length 4 cm and a height of 3 cm.

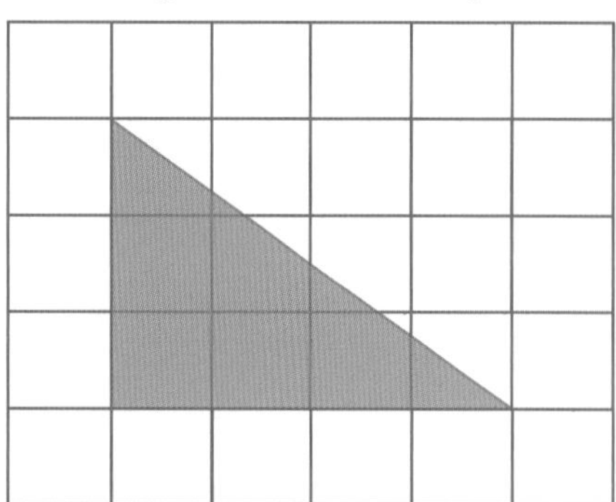

The area of the triangle is 6 square centimetres if we count the squares.
The area can also be calculated by observing that the triangle is half of a rectangle with length 4 cm and width 3 cm.
So the area of the triangle is $\frac{1}{2} \times 4 \times 3 = 6$ cm².

Formula for the area of a triangle:

Area $= \frac{1}{2} \times$ base $\times$ height

$A = \frac{1}{2}bh$

EXAMPLE 4

Find the area of each triangle:

a

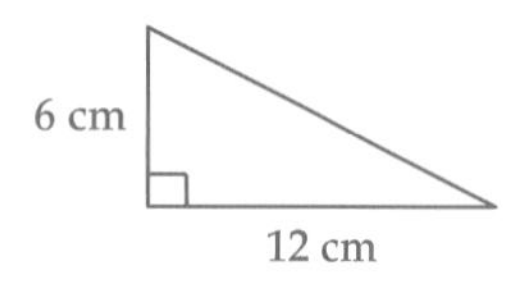

b

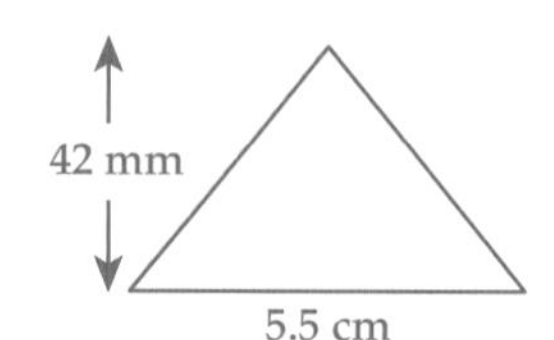

c

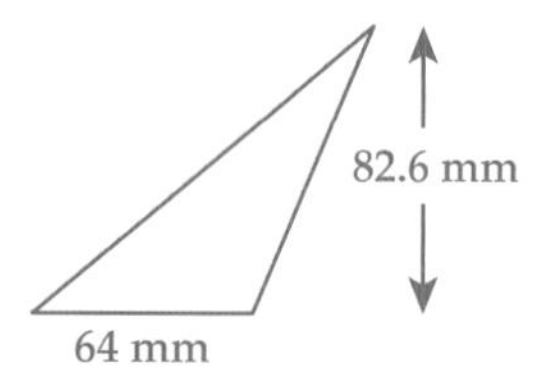

SOLUTION

a $A = \frac{1}{2}bh$

$= \frac{1}{2} \times 12 \times 6$

$= 36$ cm²

b $A = \frac{1}{2}bh$

$= \frac{1}{2} \times 5.5 \times 4.2$ ← 42 mm = 4.2 cm

$= 11.55$ cm²

All units must be converted to the same units.

c $A = \frac{1}{2}bh$

$= \frac{1}{2} \times 64 \times 82.6$

$= 2643.2$ mm²

ISBN 9780170350969

EXERCISE 13–03

1 What is the area of a triangle with base 15 cm and height 12 cm? Select the correct answer A, B, C or D.

A 180 cm^2 B 45 cm^2 C 90 cm^2 D 54 cm^2

2 What is the area of a triangle with base 8.6 m and height 4.2 m? Select A, B, C or D.

A 36.12 m^2 B 18.06 m^2 C 9.03 m^2 D 25.6 m^2

3 Find the area of each triangle below.

a
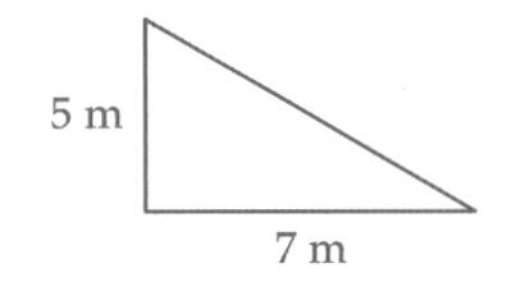

b
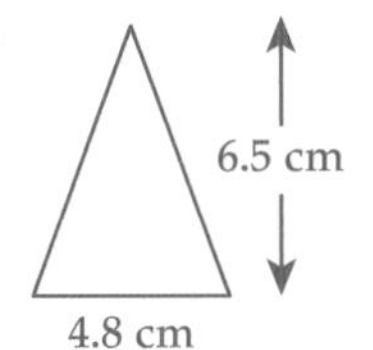

c
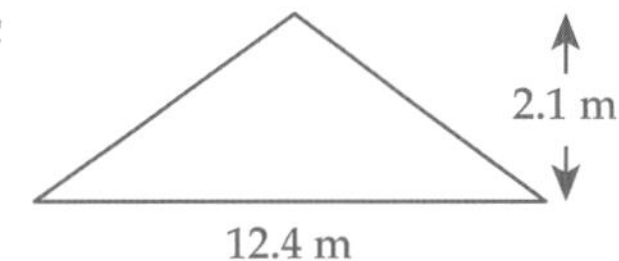

d
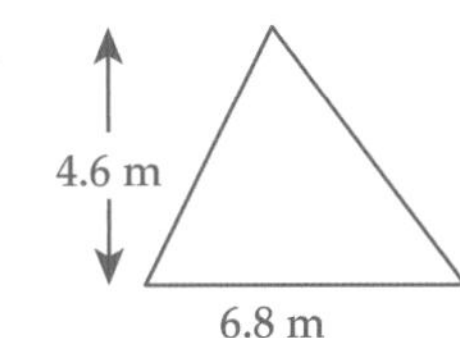

e
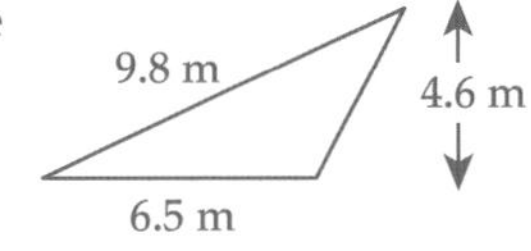

f
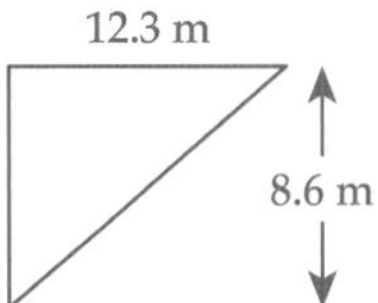

4 Find the area of each triangle.

a
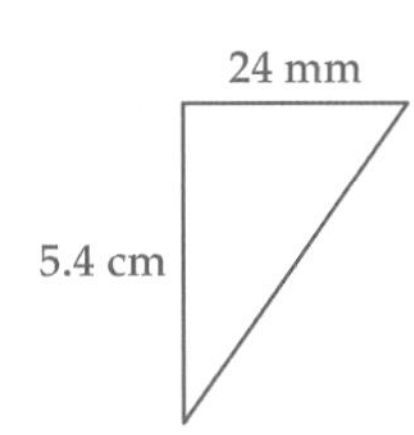

b
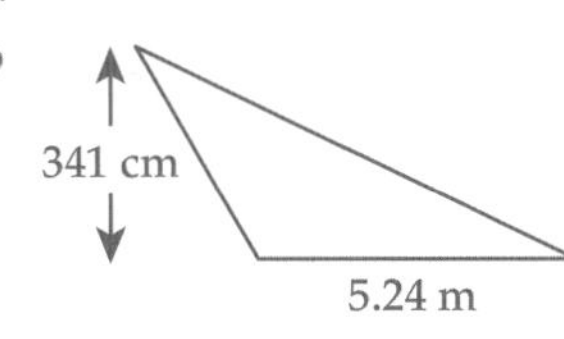

c
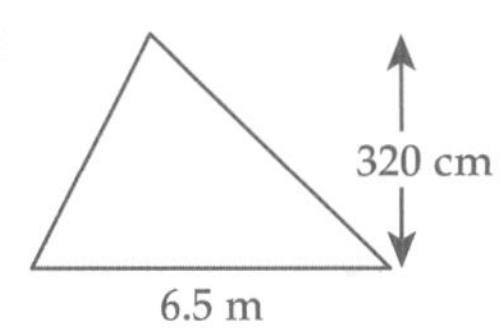

5 Find the area of each figure described below.

a A triangle with base 5.9 m and height 45 cm

b A right-angled triangle with both sides of length 4.3 mm about the right angle

c A triangle with height 4.6 mm and base 0.6 m

d An equilateral triangle with sides 82 cm and perpendicular height of 71 cm.

6 Greg was paving a triangular shaped courtyard. The base of the courtyard was 8 m and the height was 4.3 m.

a What is the area of the courtyard?

b How much will it cost Greg to pave it if the pavers cost \$23.50 per m²?

Alamy/GerryRousseau

13-04 Area of a parallelogram

This parallelogram has a base of length 3 cm and a height of 2 cm.

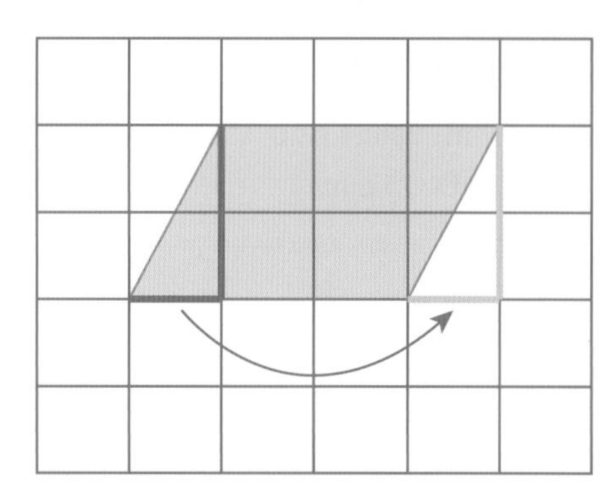

If we move the triangle outlined in dark blue to the space outlined in light blue then the parallelogram will become a rectangle and its area will be the same as the area of a rectangle with length 3 cm and width 2 cm.

So area of parallelogram $= 3 \times 2$

$= 6 \text{ cm}^2$

Formula for the area of a parallelogram:

Area = base × height

$A = bh$

EXAMPLE 5

Find the area of each parallelogram.

a

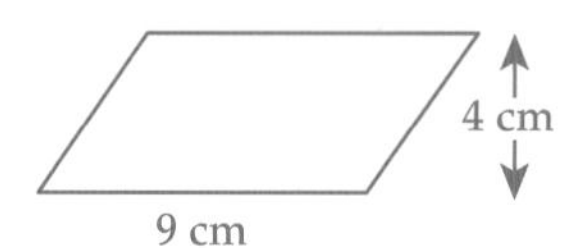

b

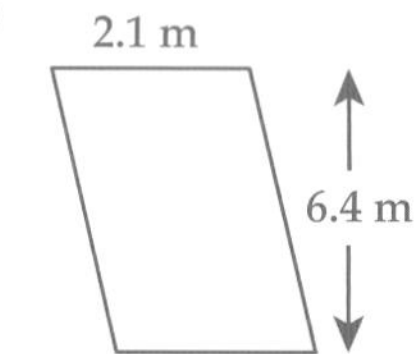

SOLUTION

a $A = bh$

$= 9 \times 4$

$= 36 \text{ cm}^2$

b $A = bh$

$= 2.1 \times 6.4$

$= 13.44 \text{ m}^2$

 ISBN 9780170350969

EXERCISE 13–04

1 Find the area of a parallelogram with base 11 cm and height 5.6 cm. Select the correct answer A, B, C or D.

A 61.6 cm^2 B 71.5 cm^2 C 30.8 cm^2 D 33.2 cm^2

2 The area of a parallelogram is 24 m^2. What is its base if its height is 8 m? Select A, B, C or D.

A 4 m B 6 m C 3 m D 12 m

3 Find the area of each parallelogram.

a
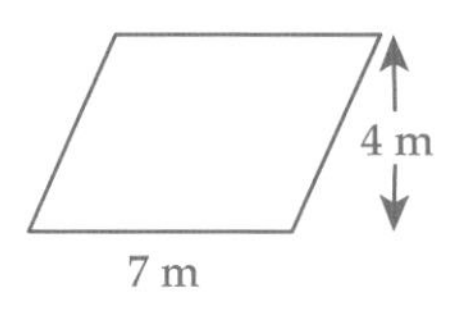

b
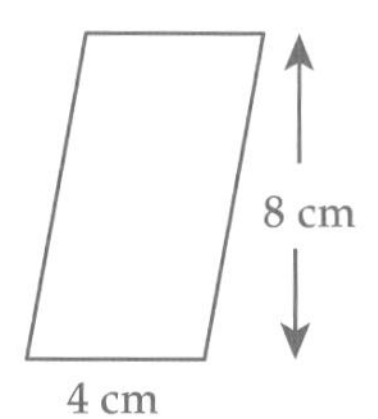

c
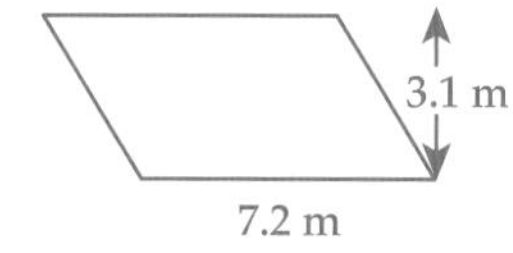

d
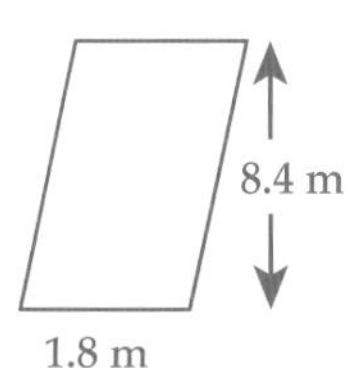

e
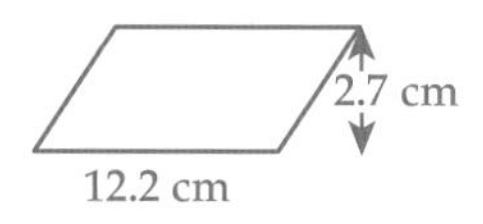

f
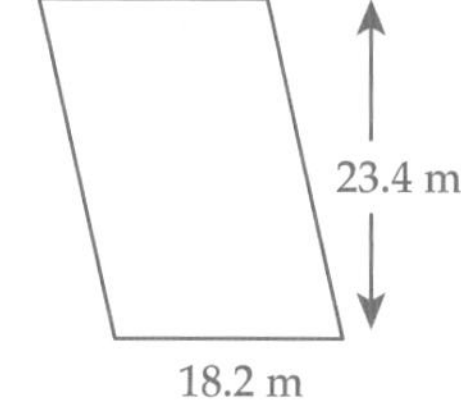

4 Find the value of the missing base or height in each of these parallelograms.

a
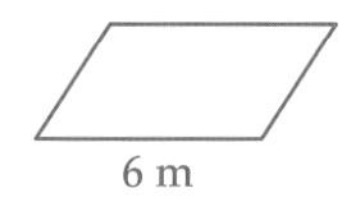

Area = 48 m^2

Height = ?

b
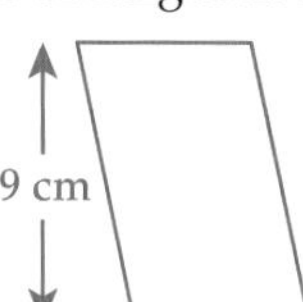
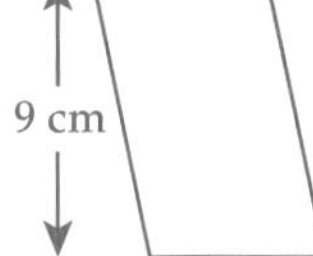

Area = 63 cm^2

Base = ?

c
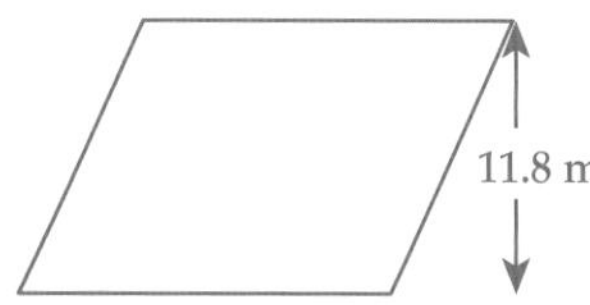
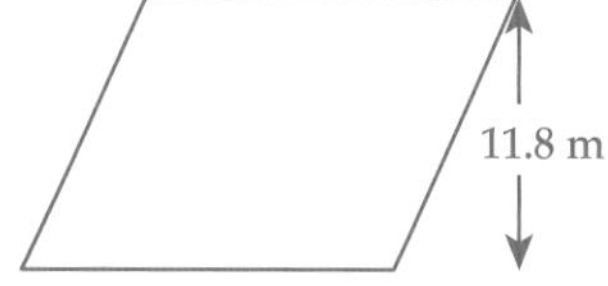

Area = 146.32 m^2

Base = ?

5 Find the size of the missing dimensions or area for each parallelogram.

a A parallelogram with base 13 cm and height 15 cm. Area = ?

b A parallelogram with area 40.5 m^2 and height 9 m. Base = ?

c A parallelogram with area 48.96 cm^2 and base 7.2 cm. Height = ?

d A parallelogram with base 4.3 mm and height 5.9 mm. Area = ?

e A parallelogram with area 114.66 m^2 and height 6.3 m. Base = ?

f A parallelogram with area 548.64 cm^2 and base 25.4 cm. Height = ?

EXAMPLE 6

Find the area of each composite shape.

a

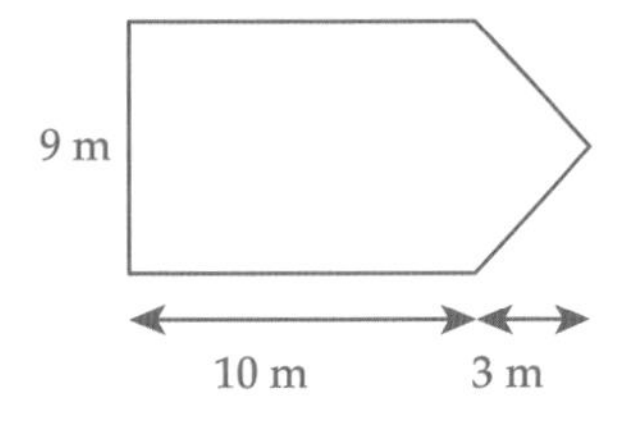

b

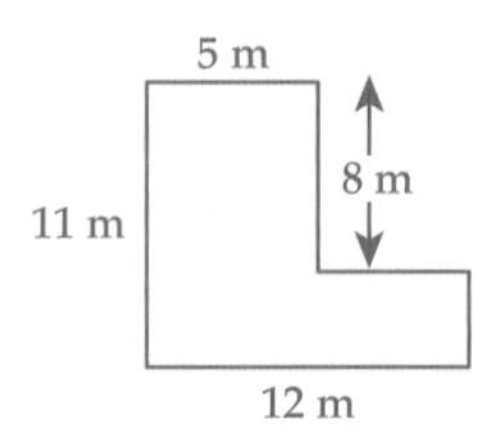

SOLUTION

Divide into standard shapes as shown with the dotted line.

a

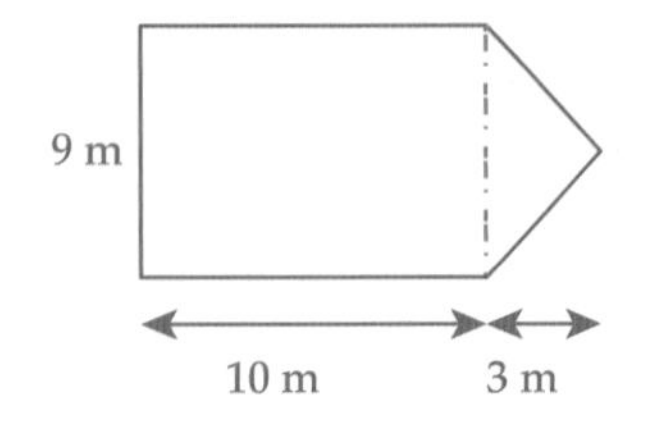

Area = Area of rectangle + Area of triangle

$= l \times w + \frac{1}{2}bh$

$= 9 \times 10 + \frac{1}{2} \times 9 \times 3$

$= 90 + 13.5$

$= 103.5 \text{ m}^2$

b

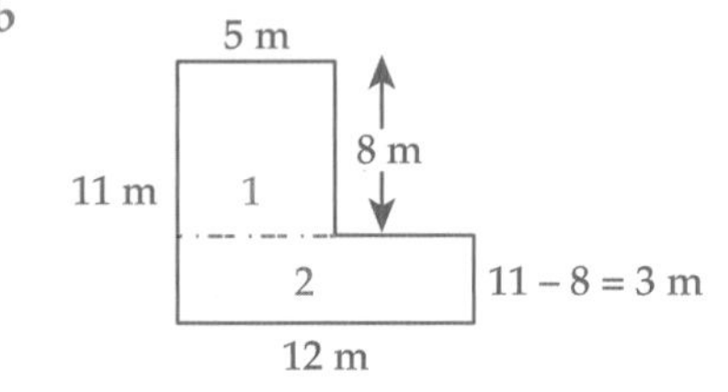

Area = Area of rectangle 1 + Area of rectangle 2

$= l \times w + l \times w$

$= 8 \times 5 + 12 \times 3$

$= 40 + 36$

$= 76 \text{ m}^2$

EXAMPLE 7

Find the shaded area of the shape below:

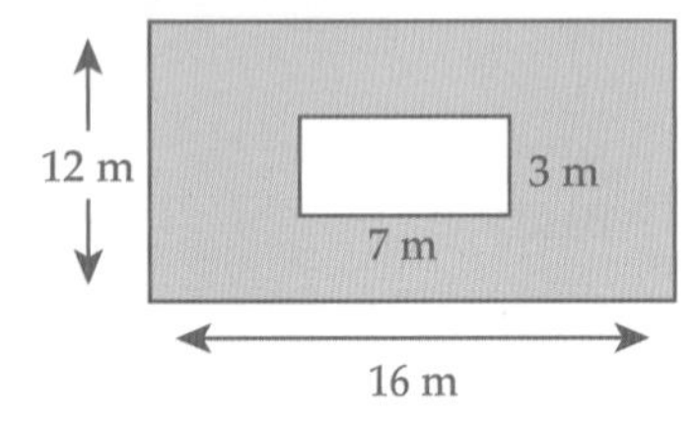

SOLUTION

Shaded area $= 12 \times 16 - 7 \times 3$

$= 171 \text{ m}^2$

ISBN 9780170350969

EXERCISE 13–05

1 A composite shape is made up of a 28 m² rectangle with a hole in the shape of a 9 m² square. What is the area of the composite shape? Select the correct answer A, B, C or D.

A 19 m² B 25 m² C 37 m² D 252 m²

2 A composite shape is made up of a 28 m² rectangle and a 16 m² triangle. What is the area of the composite shape? Select A, B, C or D.

A 34 m² B 44 m² C 12 m² D 54 m²

3 Find the area of each of the composite shapes below by adding areas of smaller shapes.

a

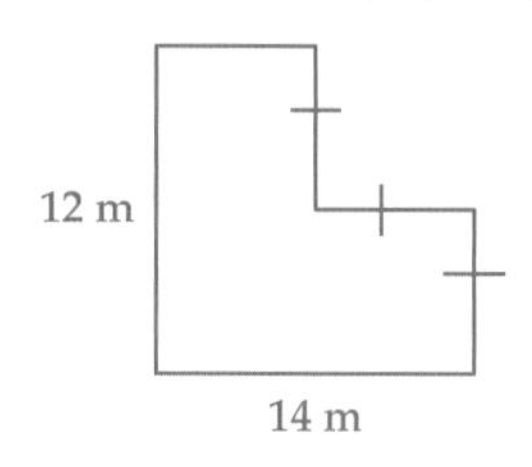

b

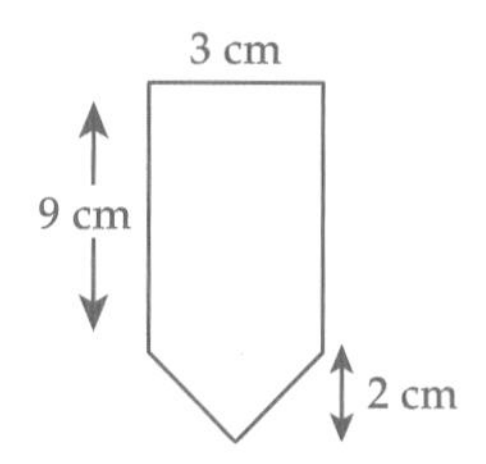

c

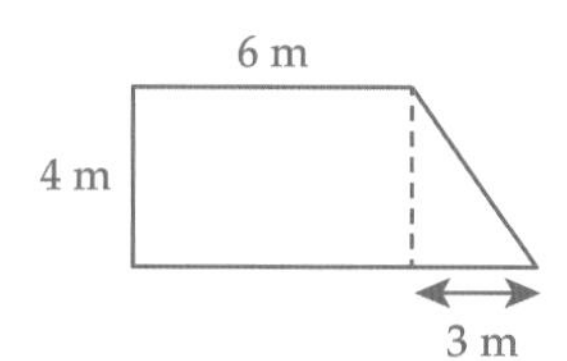

d

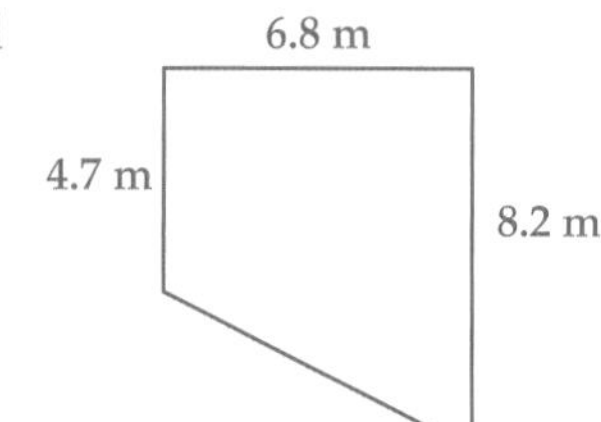

e

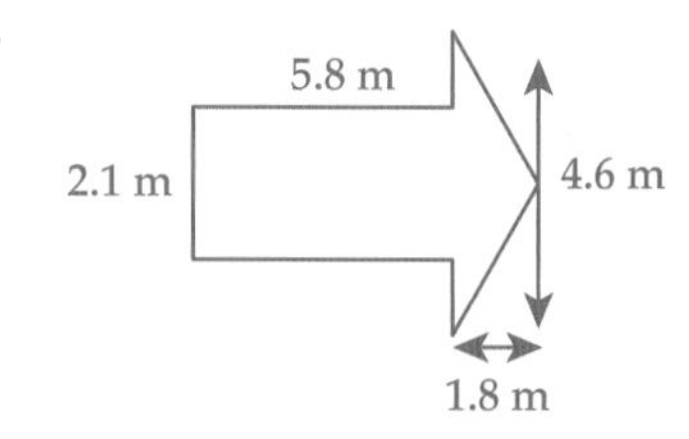

f

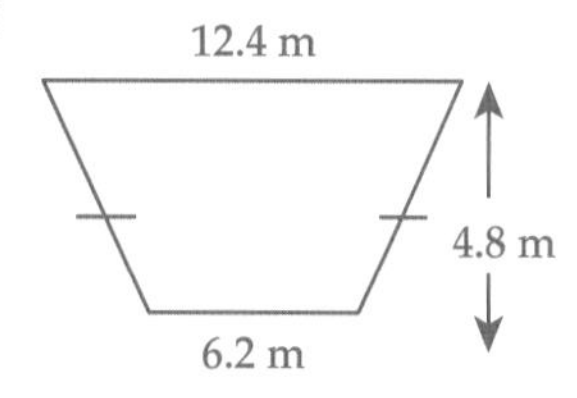

4 Question 3f can also be done by subtracting areas. Show how you will get the same answer by subtracting areas in this question.

5 Find the area of each of the composite shaded shapes below by subtracting areas.

a

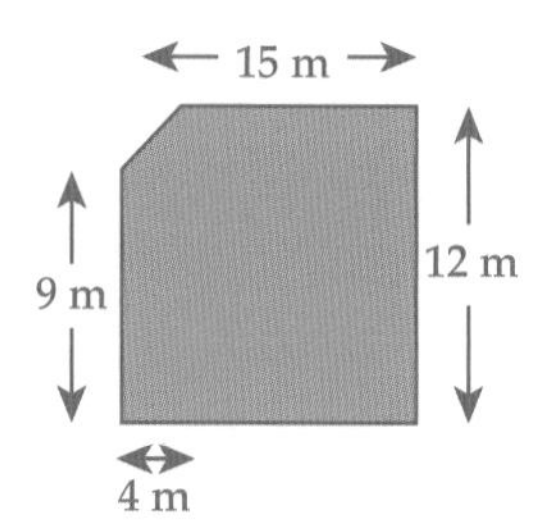

b

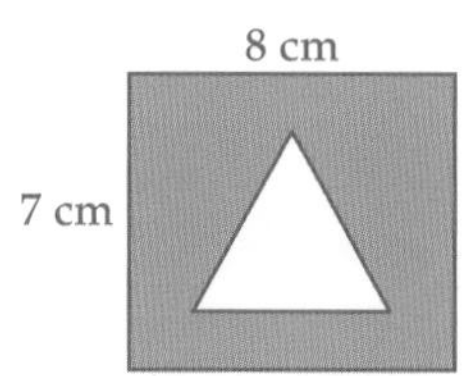

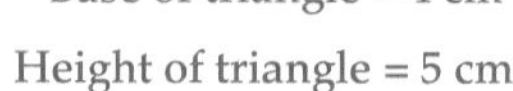

Base of triangle = 4 cm

Height of triangle = 5 cm

c

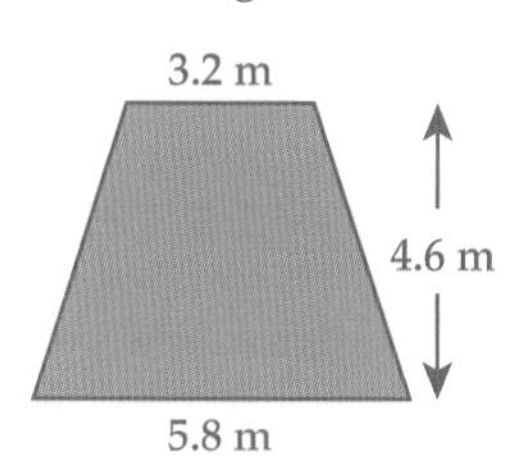

6 Draw an L-shape and show how you could find the area in two different ways.

13-06 Volume

WORDBANK

volume The amount of space inside a solid shape. Volume is measured in cubic units such as mm^3, cm^3, m^3 or km^3.

A **cubic centimetre (cm^3)** is the volume of a cube of length 1 cm

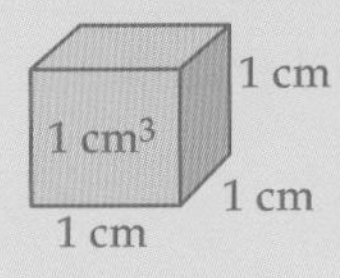

A rectangular prism has 3 dimensions: length, width and height:

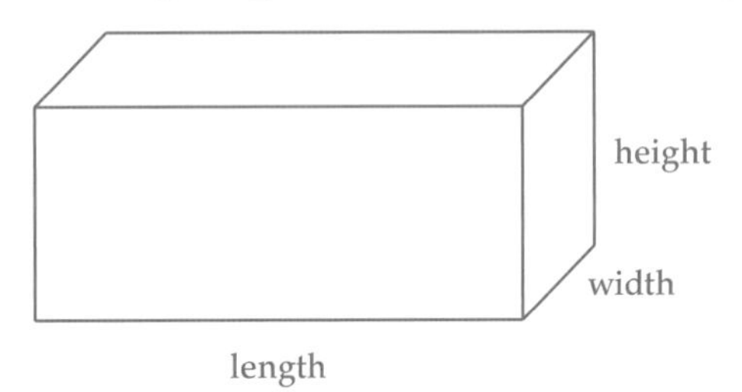

EXAMPLE 8

Find the volume of this rectangular prism.

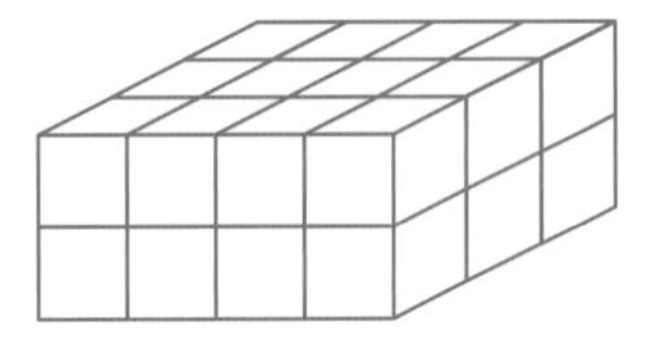

SOLUTION

Number of cubes in 1 layer: 12 ← 4 cubes by 3 cubes
Number of layers: 2
Volume = 12×2
$= 24$ units^3

To find the volume of a rectangular prism:

- count the number of cubes on one layer of the solid
- multiply by the number of layers.

ISBN 9780170350969

EXERCISE 13–06

1 What is the volume of a rectangular prism that has 16 cubes on its base and is 3 layers high? Select the correct answer A, B, C or D.

A 19 units³ B 13 units³ C 48 units³ D 32 units³

2 What is the volume of a rectangular prism with 22 cubic centimetres on its base and is 5 layers high? Select A, B, C or D.

A 27 cm³ B 110 cm³ C 105 cm³ D 115 cm³

3 Find the volume of each rectangular prism if each small cube is a cubic centimetre.

a

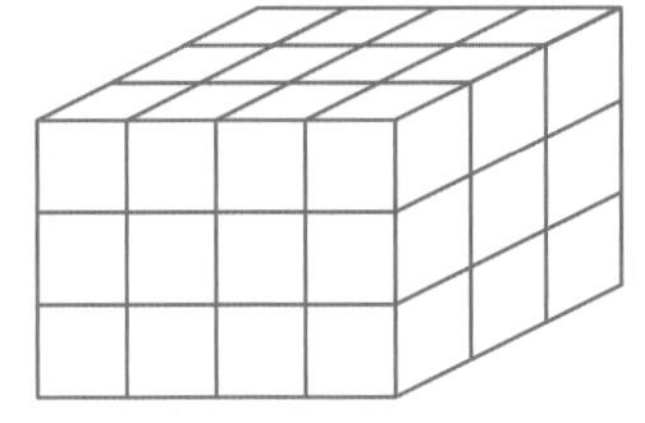

b

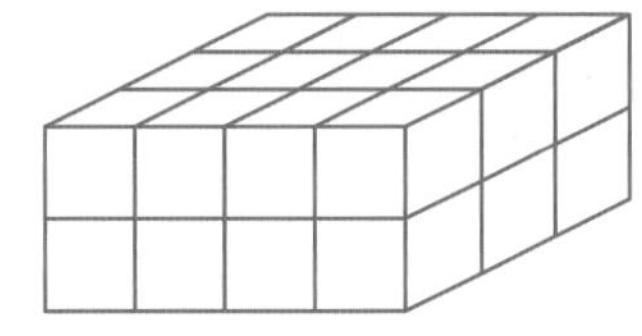

c

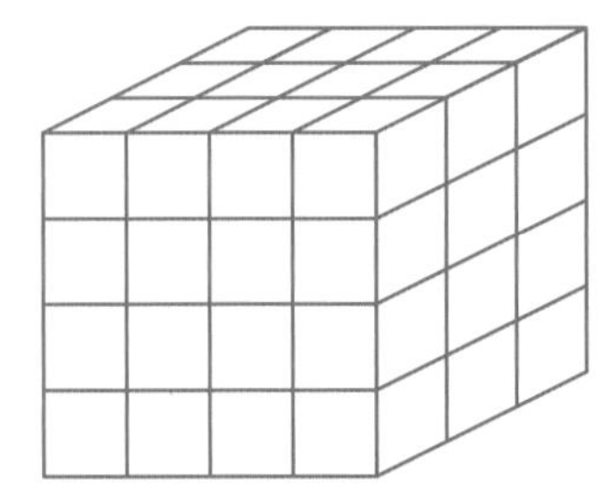

d

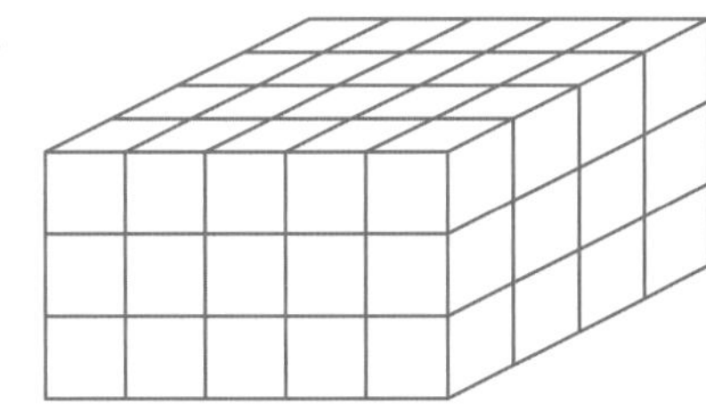

e

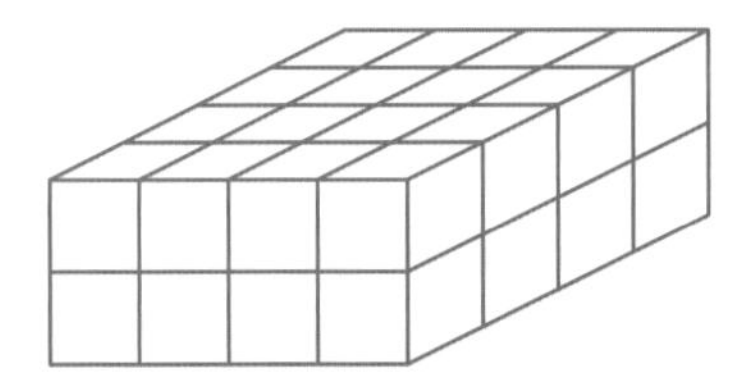

f

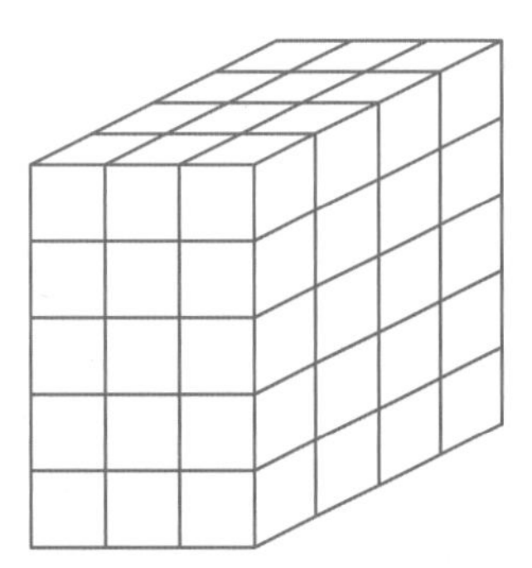

4 Copy these shapes and draw in cubes, and then count them to find the volume of each prism in cubic units.

a

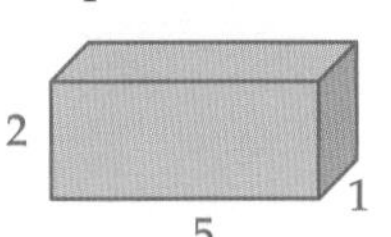

b

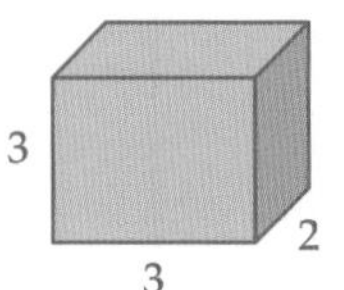

c

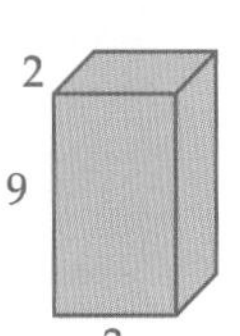

13-07 Volume of a rectangular prism

WORDBANK

rectangular prism A prism with a rectangle as its base.

A quicker way to find the volume of a rectangular prism, rather than counting cubes, is to multiply its length by its width by its the height.

Formula for the volume of a rectangular prism:
Volume = length × width × height
$V = lwh$

EXAMPLE 9

Find the volume of each rectangular prism.

a

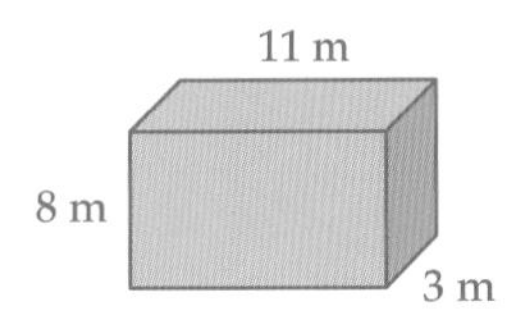

b

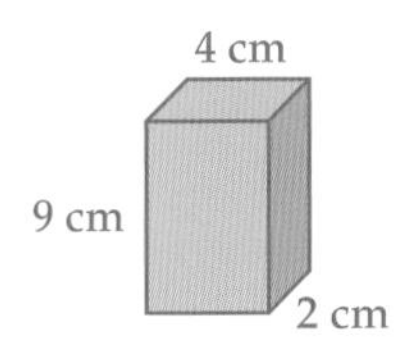

SOLUTION

a $V = lwh$
$= 11 \times 3 \times 8$
$= 264 \text{ m}^3$

b $V = lwh$
$= 4 \times 2 \times 9$
$= 72 \text{ cm}^3$

Alamy/Sean Pavone

ISBN 9780170350969

EXERCISE 13–07

1 What is the volume of a rectangular prism with length 16 m, width 5 m and height 11 m? Select the correct answer A, B, C or D.

A 1056 m^3 B 990 m^3 C 880 m^3 D 231 m^3

2 What is the volume of a rectangular prism with length 12 m, width 9 m and height 16 m? Select A, B, C or D.

A 648 m^3 B 1536 m^3 C 1944 m^3 D 1728 m^3

3 Find the volume of each prism below.

a

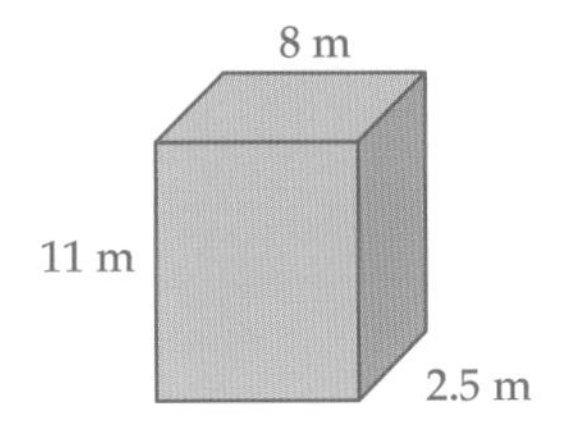

b

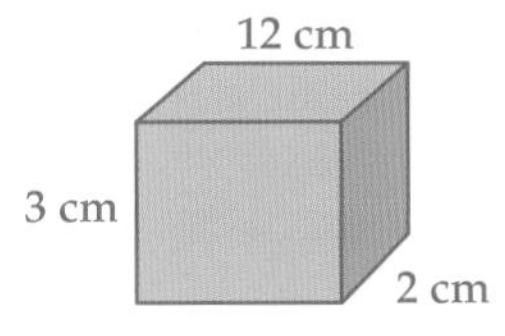

c

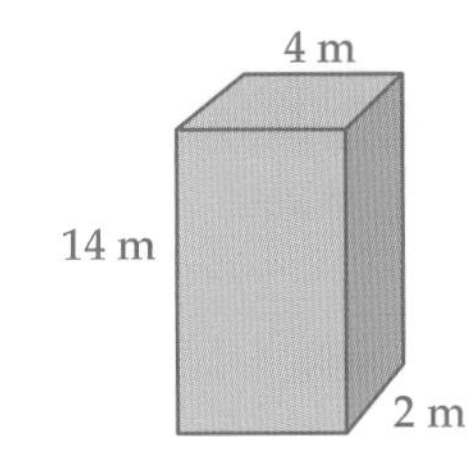

d

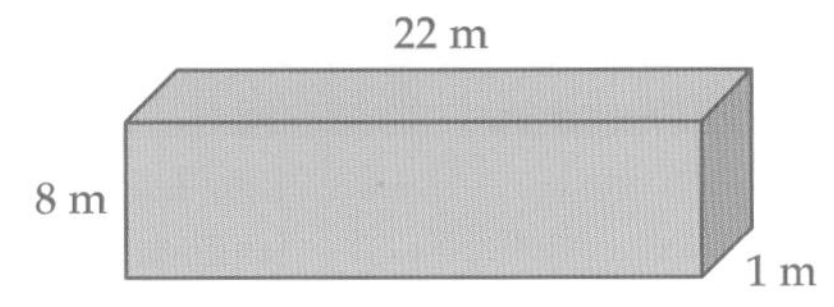

e

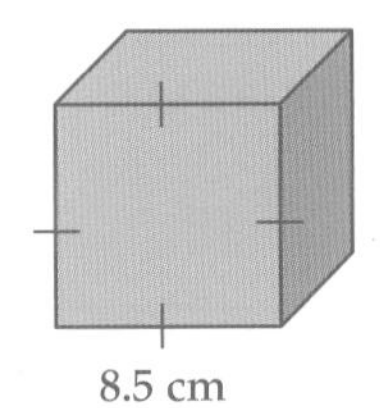

f

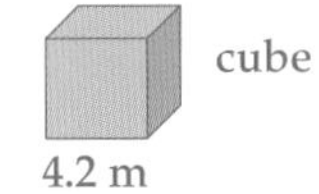

4 Find the volume of each rectangular prism described.

Solid	Dimensions
a Rectangular prism	$l = 15$ cm, $w = 13$ cm, $h = 9$ cm
b Cube	$s = 14$ m
c Rectangular prism	$l = 7.5$ cm, $w = 2.6$ cm, $h = 0.4$ cm
d Cube	$s = 3.8$ cm
e Square prism	$s = 22$ m, $h = 9$ m

5 A cube has a volume of 729 cm^3. What is its side length?

6 A rectangular prism has a volume of 192 m^3. If its length is 8 m and width is 3 m, what is its height?

13–08 Volume and capacity

WORDBANK

capacity The amount of liquid or material that a container can hold, measured in millilitres (mL), litres (L) and kilolitres (kL).

The units for volume and capacity are related. A cubic centimetre holds 1 mL, while a cubic metre holds 1000 L or 1 kL.

VOLUME AND CAPACITY

1 cm^3 contains 1 mL

1 m^3 contains 1 kL or 1000 L

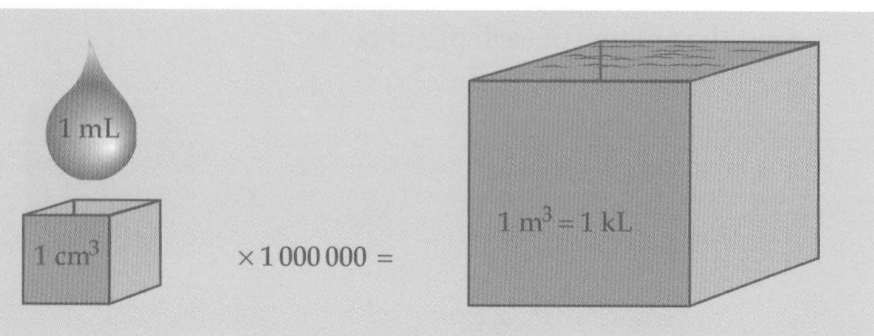

Unit	Abbreviation	Size	Example
millilitre	mL	1 mL = 1 cm^3 1000 mL = 1 L	A large drop of water
litre	L	base unit 1000 L = 1 m^3	A carton of milk
kilolitre	kL	1 kL = 1 m^3 1000 L = 1 kL	Volume of water in a spa

EXAMPLE 10

What unit of measure would you use to find the capacity of:

a a cup? b a bucket?

SOLUTION:

a millilitres b litres

EXAMPLE 11

Convert a volume of 0.004 m^3 to L.

SOLUTION

$0.004\ m^3 = 0.004\ kL$ ← $1\ m^3 = 1\ kL$

$= 0.004 \times 1000\ L$ ← LOSM: $1\ kL = 1000\ L$

$= 4\ L$

ISBN 9780170350969

EXERCISE 13–08

1 What is the unit used to measure the capacity of a kitchen sink? Select the correct answer A, B, C or D.

A millilitres B kilolitres C metres^3 D litres

2 What is the approximate capacity of a bathtub? Select A, B, C or D.

A 150 mL B 1500 kL C 1500 m^3 D 150 L

3 What unit of measure would you use to measure the capacity of these containers?

a a bowl b a dam c a small water bottle
d a soft-drink can e a lake f a large milk carton

4 Use a measuring jug marked with mL and L to find the capacity of:

a a teaspoon b a cup
c a tablespoon d a bucket
e a bowl f a vase

Shutterstock.com/Christian Draghici

Shutterstock.com/Seleznov Oleksandr

Shutterstock.com/KPG_Payless

5 Copy and complete:

a $6 \text{ cm}^3 = \_\_\_\_ \text{ mL}$ b $45 \text{ mL} = \_\_\_\_ \text{ cm}^3$
c $8000 \text{ cm}^3 = \_\_\_\_ \text{ L}$ d $9.4 \text{ L} = \_\_\_\_ \text{ cm}^3$
e $7654 \text{ mL} = \_\_\_\_ \text{ L}$ f $82 \text{ L} = \_\_\_\_ \text{ mL}$
g $450 \text{ L} = \_\_\_\_ \text{ m}^3$ h $4.56 \text{ kL} = \_\_\_\_ \text{ m}^3$

6 A bottle of medicine contains 260 mL. David is told to take 10 mL twice a day. How long will the medicine bottle last?

7 Find the volume of each rectangular prism and its capacity in litres.

a

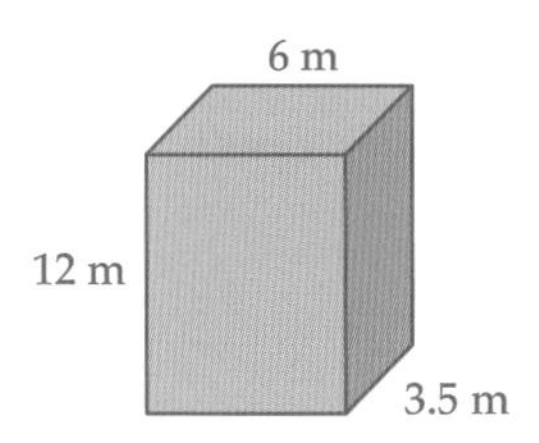

b

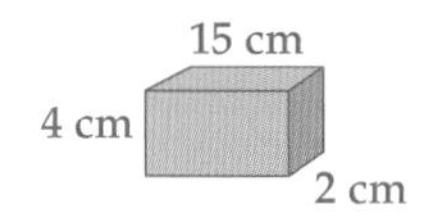

c

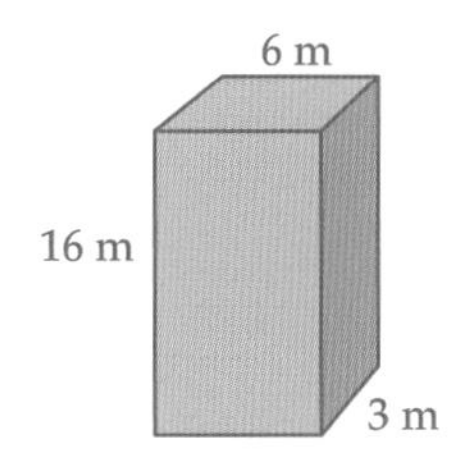

CODE PUZZLE

What do you call a numerical genius from another planet?

To decode the answer, find the area of each shape, then match the letter inside the shape to the correct answer at the bottom of the page.

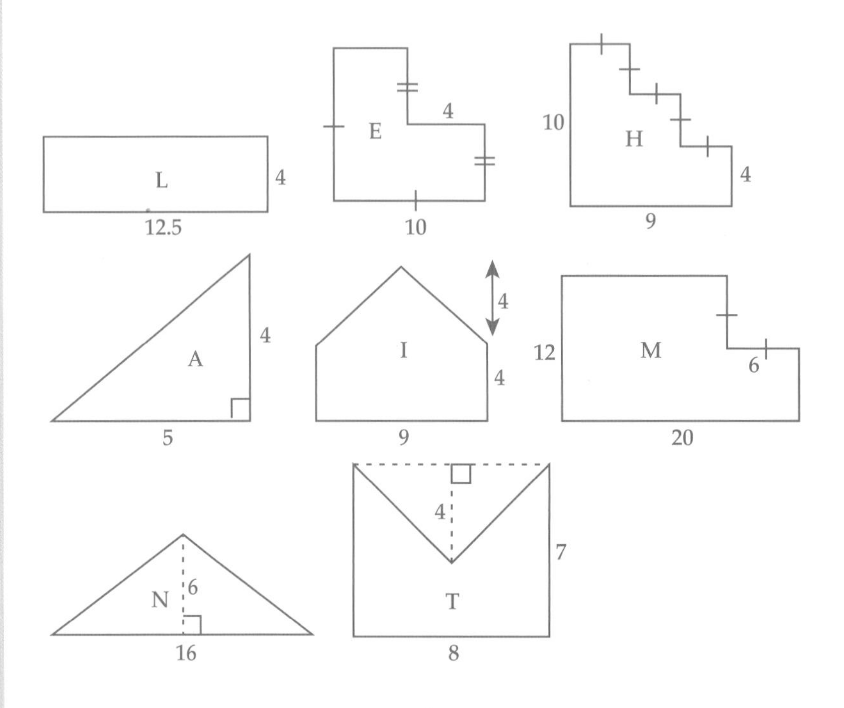

10 204–10–40–63–10–204–10–50–54–80–48

ISBN 9780170350969

PRACTICE TEST 13

Part A General topics

Calculators are not allowed.

1 Copy and complete: 250 cm = ___ m

2 Round 3516 to the nearest hundred.

3 Find the perimeter of a rectangle 6 m long and 4 m wide.

4 Find a and b.

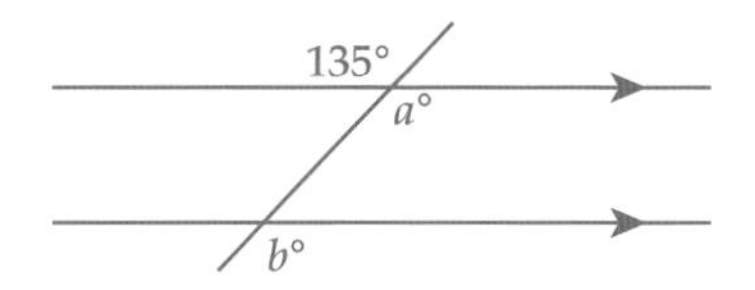

5 Evaluate 3 – (–4).

6 Solve $2c - 21 = 7$.

7 Evaluate 3^3.

8 Draw a triangular prism.

9 Evaluate 5×12.

10 Find the highest common factor of 16 and 20.

Part B Area and volume

Calculators are allowed.

13-01 Area

11 Which units could we use to measure area? Select the correct answer A, B, C or D.

A metres B metre^2 C metre^3 D none of these

13-02 Area of a rectangle

12 Calculate the area of a rectangle with length 16 m and width 12 m. Select A, B, C or D.

A 28 m^2 B 96 m^2 C 192 m^2 D 384 m^2

13 The area of a rectangle is 84 m^2 and its length is 16 m. What is its width? Select A, B, C or D.

A 1344 m B 12 m C 10.5 m D 5.25 m

13-03 Area of a triangle

14 What is the area of the triangle below? Select A, B, C or D.

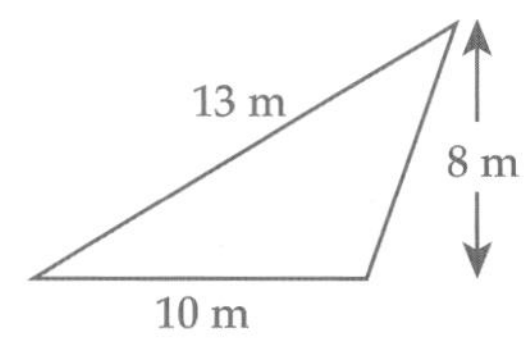

A 40 m^2 B 130 m^2 C 65 m^2 D 52 m^2

15 What is the area of a triangle with a base of 14 m and a height of 6 m?

13-04 Area of a parallelogram

16 Find the area of a parallelogram with a base of 20 cm and a height of 12 cm.

17 If the area of a parallelogram is 56 m² and its height is 4 m, what is its base?

13-05 Area of composite shapes

18 Calculate the area of the shape below.

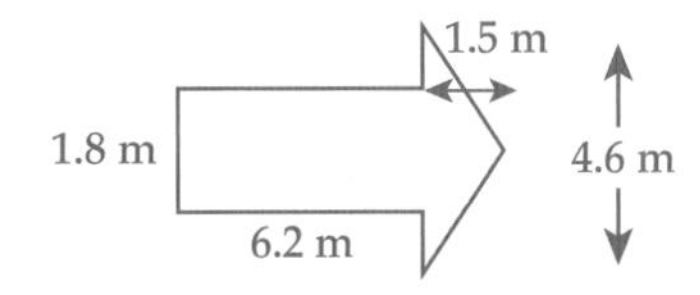

19 In which line is an error first made when finding the area of the shape?

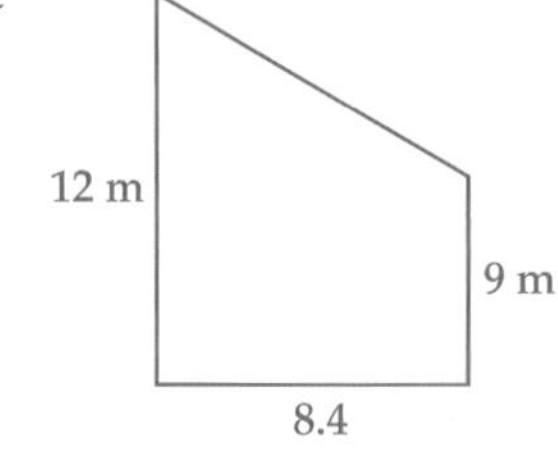

Area = Area of rectangle + Area of triangle — Line 1

$= lw + \frac{1}{2}bh$ — Line 2

$= 8.4 \times 9 + \frac{1}{2} \times 8.4 \times 3$ — Line 3

$= 120.6 \text{ m}^2$ — Line 4

13-06 Volume

20 Which units could we use to measure volume? Select A, B, C or D.

A metres　B metre²　C metre³　D all of these

13-07 Volume of a rectangular prism

21 Find the volume of a rectangular prism 18 m long, 9.2 m wide and 4 m high.

13-08 Volume and capacity

22 What is the capacity in kL of a swimming pool in the shape of a rectangular prism 23 m long, 15 m wide and 2 m deep?

ISBN 9780170350969

STATISTICAL GRAPHS

14

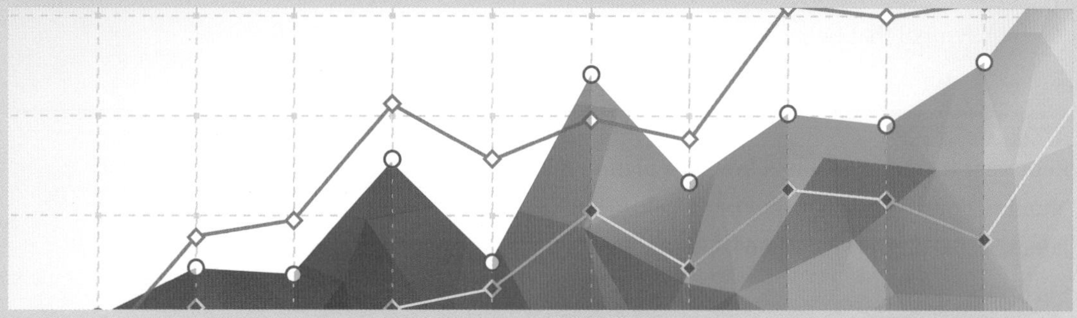

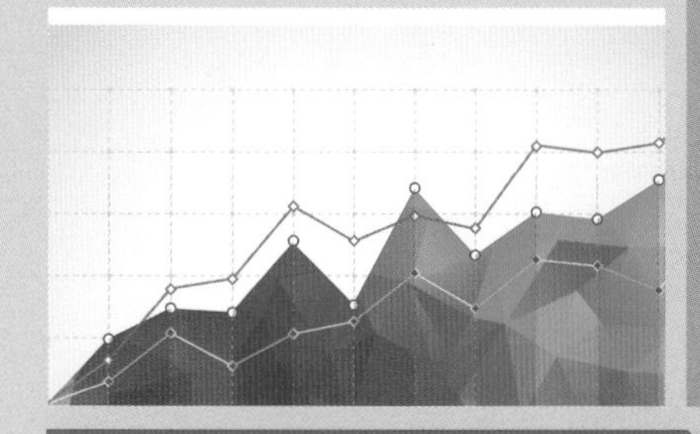

WHAT'S IN CHAPTER 14?

14-01 Picture graphs
14-02 Column graphs
14-03 Line graphs
14-04 Divided bar graphs
14-05 Sector graphs
14-06 Misleading graphs

IN THIS CHAPTER YOU WILL:

- read, interpret and construct different types of statistical graphs: picture graphs, column graphs, line graphs, divided bar graphs and sector graphs
- identify graphs that are incorrect and misleading

Shutterstock.com

14-01 Picture graphs

WORDBANK

picture graph A graph where pictures or symbols are drawn to represent the data.

key A symbol that helps to read a graph. For example 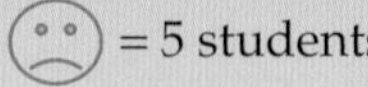 = 5 students

EXAMPLE 1

Represent the information below on a picture graph.

Number of text messages sent per week:

James 56; Sal 32; Tori 24; Ben 64; George 16

SOLUTION

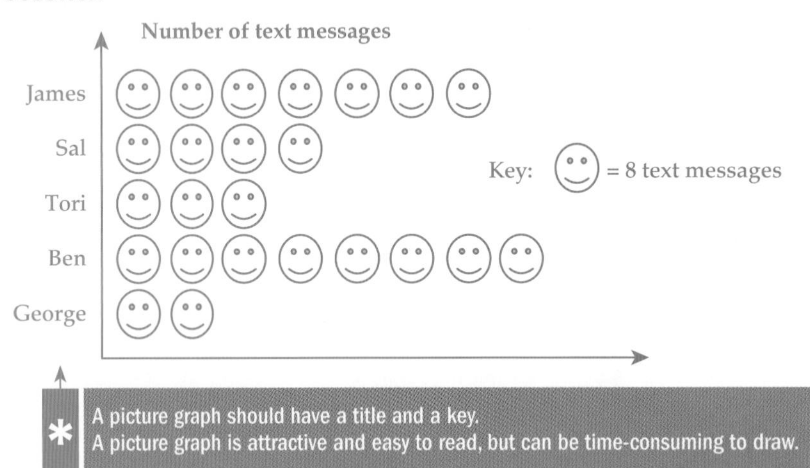

*
A picture graph should have a title and a key.
A picture graph is attractive and easy to read, but can be time-consuming to draw.

Shutterstock.com/Pavels Rumme

ISBN 9780170350969

EXERCISE 14–01

1 This picture graph shows the sales of used cars at a car yard.

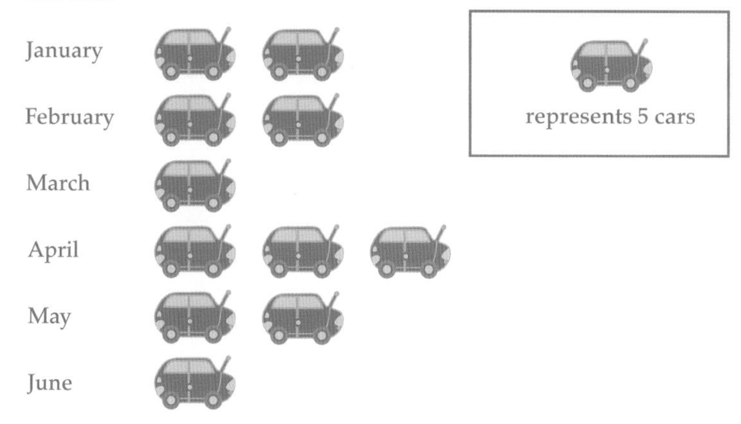

a How many used cars were sold in January?

b How many cars were sold in June?

c In which month were the most cars sold?

d How many cars were sold that month?

e In which months were the fewest cars sold?

f How many cars were sold in those months?

2 Draw a picture graph to represent the data below. You may choose your own symbol and key to use in the graph.

Favourite ice-cream flavour: Chocolate 12; Strawberry 8; Vanilla 6; Choc chip 4; Blueberry 4; Mocha 10

3 Answer the questions about the graph below:

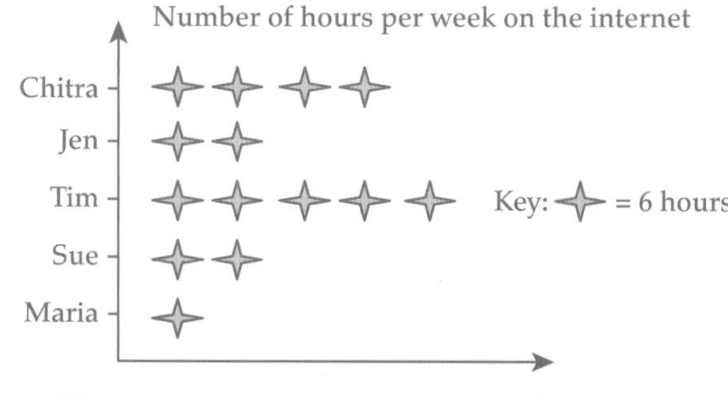

a Who spent the most hours per week on the internet?

b Who spent the fewest hours on the internet?

c What was the total number of hours spent by all students on the internet in 1 week?

d Who spent more than 14 hours per week on the internet?

4 Represent the information below on a picture graph.

Method of travel to school: Walking 16; Bus 20; Train 12; Car 8; Bike 6; Two or more methods 14

5 Write down two advantages and two disadvantages of using a picture graph to represent data.

ISBN 9780170350969

14-02 Column graphs

WORDBANK

column graph A graph that uses columns of different heights to represent data.

axes Plural of axis. The two number lines that form the edges of a graph: the horizontal axis and the vertical axis.

EXAMPLE 2

a Draw a column graph illustrating the data below.
Number of sunny days in May:
Monday 4; Tuesday 4; Wednesday 5; Thursday 5; Friday 2; Saturday 4; Sunday 2

b How many days in May were not sunny?

SOLUTION

a

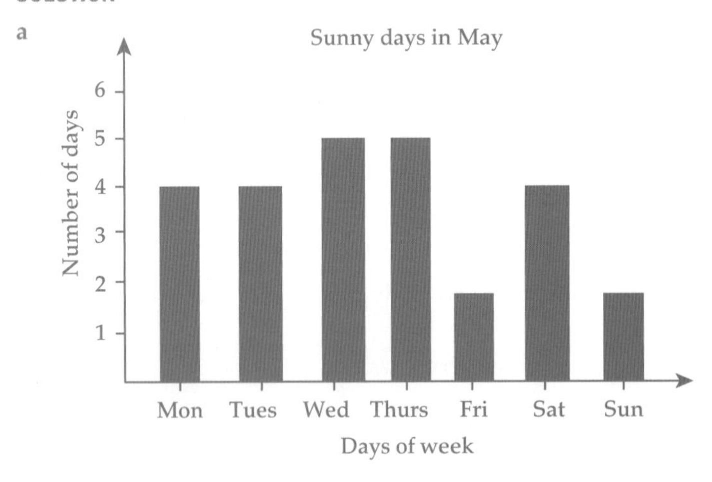

b There were 26 sunny days in May and 31 days in May altogether, so 5 days were not sunny.

Shutterstock.com/Dmitrijs Bindemanis

ISBN 9780170350969

EXERCISE 14–02

1 When drawing a column graph the columns must be ______. Select the correct answer A, B, C or D.

A the same width B the same height C joined D different widths

2 This graph shows the number of children in different families living in Chalmers Street.

a Which family had the smallest number of children?

b How many children were in the Chu family?

c What was the greatest number of children per family?

d Which families had fewer than 5 children?

e How many families had more than 4 children?

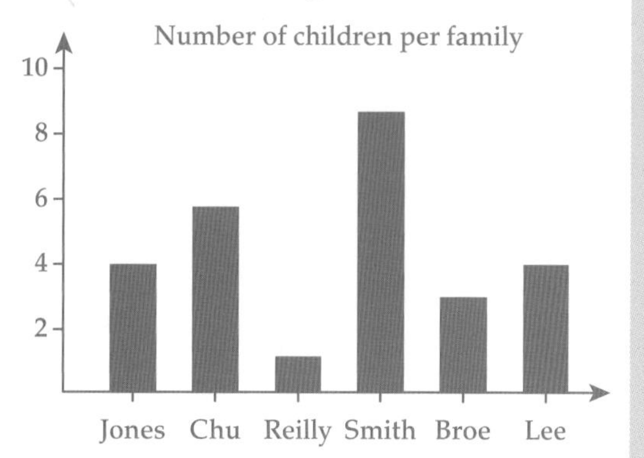

3 a What type of graph is shown on the right?

b Who scored 8 or more on the quiz?

c If 50% was a pass and the quiz was out of 12, did everyone pass?

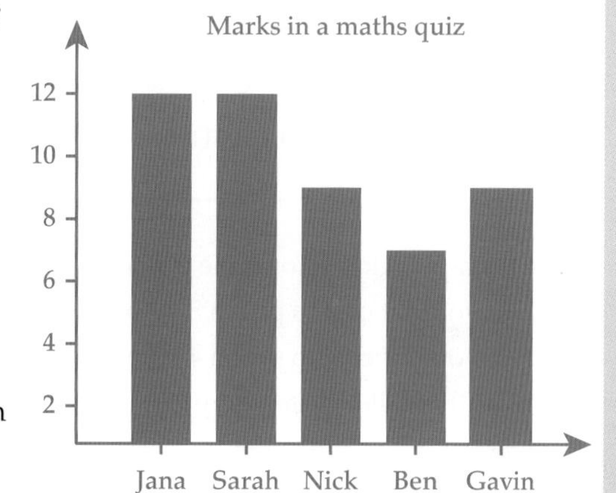

4 a Draw a column graph displaying the data below:

Number of kilometres from the beach:
Salmah 6; Julie 2; Connor 3.5; Brad 5; Sharelle 4

b How many students lived less than 4 km from the beach?

5 This column graph shows Ryan's exam marks in different subjects.

a What is Ryan's best subject?

b What is his worst subject?

c How many marks did Ryan get for Computing?

d What was his mark in English?

e If Ryan's aim was a mark of over 50, did he reach this in every subject?

f Name the subjects in which he reached his aim.

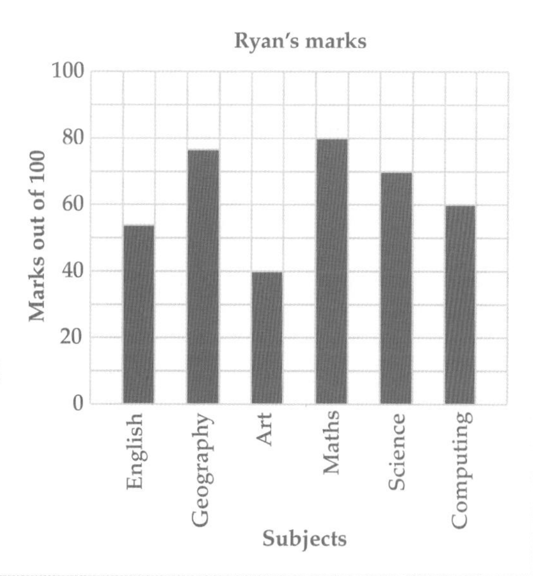

ISBN 9780170350969

14–03 Line graphs

A **line graph** uses lines to represent data.
Two special types of line graphs are **conversion graphs** and **travel graphs**.
A **step graph** is a flat line graph that goes up or down in steps.

EXAMPLE 3

This conversion graph converts between A$ (Australian dollars) and US$ (US dollars).

a Use the graph to convert A$100 to US$.

b How many A$ is US$80?

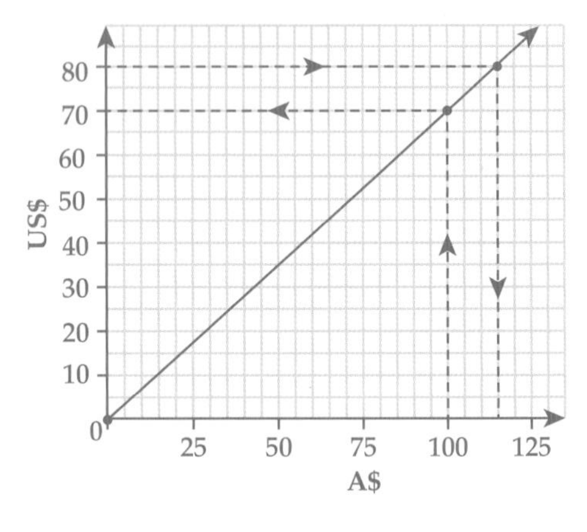

SOLUTION

a From the graph A$100 = US$70 ← Follow the dotted line up and across.

b From the graph US$80 = A$115 ← Follow the dotted line across and down.

EXAMPLE 4

The step graph below shows the postal charges for parcels.

a What would it cost to post a 5 kg parcel?

b What would it cost to post a 4 kg parcel?

c Why is this called a step graph?

SOLUTION

a A 5 kg parcel would cost $20 to post.

b A 4 kg parcel would cost $17 to post.

c It is called a step graph because it goes up in steps for certain masses of parcels.

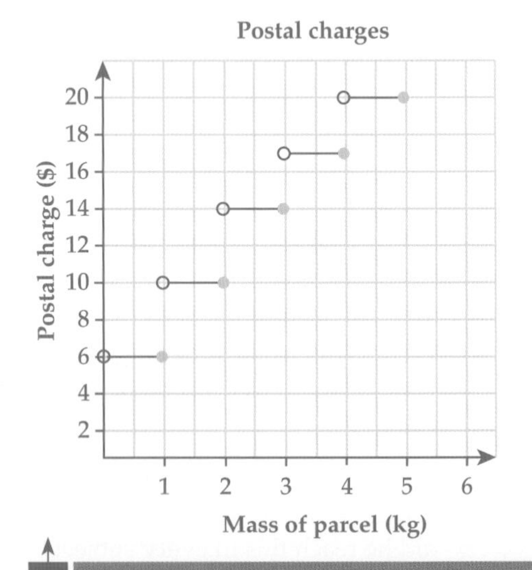

* On a step graph, read the filled-in circle, as the blank circle is not included. A 4 kg parcel would cost $17, not $20.

 ISBN 9780170350969

EXERCISE 14–03

1 The step graph below shows parking fees for a city car park.

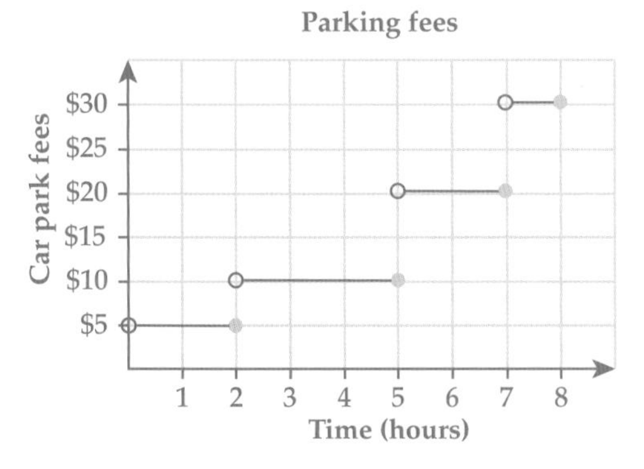

a How much does it cost to park for 1 hour?

b How much does it cost to park for 5 hours?

c How much does it cost to park all day (8 hours)?

Shutterstock.com/harper kt

2 Investigate the current postage rates for parcels at Australia Post and draw a step graph to illustrate this.

3 Describe what is being shown on this line graph.

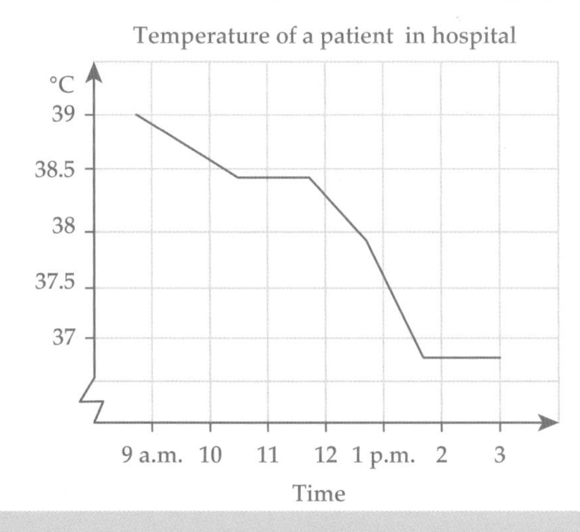

ISBN 9780170350969

EXERCISE 14–03

4 This conversion graph converts NZ$ (New Zealand dollars) to A$ (Australian dollars).

a Use the graph to convert NZ$50 to A$.

b How many A$ equals NZ$125?

c Convert A$90 to NZ$.

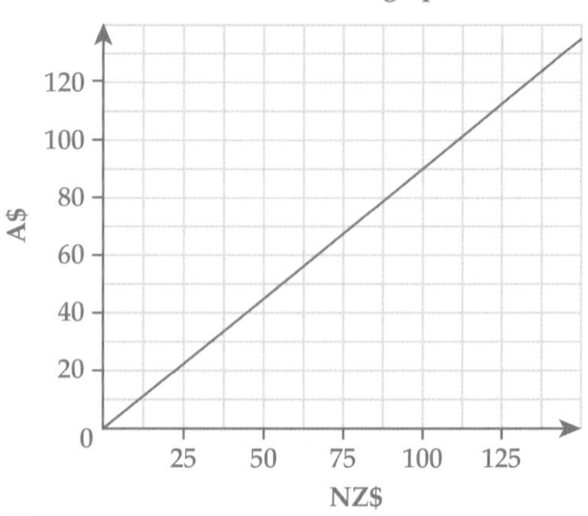

5 The line graph below shows the bicycle trip made by a group of students in Year 7 as part of their Activity Week.

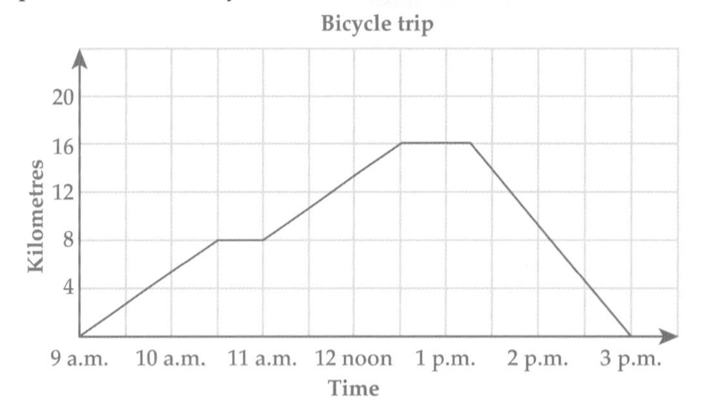

a What time did they leave on their trip?

b How far did they travel altogether?

c How many times did they stop and rest?

d What time did they start their journey home?

e Find their speed in km/h for their trip home.

ISBN 9780170350969

EXERCISE 14–03

6 The line graph below shows temperatures over an 8-hour period in a country town.

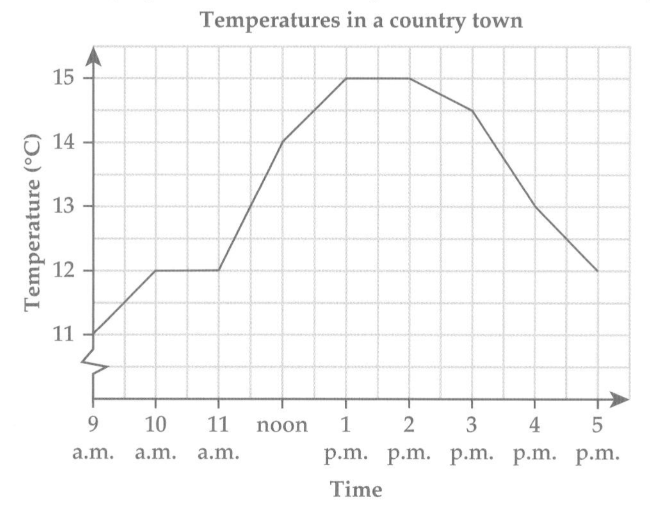

a Write the temperature at:

i 9 a.m. ii 5 p.m.

b At what times was the temperature 14°C?

c Between what times was the largest hourly increase in temperature?

d Write this increase as a rate in °C/h.

e Between which two hours did the temperature remain the same?

Getty Images/Robin Smith

14–04 Divided bar graphs

A **divided bar graph** is a graph in which a rectangle (or bar) is divided into parts to represent data.

EXAMPLE 5

This table shows the hours worked by 30 employees at a supermarket last weekend.
Display the data on a divided bar graph.

Number of hours worked	10	12	14	16	18	20
Number of workers	2	3	5	11	7	2

SOLUTION

As there are 30 workers altogether, draw a rectangle 12 cm long as this will fit on the page and it is divisible by 30.

12 cm = 120 mm and represents 30 workers, so 1 worker is represented by 120 mm ÷ 30 = 4 mm.

Each section can be calculated: 10 h, 2 workers, 2×4 mm = 8 mm

12 h, 3 workers, 3×4 mm = 12 mm

14 h, 5 workers, 5×4 mm = 20 mm, and so on.

Hours worked

10 h	12 h	14 h	16 h	18 h	20 h

4 mm = 1 worker

Shutterstock.com/joyfull

ISBN 9780170350969

EXERCISE 14–04

1 What shape do we divide up for a bar graph? Select the correct answer A, B, C or D.

A a square B a circle C a triangle D a rectangle

2 Each section of a divided bar graph should be ______. Select A, B, C or D.

A equal B increasing
C different sizes D decreasing

3 Andrew worked at a timber yard during the weekend and earned $160. The divided bar graph shows how he spent his money.

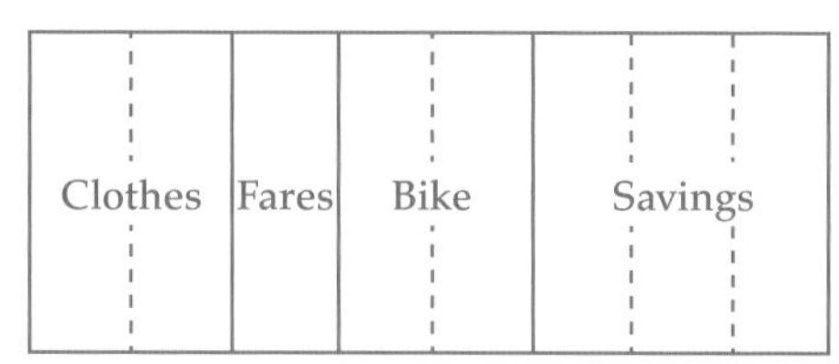

a What does each unit represent on this bar graph?
b What did Andrew spend on clothes?
c How much did he spend on fares?
d How much did he spend on his bike?
e How much did he save?
f Do you think Andrew was careful with his money?

4 Draw a divided bar graph, using a suitable scale, for the information below:

Number of food items sold at the school canteen on Friday: Chicken nuggets 41, Pies 30, Sandwiches 25, Sushi 40, Hot dogs 46, Pasta 18.

5 This divided bar graph shows how much of every dollar of its yearly budget Clifford City Council spends on different areas. Each unit represents 10 cents.

Council spending

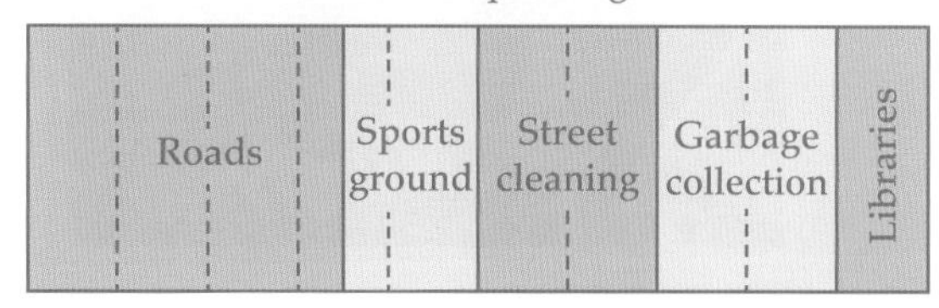

a How much of every dollar is spent on garbage collection?
b How much of every dollar is spent on roads?
c Twice as much is spent on roads as is spent on sports grounds. True or false?

6 Survey your class on their eye colour and show the information on a divided bar graph.

ISBN 9780170350969

14-05 Sector graphs

A **sector graph** is a graph in which a circle is divided into parts (called sectors) to represent data. A sector has the shape of a pizza slice.

To draw a sector graph:

- draw a circle
- calculate the total of the data
- for each category, calculate the sector angle by multiplying a fraction by 360°
- draw in sectors to represent each part of the data.

EXAMPLE 6

Eleni's wage is \$1080 per week. She spends \$324 on rent, \$108 on food, \$216 on clothes and entertainment and saves the rest.

Draw a sector graph to represent how Eleni's weekly wage is divided.

SOLUTION

Work out each part of the data as a fraction of 360° to calculate the angle at the centre.

$\text{Rent} = \frac{324}{1080} \times 360° = 108°$ $\text{Food} = \frac{108}{1080} \times 360° = 36°$

$\text{Clothes} = \frac{216}{1080} \times 360° = 72°$ $\text{Savings} = \frac{432}{1080} \times 360° = 144°$

Check that the angles all add up to 360°.

108 + 36 + 72 + 144 = 360

EXERCISE 14-05

1 What size sector in a sector graph would represent $\frac{1}{5}$ of the data? Select the correct answer A, B, C or D.

A 60° B 120° C 72° D 20°

2 What size sector in a sector graph would represent $\frac{2}{3}$ of the data? Select A, B, C or D.

A 120° B 90° C 200° D 240°

 ISBN 9780170350969

EXERCISE 14–05

3 a One-third of Amanda's day is spent sleeping. How many hours is this?

b What fraction of the day is spent at school?

c How many hours are spent at school?

d How many hours of homework does Amanda do?

e How much free time does Amanda have?

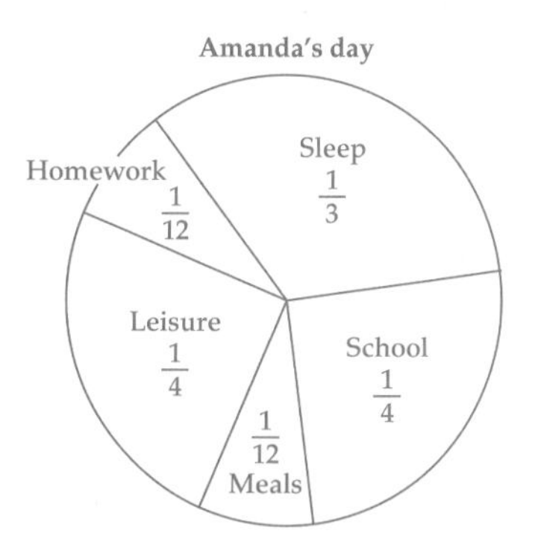

4 Ms Santana budgets her weekly income of $640. This sector graph shows how she spends her money.

a What is most of her weekly income spent on?

b How much of her income is spent on rent?

c How much does it cost her to run her car each week?

d Can you tell from the graph exactly how much Ms Santana spends on entertainment each week?

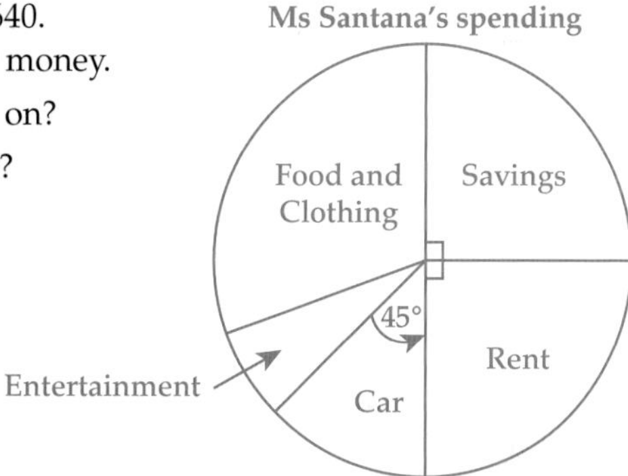

5 a Calculate the size of each sector required to place the data below on a sector graph.

Tien's wage: $288 on rent, $240 on food, $96 on entertainment, $144 on clothes, $192 on savings.

b Draw the sector graph.

6 The sector graph shows how Mr Milosevski spends his weekly income of $840.

a What is most of his weekly income spent on?

b What fraction of his income is spent on food and clothing?

c How much does he spend on entertainment and savings?

d What do car expenses come to?

ISBN 9780170350969

14-06 Misleading graphs

Graphs can be misleading if they are not drawn carefully and accurately. This means if you look at the graphs they can give you a false impression.

Some features that can make a graph misleading are:

- incorrect picture size
- no scale or uneven scale on one axis
- no title.

EXAMPLE 7

Investigate the graph below if it is meant to represent 'Sales of books have doubled every month for the past 2 months.'

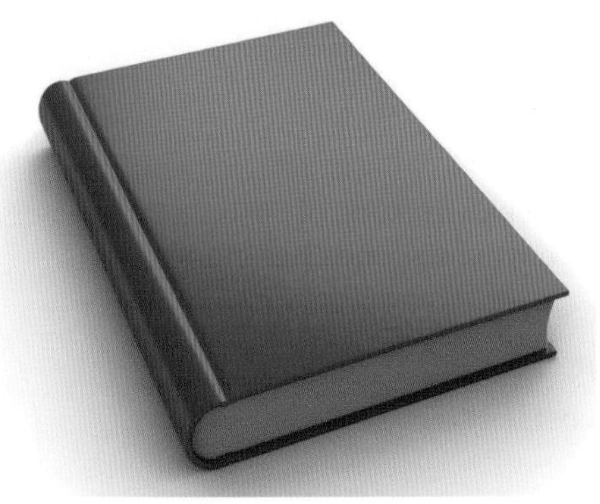

iStockphoto/pavlen

SOLUTION

The graph is misleading because the large picture is more than double in volume since the length, width and thickness of the book have been doubled. This makes the book 8 times bigger.

EXAMPLE 8

Describe any problems with the graph below.

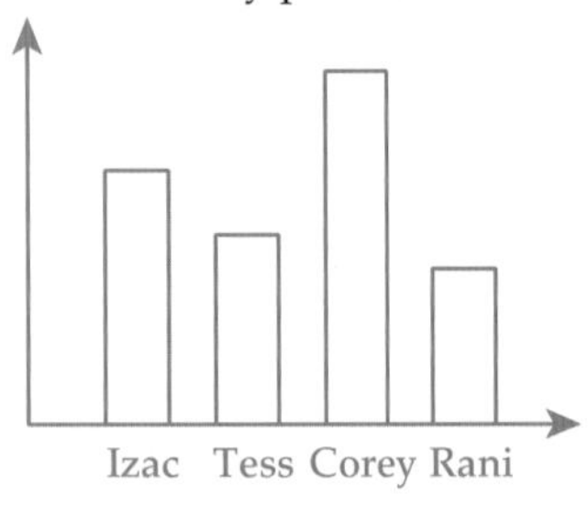

SOLUTION

There is no title so we don't know what the data represents.

There is no scale on the vertical axis so we don't know the height of each column.

 ISBN 9780170350969

EXERCISE 14–06

1 What can make a graph misleading? Select the correct answer A, B, C or D.

A a long title

B different heights of columns

C lots of pictures

D a missing scale

2 Investigate the graph below and explain if it is misleading.

'Sales of canned soup have tripled in the past month'

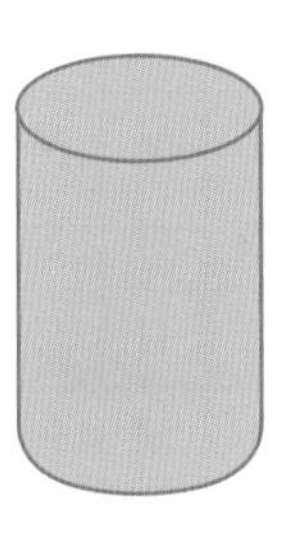

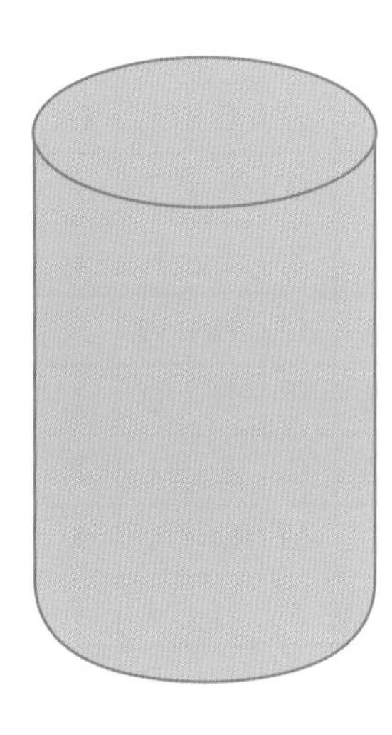

3 Draw a more accurate graph for Question 2.

4 What is wrong with this graph? Name as many things as possible.

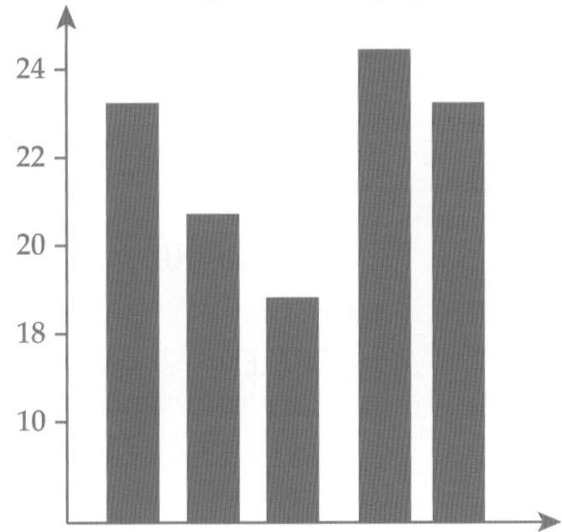

5 a Draw a more accurate graph for Question 4.

b Can you complete all details in this graph?

6 What is misleading about the graph below? Correct it.

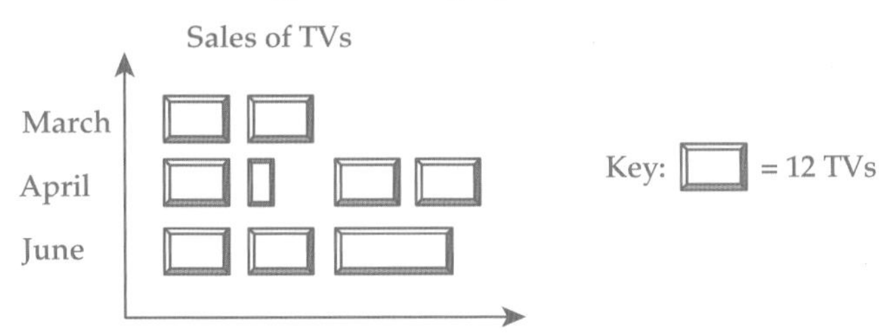

LANGUAGE ACTIVITY

FIND-A-WORD PUZZLE

Make a copy of this page, then find the words below in the grid of letters.

E	D	A	T	A	S	I	X	E	I	G	Y	H	C	X
L	G	Z	G	Y	C	N	I	T	V	V	O	O	T	G
A	C	O	N	T	I	N	U	O	U	S	L	Z	N	D
C	R	Y	E	K	T	U	P	C	H	U	I	I	Q	K
S	W	L	H	W	S	V	X	I	M	P	D	C	B	O
D	S	B	U	D	I	Y	H	N	C	A	A	F	E	L
A	A	O	E	I	T	J	S	G	E	T	X	R	J	I
R	L	Y	M	V	A	E	R	L	E	Z	U	E	G	N
F	B	X	H	I	T	W	S	V	T	V	M	R	S	E
N	H	S	P	D	S	I	R	O	T	C	E	S	E	E
S	M	I	U	E	M	I	F	K	A	L	P	V	W	D
A	P	P	D	D	U	D	Q	J	T	I	H	D	B	B
S	D	L	V	D	D	U	Y	I	M	J	Z	D	R	O
E	H	F	E	T	O	Q	T	L	N	V	J	Z	E	J
R	O	Z	P	H	C	T	K	P	X	B	L	W	H	R

AXES	BAR	COLUMN	CONTINUOUS
DATA	DIVIDED	GRAPH	KEY
LINE	MISLEADING	PICTURE	SCALE
SECTOR	STATISTICS	TITLE	

ISBN 9780170350969

PRACTICE TEST 14

Part A General topics

Calculators are not allowed.

1 Simplify $7b - b$.

2 Evaluate $99 - 3 \times 8$.

3 Construct a 124° angle.

4 Write 8:28 p.m. in 24-hour time.

5 Evaluate $0.024 \div 0.8$.

6 Evaluate 3.851×100.

7 Is 23 583 divisible by 6?

8 What do the angles in a quadrilateral add up to?

9 What is a shape with 5 sides called?

10 Find the perimeter of this parallelogram.

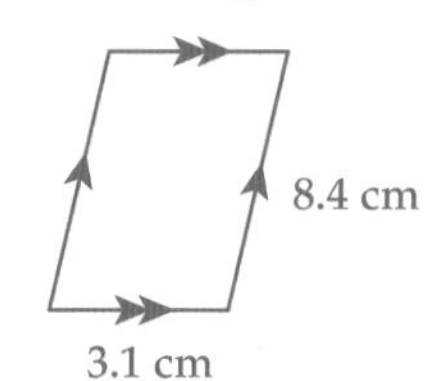

Part B Statistical graphs

Calculators are allowed.

14–01 Picture graphs

This picture graph represents three families landing at Alice Springs airport.

	Number in family	Total luggage
Stone	☺ ☺ ☺	62 kg
Cowan	☺ ☺ ☺ ☺	85 kg
Boyd	☺ ☺	40 kg

11 The airline allows up to 21 kg luggage per person. Which family had more than the luggage allowance?

12 Which family had the least luggage per person?

14–02 Column graphs

13 A column graph uses columns which are ______. Select the correct answer A, B, C or D.

A equally spaced   B coloured   C horizontal   D the same height

ISBN 9780170350969

14–03 Line graphs

14 The graph below shows the time, in minutes, a certain brand of paint takes to dry at various temperatures.

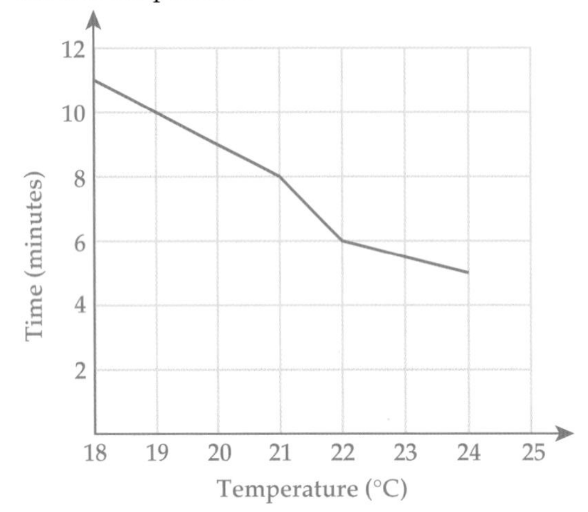

a What is the drying time at 20°C?

b If the drying time is 5 minutes, what is the temperature?

14–04 Divided bar graphs

15 How do you draw a divided bar graph?

14–05 Sector graphs

16 Display the information below on a sector graph:

Brian's working holiday: Work for 8 days; Hiking for 3 days; Driving for 5 days; Relaxing for 4 days

17 For the sector graph in Question 16 how many degrees is the sector for 'Relaxing'?

14–06 Misleading graphs

18 Write down two ways a graph can be misleading.

 ISBN 9780170350969

ANALYSING DATA

15

WHAT'S IN CHAPTER 15?

15-01 Frequency tables
15-02 Frequency histograms and polygons
15-03 Dot plots
15-04 Stem-and-leaf plots
15-05 The mean and mode
15-06 The median and range
15-07 Analysing dot plots and stem-and-leaf plots

IN THIS CHAPTER YOU WILL:

- organise data (information) into frequency tables
- read and construct column graphs called frequency histograms
- read and construct line graphs called frequency polygons
- read and construct dot plots
- read and construct stem-and-leaf plots
- find the mean and mode of a list of scores
- find the median and range of a list of scores
- find the mean, mode, median and range of scores presented in dot plots and stem-and-leaf plots

Shutterstock.com/Jirsak

15-01 Frequency tables

WORDBANK

data Statistical information, a collection of facts.

frequency The number of times a score occurs in a set of data.

frequency table A table that shows the frequency of each score in a data set.

tally marks Marks used to count the frequency of a score.

| | || ||| |||| ~~||||~~ ~~||||~~| ← Tally marks
1 2 3 4 5 6

EXAMPLE 1

The number of mistakes made by 50 students in a spelling test are listed below.

0, 1, 5, 2, 5, 4, 3, 3, 2, 0

0, 1, 3, 3, 3, 2, 2, 4, 1, 1

3, 2, 3, 1, 3, 3, 2, 3, 2, 0

4, 4, 3, 2, 3, 3, 3, 3, 1, 0

3, 3, 3, 1, 2, 2, 1, 2, 2, 5

Arrange this data into a frequency table.

SOLUTION

The 50 scores run from 0 to 5, so use a frequency table to count how many of each value.

Write a tally mark for each time a score appears in the set of data. Count the tally marks for each score to find its frequency. The total frequency should be 50 because there are 50 scores.

Number of mistakes (score)	Tally	Frequency
0	~~||||~~	5
1	~~||||~~ |||	8
2	~~||||~~ ~~||||~~ ||	12
3	~~||||~~ ~~||||~~ ~~||||~~ |||	18
4						4
5					3
	Total	50

ISBN 9780170350969

EXERCISE 15–01

1 What does the score column represent in a frequency table? Select the correct answer A, B, C or D.

A the frequency B the tally marks C the data D all of these

2 What does the frequency represent? Select A, B, C or D.

A the number of scores B how often each score occurs

C the data D none of these

3 A random sample of 50 matchboxes was taken and the contents of each box counted. Show the data in a frequency table.

50, 49, 52, 50, 51, 50, 52, 49, 48, 53, 48, 52, 49, 50, 49, 48, 47,
51, 52, 51, 51, 51, 49, 48, 47, 52, 51, 50, 50, 49, 47, 50, 50, 48,
49, 53, 52, 50, 48, 50, 50, 50, 49, 51, 53, 48, 47, 50, 50, 52

iStockphoto/janka3147 fdfl

iStockphoto/DmitriMaruta

4 In a quality-control test for tyres, a sample of 30 tyres was run continuously until they were worn out. The results are shown in thousands of kilometres. Express these scores in a frequency table.

23, 26, 24, 28, 27, 23, 26, 28, 29, 23, 22, 21, 25, 23, 23,
22, 21, 23, 25, 26, 28, 21, 23, 27, 26, 23, 23, 26, 24, 23

5 The sizes of shoes sold by a store over a weekend are shown below. Represent the data in a frequency table.

5, 6, 7, 7, 7, 6, 5, 8, 4, 3, 7, 7, 8, 9, 4, 3, 7, 3, 7, 9, 2, 9, 7, 6, 6, 7, 7, 7, 6, 7

6 A group of 50 students was given a spelling test and the number of mistakes for each student was recorded. Express the data in a frequency table.

3, 4, 0, 0, 2, 1, 3, 3, 4, 1, 2, 2, 1, 3, 3, 2, 2, 2, 1, 2, 2, 3, 4, 5,
2, 2, 2, 3, 4, 3, 3, 2, 2, 1, 0, 0, 0, 2, 3, 3, 2, 2, 3, 2, 2, 4, 5, 2

7 Survey the students in your class on the number of children in each student's family. Organise this data in a frequency table.

8 Survey the students in your class on the number of bedrooms in each student's home. Organise this data into a frequency table and discuss the results with your class.

15–02 Frequency histograms and polygons

WORDBANK

frequency histogram A column graph that shows the frequency of each score. The columns are joined together.

frequency polygon A line graph that shows the frequency of each score, drawn by joining the top of each column in a histogram.

EXAMPLE 2

This frequency table shows the number of toothpicks in a sample of toothpick boxes advertised as containing 50 toothpicks. Draw a frequency histogram and polygon for this data.

Number of toothpicks	Frequency
48	2
49	4
50	10
51	7
52	3

SOLUTION

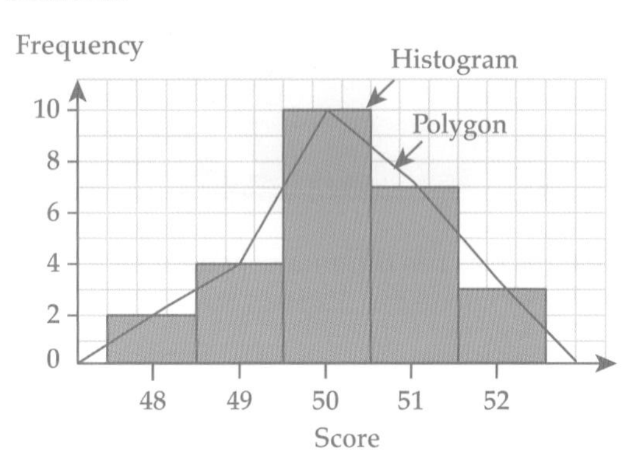

- The **frequency histogram** has columns of the same width, centred on each score
- The **frequency polygon** is formed by joining the middle of the top of each column, and starts and ends on the horizontal axis.

✱ It is called a polygon because the line graph has the shape of a many-sided figure.

ISBN 9780170350969

EXERCISE 15–02

1 A frequency histogram compares which two things? Select the correct answer A, B, C or D.

A tally to frequency
B scores to frequency
C data to scores
D frequency to information

2 What type of graph is a frequency polygon? Select A, B, C or D.

A column graph
B divided bar graph
C sector graph
D line graph

3 Draw a frequency histogram for each frequency table.

a

Score	Frequency
0	3
1	6
2	9
3	4
4	3

b

Score	Frequency
12	4
13	9
14	12
15	7
16	5
17	3

4 Draw a frequency polygon on each histogram in Question 3.

5 The ages of the children at the cinema are shown below.

6, 8, 10, 9, 8, 7, 8, 6, 8, 7
5, 8, 9, 7, 6, 8, 7, 10, 8, 9
7, 8, 10, 11, 9, 7, 8, 5, 6, 8

a Organise this data in a frequency table.
b Draw a frequency histogram illustrating this data.
c Add a frequency polygon to your diagram in part b.
d What was the most common age?
e How many children were under 7 years old?

6 The number of ships entering Port Kembla harbour each week over six months is shown.

6, 6, 8, 7, 7, 9, 11, 12, 6, 11, 7, 8, 8, 9, 10, 9, 11, 12, 9, 10, 8, 8, 8, 7

a Express the data in a frequency table.
b Draw a frequency polygon for this data.
c What is the most common number of ships visiting the harbour?
d What was the highest number of ships in Port Kembla harbour?

15–03 Dot plots

WORDBANK

dot plot A diagram showing frequency of data scores using dots.

outlier An extreme score that is much higher or lower than the other scores in a data set.

cluster A group of scores that are bunched close together.

EXAMPLE 3

The data below shows the amount of pocket money in dollars a group of students is given for a week.

6, 10, 8, 12, 8, 10, 12, 5, 6, 8, 7, 5, 10, 12, 11, 9, 12, 10, 15, 7

10, 14, 12, 13, 9, 14, 10, 6, 10, 8

a Plot this data on a dot plot.

b What was the largest amount of pocket money given?

c How many students were given more than $10?

d How many students were given more than $7 but less than $11?

e What amount of pocket money is given most often?

SOLUTION

a

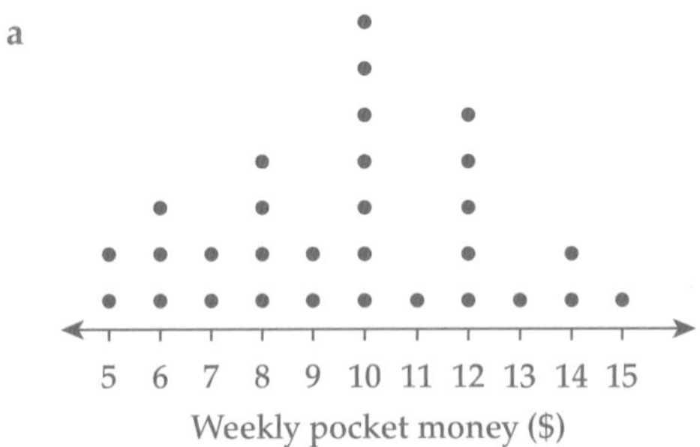

b $15 c 10 d 13 e $10

Alamy/allesalltag

 ISBN 9780170350969

EXERCISE 15–03

1 Besides dots, a dot plot must have what? Select the correct answer A, B, C or D.

A a horizontal axis
B a vertical axis
C a key
D lines

2 In a dot plot, what does the number of dots above each score represent? Select A, B, C or D.

A the vertical axis
B the scale
C the data
D the frequency

3 a Draw a dot plot to represent these marks out of 10 for a maths quiz:

8, 9, 5, 6, 7, 8, 9, 4, 6
7, 6, 7, 8, 4, 8, 10, 7, 5
8, 7, 6, 5, 7, 4, 7, 10, 7

b What was the highest score?

c What was the lowest score?

d What was the most common score?

e If 6 out of 10 was considered a pass, how many students passed?

4 a Draw a dot plot to display the number of text messages Allan receives each day:

12, 18, 14, 12, 13, 14, 12, 4, 13
13, 12, 12, 12, 14, 14, 12, 17, 12
12, 13, 17, 13, 12, 12, 17, 14, 12

b What was the greatest number of messages Allan received in one day?

c What was the smallest number of messages Allan received?

d What was the most common number of messages Allan received?

e When there is a score that is much lower or higher than all the other scores, it is called an **outlier**. Is there an outlier here?

f If scores are grouped close together they are called a **cluster**. Is there a cluster of scores here?

5 The amounts below are Effie's part-time weekly wages for 20 weeks.

190, 220, 280, 200, 250, 240, 230, 220, 180, 200
270, 280, 260, 250, 240, 250, 230, 250, 180, 250

a Organise the wages into a dot plot to answer the questions.

b What was the greatest amount that Effie earned for a week?

c What was the amount that she earned most often?

d If she saved everything she earned above $200/week, how much did Effie save in this time?

15–04 Stem-and-leaf plots

WORDBANK

stem-and-leaf plot A table listing data scores where the tens digits are in the stem and the units digits are in the leaf.

EXAMPLE 4

A group of 30 shoppers were surveyed on the number of hours spent each week shopping. Illustrate the results in a stem-and-leaf plot.

22, 13, 14, 15, 11, 8, 10, 12, 9, 10, 14, 21, 17, 6, 12, 27, 8, 17, 14, 26

16, 19, 24, 5, 18, 23, 6, 18, 28, 3

Alamy/Ilene MacDonald

SOLUTION

The scores range from 3 to 28, so the stems (the tens value) will range from 0 to 2.

List the scores in order from left to right:

Stem	Leaf
0	8 9 6 8 5 6 3
1	3 4 5 1 0 2 0 4 7 2 7 4 6 9 8 8
2	2 1 7 6 4 3 8

This is an **unordered stem-and-leaf plot** as the leaves are not in numerical order. We now need to place them in order from low to high:

Stem	Leaf
0	3 5 6 6 8 8 9
1	0 0 1 2 2 3 4 4 4 5 6 7 7 8 8 9
2	1 2 3 4 6 7 8

This is an ordered stem-and-leaf plot.

It is easy to notice trends (patterns) in stem-and-leaf plots, such as the most common number of hours per week shopping is between 10 and 19 with a stem of 1. The lowest number of shopping hours is 3 and the highest is 28 per week. The most common number of shopping hours per week is 14, which occurs three times.

ISBN 9780170350969

EXERCISE 15–04

1 What does the stem in a stem-and-leaf plot have? Select the correct answer A, B, C or D.

A the units digit B from 0 to 2 C the tens digit D none of these

2 The stem-and-leaf plot represents the number of pies sold per day by a bakery during June.

Stem	Leaf
1	2 3 5 6 6
2	1 2 2 3 4 5 7 7 7 8 9
3	1 2 3 3 4 6 8 9
4	2 3 3 5 6 7

a What was the lowest number of pies sold in a day?

b What was the highest number of pies sold?

c On how many days were more than 35 pies sold?

d On how many days were fewer than 15 pies sold?

e What was the most common number of pies sold?

3 The data below represents the marks out of 50 for a maths test.

23, 32, 45, 28, 36, 19, 48, 33, 29, 35
45, 39, 44, 15, 47, 38, 33, 24, 35, 37
49, 33, 42, 27, 33, 46, 25, 33, 27, 42

a Record the data in an ordered stem-and-leaf plot.

b Use the plot to find the highest, lowest and most common score.

4 a The number of toothpicks in different packets of toothpicks are shown here. Represent the data in an ordered stem-and-leaf plot.

49, 52, 47, 58, 42, 55, 39, 47, 52, 62
48, 51, 53, 52, 55, 52, 61, 48, 52, 58
51, 52, 61, 55, 52, 59, 52, 54, 57, 52

b If you bought a packet at random, how many toothpicks would you be most likely to get?

5 The daily sales figures in dollars for two local hamburger stores are listed below.

McDavids

342, 280, 295, 314, 327, 345, 299, 307, 318, 326, 317

Burger Prince

306, 348, 359, 367, 298, 356, 345, 355, 366, 354, 342

a Draw an ordered stem-and-leaf plot for each set of data, using stems of 28, 29, 30, ...

b Which store had the greatest sales for one day?

c Which store sold the least in one day?

d Which store had the greater sales over the 11 days? Show all working.

15–05 The mean and mode

WORDBANK

mean The average of a set of data scores, with symbol $\bar{x}$, calculated by adding the scores and dividing by the number of scores.

mode The most popular score(s) in a data set, the score that occurs most often.

In statistics, there are three **measures of location** that measure the central or middle position of a set of data. They are called the **mean**, **mode** and **median**.

$$\text{Mean} = \bar{x} = \frac{\text{sum of scores}}{\text{number of scores}}$$

EXAMPLE 5

Find the mean of these scores:

8, 5, 4, 7, 6, 10, 9, 3

SOLUTION

$$\begin{aligned}\text{Mean } \bar{x} &= \frac{\text{sum of scores}}{\text{number of scores}}\\ &= \frac{8+5+4+7+6+10+9+3}{8}\\ &= \frac{52}{8}\\ &= 6.5\end{aligned}$$

 Note that the mean is around the centre of the set of scores.

Mode = the most frequently occurring score(s).
A set of data can have more than one mode, or no mode at all.

EXAMPLE 6

For each set of data, find the mode.

a 3, 5, 3, 6, 4, 5, 7, 6, 3, 4, 8, 3, 6

b 12, 11, 13, 14, 12, 13, 16, 14, 13, 12, 18, 13

c 8, 9, 7, 6, 8, 7, 9, 5, 8, 7, 4, 10, 3

SOLUTION

a The mode is 3. ← 3 occurs four times and the other scores occur less often.

b The mode is 13. ← 13 occurs more often than the other scores.

c There are 2 modes: 7 and 8.

ISBN 9780170350969

EXERCISE 15–05

1 What is the statistical term for the score that occurs most often in a set of data? Select the correct answer A, B, C or D.

A the mean B the frequency C the average D the mode

2 Find the mean of the scores 5, 7, 2, 7, 9, 6. Select A, B, C or D.

A 4 B 6 C 7 D 5

3 For each set of data, find the mean and the mode:

a 7, 8, 9, 5, 6, 5, 7, 8, 7

b 4, 6, 7, 3, 4, 5, 4, 6, 8

c 11, 12, 11, 14, 13, 16, 13, 15, 13

d 21, 28, 25, 28, 26, 28, 24, 28, 27

e 52, 55, 56, 54, 55, 54, 51, 58, 57

4 What do you notice about the mode of the set of scores in Question 3e?

5 The following scores are the heights in centimetres of members of the Knights rugby league team:

175, 186, 181, 190, 165, 172, 186, 184
190, 168, 189, 178, 179, 184, 177, 186

a Calculate the mean height of the players.

b Is there a mode? If so, find it.

Fairfax/Tim Clayton

6 The following data shows the number of races each Year 7 student ran in at the school athletics carnival:

8,4, 7, 5, 6, 8, 3, 1, 2, 7, 2, 3, 4, 6, 2, 5, 2, 3
5, 4, 2, 3, 2, 9, 3, 2, 4

a Calculate the mean and the mode for this data.

b The students were given a trip to the fun park if the average participation was greater than 6 races. Were they given a trip?

15–06 The median and range

WORDBANK

median The middle score when the scores are ordered from lowest to highest.

range The highest score minus the lowest score.

To find the median:

- order the scores from lowest to highest
- if there is an odd number of scores, **median = middle score**
- if there is an even number of scores, **median = average of the two middle scores**

Range = highest score − lowest score

The median is another **measure of location** while the range is a **measure of spread**.

EXAMPLE 7

Find the median and range of each set of data.

a 8, 5, 7, 3, 4, 5, 9

b 11, 18, 13, 16, 12, 14, 19, 17

SOLUTION

a Write the scores from lowest to highest: 3, 4, 5, 5, 7, 8, 9

There is an odd number of scores so the median will be the one in the middle:

3, 4, 5, 5, 7, 8, 9 ← There are 3 scores above 5 and 3 scores below 5

Median = 5.

Range = 9 − 3 ← Highest − lowest

= 6

b Write the scores from lowest to highest: 11, 12, 13, 14, 16, 17, 18, 19

There is an even number of scores so the median will be the average of the 2 middle scores:

11, 12, 13, 14, 16, 17, 18, 19 ← The 2 middle scores are 14 and 16

$$\text{Median} = \frac{14+16}{2}$$

= 15 ← The number halfway between 14 and 16

Range = 19 − 11

= 8

ISBN 9780170350969

EXERCISE 15–06

1 What does the median describe about a set of data? Select the correct answer A, B, C or D.

A all the data   B the most common   C the frequency   D the middle

2 Find the median of 4, 8, 3, 5, 9, 7. Select A, B, C or D.

A 6   B 4   C 5   D 7

3 Find the median and the range for each set of data.

a 6, 8, 4, 7, 9, 2, 5, 9, 5

b 11, 16, 12, 14, 18, 14, 13

c 22, 28, 26, 24, 23, 29, 22, 27, 21

d 54, 58, 57, 56, 54, 52

e 102, 110, 109, 105, 115, 118, 121, 104, 115

f 65, 63, 34, 89, 52, 73, 45, 92

4 The data below represents the number of people attending a cinema each day in September.

Mon	Tues	Wed	Thurs	Fri	Sat	Sun
45	82	62	23	74	98	86
35	86	45	32	75	96	84
38	84	48	37	82	98	78
47	86	36	42	76	98	82
38	88					

a Find the median and the range.

b Find the mean and the mode.

c Which weekday (not weekend) has the highest number of movie-goers? Why might that be?

d On what other days is the attendance high? Why would this be?

5 The number of goals scored by a hockey team in one season is listed below:

10, 12, 11, 15, 10, 12, 12, 11, 12, 14, 16, 12, 12, 13, 14, 11, 12, 13, 12, 15

a What is the range?

b What is the mode?

c Find the mean.

d Find the median.

e Which measure would be most useful for the coach if she wanted to find their average number of goals?

6 Find the range and median of these scores, correct to one decimal place:

21.5, 13.6, 5.7, 8.9, 11.1.

15–07 Analysing dot plots and stem-and-leaf plots

EXAMPLE 8

This dot plot shows the number of hours each student in a Year 7 class used a computer on the weekend.

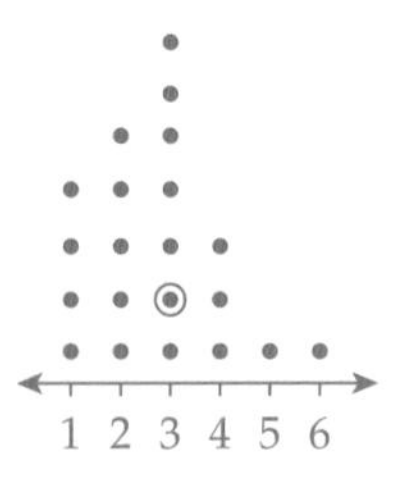

Find:

a the mean correct to two decimal places

b the mode c the range d the median.

SOLUTION

a $\text{Mean} = \frac{4 \times 1 + 5 \times 2 + 7 \times 3 + 3 \times 4 + 5 + 6}{21}$ $\frac{\text{sum of scores}}{\text{number of scores}}$

$= \frac{58}{21}$

$= 2.7619$

≈ 2.76

b Mode = 3 ⟵ the score with the most dots

c Range = 6 – 1 ⟵ highest – lowest

= 5

d Median = middle score ⟵ 11th dot, circled

= 3

✱ When counting the middle score for the median, count the dots in columns going up.

EXAMPLE 9

This stem-and-leaf plot shows the daily number of coffees sold at a café over two weeks.

Stem	Leaf
3	2 5 8 8
4	3 3 3 6 7 8 9
5	3 4 9

a Find the mean number of coffees ordered per day, rounded to two decimal places.

b What is the mode?

c Find the range.

d Find the median.

SOLUTION

a $\text{Mean} = \frac{32 + 35 + 2 \times 38 + 3 \times 43 + 46 + 47 + 48 + 49 + 53 + 54 + 59}{14}$

$= \frac{628}{14}$

$= 44.8571...$

≈ 44.86

b Mode = 43 ⟵ The most common score

c Range = 59 – 32 = 27

d $\text{Median} = \frac{43 + 46}{2}$ ⟵ Average of 7th and 8th scores, the middle two scores

$= 44.5$

✱ When counting the middle scores for the median, count the leaves in rows from left to right

ISBN 9780170350969

EXERCISE 15–07

1 For the scores 3.2, 4.8, 7.2, 1.9, 2.5, 7.5, 6.4, what is the range? Select the correct answer A, B, C or D.

A 6.6 B 5.6 C 5.4 D 6.4

2 For the scores in Question 1, what is the median? Select A, B, C or D.

A 4.8 B 3.2 C 6.4 D 4

3 These marks were scored out of 10 in a maths quiz.

5, 6, 9, 4, 2, 3, 5, 7, 8, 4, 5, 6, 2, 5, 8, 4, 5, 9, 3, 5, 7, 2, 4, 5, 8, 5, 6

a Represent the marks on a dot plot.

b Find the mean correct to one decimal place.

c Find the mode, the range and the median.

d If the teacher was interested in the average for the test, which measure would be most useful?

e If the teacher was interested in the most common mark scored in the test, which measure would be most useful?

4 This stem-and-leaf plot shows the number of ice-creams sold per day over 3 weeks.

Stem	Leaf
4	2 3 6 8
5	1 4 4 7 8 9
6	2 3 3 3 6 7 8
7	3 4 5 7

a Find the mean correct to two decimal places.

b Find the mode, the range and the median.

c If you were ordering ice-cream, which measure would you use to predict the most likely number of ice-creams to be sold the next day?

5 The data below shows the number of students attending the art gallery each day over 2 weeks.

125, 132, 156, 121, 148, 146, 135, 152, 168, 128, 135, 142, 158, 163

a Show the data on a stem-and-leaf plot.

b Find the mean correct to one decimal place.

c Find the mode, the range and the median.

6 A back-to-back stem-and-leaf plot can be drawn for 2 sets of scores you wish to compare.
The scores below are for two students who played 12 rounds of a computer game.

Anitha's scores		Joe's scores
	5	4 4 8
9	6	5 7 8 8
8 5 2	7	2 4 7 8
6 6 5 4 2	8	4
8 3 1	9	

a Who had the highest score in a round?

b Who had the lowest score?

c Who performed better over all the rounds?
How can you tell from the stem-and-leaf plot?

ISBN 9780170350969

CROSSWORD PUZZLE

Make a copy of this page and complete the crossword below using these words:

ANALYSIS	AVERAGE	DATA	FREQUENCY
GRAPH	HISTOGRAM	MEAN	MEDIAN
MODE	POLYGON	RANGE	SCORE
STATISTICS	STEM-AND-LEAF PLOT	TALLY	

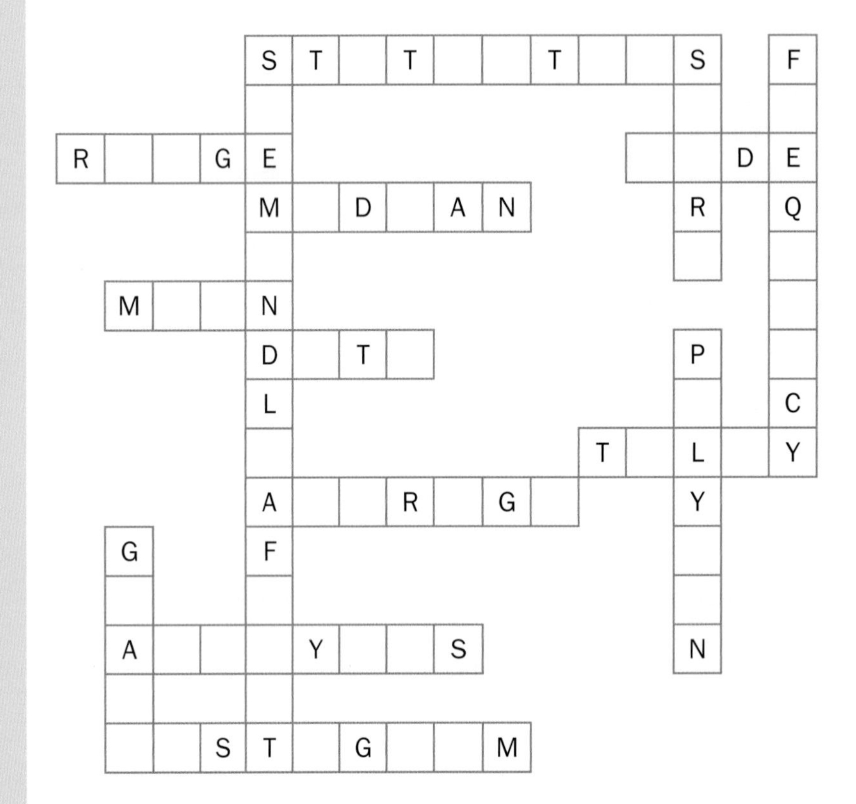

 ISBN 9780170350969

PRACTICE TEST 15

Part A General topics

Calculators are not allowed.

1 Draw a hexagon.

2 Convert $\frac{1}{8}$ to a decimal.

3 Evaluate $-6 \times (-5) + (-4)$.

4 Write an algebraic expression for the number that is 10 less than y.

5 Name the quadrilateral with one pair of parallel sides.

6 Use a factor tree to write 90 as a product of prime factors.

7 Evaluate $\frac{2}{5} \times \frac{3}{4}$.

8 Find p.

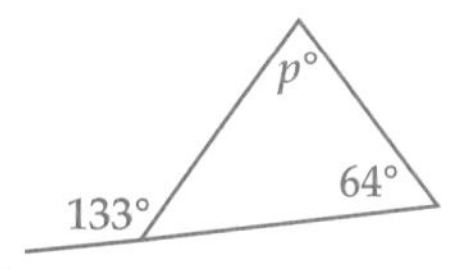

9 Evaluate $7 \times 2 \times 50$.

10 True or False? For any two numbers a and b, $a - b = b - a$.

Part B Analysing data

Calculators are allowed.

15-01 Frequency tables

11 Which two columns represent the same information in a frequency table? Select the correct answer A, B, C or D.

A score and frequency
B score and tally
C tally and frequency
D frequency and mode

15-02 Frequency histograms and polygons

12 What type of graph is a frequency histogram? Select A, B, C or D.

A line graph
B picture graph
C sector graph
D column graph

13 Which column would be highest if the data below was drawn on a frequency histogram? Select A, B, C or D.

21, 26, 22, 24, 22, 23, 26, 24, 26, 25, 27, 26

A 22
B 26
C 24
D 25

15-03 Dot plots

14 Copy and complete: The largest number of dots on a dot ______ represents the ______.

15 What is the statistical name for an extreme score that is much different to the other scores in a data set?

15-04 Stem-and-leaf plots

16 This stem-and-leaf plot shows the number of drinks sold per day at a petrol station.

Stem	Leaf
4	2 3 5 5 6
5	1 2 3 3 3 7 8
6	3 4 5 6 6 8 9 9

Find

a the range b the mode c the mean

17 If you managed the petrol station in Question 16, and you were ordering drinks for the next day, which measure would you use to predict your order: the range, the mode or the mean?

15-05 The mean and mode

18 Find the mean of these scores. Select A, B, C or D.

12, 14, 9, 6, 11, 8, 10, 14

A 10 B 10.5 C 11 D 11.5

19 What is the mode of the scores in Question 18?

15-06 The median and range

20 What is the range of these scores?

15, 17, 22, 13, 17, 45, 32, 12

21 Find the median of the scores in Question 20.

15-07 Analysing dot plots and stem-and-leaf plots

22 This dot plot shows the number of hours each student spends watching TV each night.

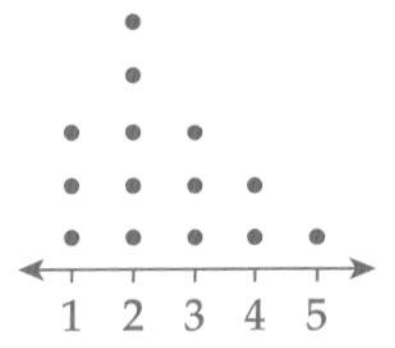

a Find the mean number of hours spent watching TV.

b Find the mode.

ISBN 9780170350969

PROBABILITY

16

WHAT'S IN CHAPTER 16?

16-01 The language of chance
16-02 Sample spaces
16-03 Probability
16-04 The range of probability
16-05 Experimental probability

IN THIS CHAPTER YOU WILL:

- understand probability and words related to chance
- list all of the possible outcomes of a situation (chance experiment)
- calculate probability as a fraction, decimal or percentage
- calculate probability based on experiments and past records

Shutterstock.com/Doug James

16-01 The language of chance

WORDBANK

Probability The chance of an event occurring, written as a fraction between 0 and 1.

Outcome A result of a situation involving chance.

Event One or more outcomes of an experiment.

The chance of an event occurring is often discussed in everyday conversation. We say: 'What is the chance that it will be a sunny day tomorrow?' or 'What chance have I got of winning the race?' and so on.

Here are some words that we use to describe these chances:

- An **impossible** event means the event cannot occur. For example: it will snow in Alice Springs today.
- An **unlikely** event means the event will probably not occur. For example: winning a lotto prize.
- An **even chance** or **50-50 chance** means the event has an equal chance of occurring or not occurring. For example: getting a head when you toss a coin.
- A **likely event** means the event will probably occur. It is more likely to occur than not occur. For example: the next vehicle to pass the school is a car.
- A **certain event** means the event must occur. For example: it will get dark tonight.

The chance of an impossible event occurring is 0, while the chance of a certain event occurring is 1. An event that has an even chance is placed halfway between 0 and 1. Likely and unlikely events can be ordered on the number line as follows:

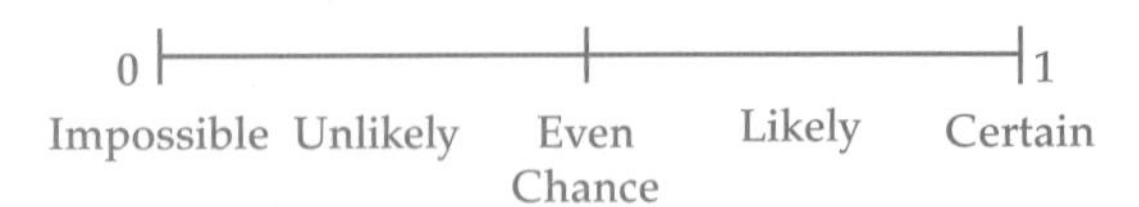

EXAMPLE 1

Describe in words the chance of each event listed and then mark their position on a number line.

a Asher choosing a red card from 10 red cards.

b Sophie winning a raffle if she has no ticket.

c A newborn baby being a girl.

d Rolling a 2 on a die.

SOLUTION

a Certain b Impossible

c Even chance d Unlikely

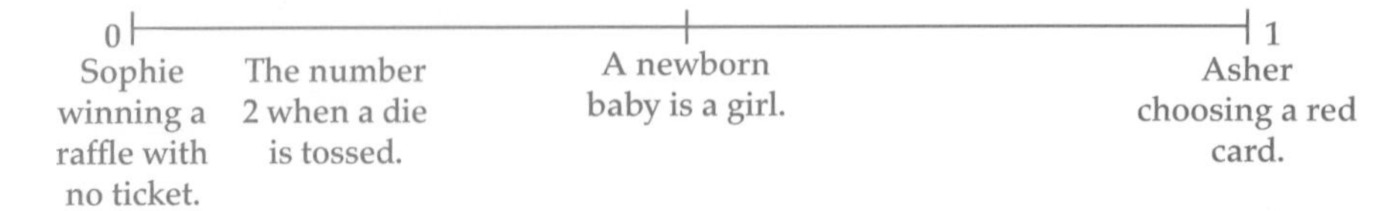

 ISBN 9780170350969

EXERCISE 16–01

1 How would you describe the chance of choosing a red ball from a bag containing 1 red and the rest blue balls? Select the correct answer A, B, C or D.

A Certain B Likely C Unlikely D Impossible

2 How would you describe the chance that it will be cold on a day in winter? Select A, B, C or D.

A Certain B Likely C Unlikely D Impossible

3 Describe the events below using a suitable word.

a A day in summer being hot.
b A newborn baby being a boy.
c Choosing a white marble from a bag containing red and black marbles.
d Throwing a tail when a coin is tossed.
e Choosing a white sock from a drawer with 20 white socks.
f Rolling a 5 when a die is rolled.
g Choosing a black marble from a bag containing 5 black marbles and 1 red marble.
h Rolling an even number when a die is tossed.
i Throwing a head or a tail when a coin is tossed.
j A sunny day in Perth on a December day.

Shutterstock.com/Simon Krzic

4 Place the events 3a–j on a number line between 0 and 1.

5 Is each statement true or false?

a The chance of it being hot on a summer's day is likely.
b It is certain that the sun will rise tomorrow morning.
c It is impossible to roll a number less than 2 on a die.
d There is an even chance that the next baby born will be a boy.
e It is unlikely to roll a number less than 5 on a die.
f It is likely that a day in February in Queensland will be cold.

6 Describe an event that matches each probability word.

a Even chance b Impossible c Likely
d Unlikely e Certain

16–02 Sample spaces

WORDBANK

sample space The set of all possible outcomes in a chance situation. For example, when tossing a coin the sample space is {head, tail}.

equally likely Having exactly the same chance of occurring. For example, head or tail are equally likely when tossing a coin.

EXAMPLE 2

A die is rolled.

✱ A die is the singular of dice: one die, two dice.

a What is the sample space?

b Are the outcomes equally likely?

c Describe the chance of rolling an odd number on the die.

SOLUTION

a The sample space = {1, 2, 3, 4, 5, 6}.

b Yes, as the die is equally likely to land on each of these numbers.

c There is an even chance of rolling an odd number as there are 3 odd and 3 even numbers.

EXAMPLE 3

A letter of the alphabet from C to G is chosen at random.

a What is the sample space?

b How many outcomes are in the sample space?

SOLUTION

a {C, D, E, F, G}

b Number of outcomes = 5

iStockphoto/scibak

ISBN 9780170350969

EXERCISE 16–02

1 A number from 12 to 23 is selected at random. How many outcomes are possible? Select the correct answer A, B, C or D.

A 12 B 11 C 13 D 23

2 Write down the sample space for each chance situation:

a tossing a coin
b choosing a letter of the alphabet
c choosing a number from 4 to 10
d rolling 2 dice
e choosing a note in the Australian currency
f choosing a vowel from the possible vowels
g tossing two coins
h choosing an even number from 7 to 17
i picking a coin from the Australian currency
j choosing a prime number from the numbers 1 to 10

3 Write down the number of outcomes for each sample space in Question 2.

4 In a bag of marbles there are 12 red, 4 blue and 6 green marbles. One marble is chosen and its colour is noted.

a List the sample space and count the number of possible outcomes in this sample space.
b Are the outcomes equally likely?

5 a What is the sample space if I spin the arrow?
b Are all outcomes equally likely?
c Is there an even chance of spinning red?
d Is there an even chance of spinning blue or red?

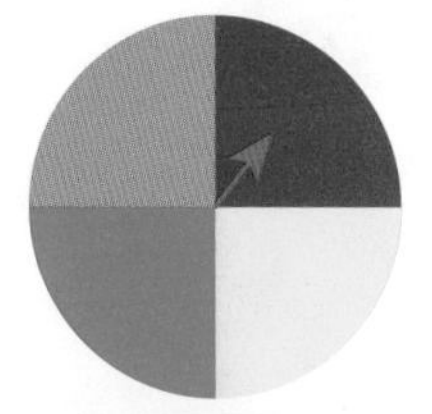

6 a What is the sample space if I spin the arrow?
b Are all outcomes equally likely?
c Is there an even chance of spinning purple?
d What is the chance of spinning green?
e Is there an even chance of spinning orange?
f Describe the chance of spinning purple or green.

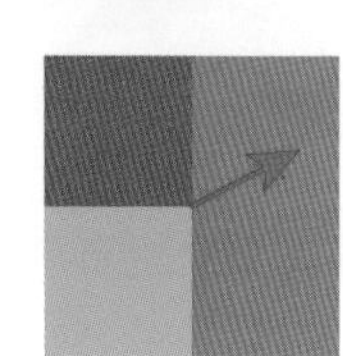

7 a Draw a spinner that has outcomes that are equally likely.
b Draw a spinner where the outcomes are not equally likely.

16–03 Probability

If we roll a die, what is the probability of getting a number greater than 4?
Sample space = {1, 2, 3, 4, 5, 6}
Two of these numbers—5 and 6—are greater than 4.
So there are 2 chances out of 6 of rolling a number greater than 4.

> The **probability of an event** has the abbreviation $P(E)$. If all outcomes are equally likely, then:
>
> $$P(E) = \frac{\text{number of outcomes in the event}}{\text{number of outcomes in the sample space}}$$

EXAMPLE 4

Find the probability of choosing a white ball from a bag containing 3 white, 5 blue and 6 red balls.

SOLUTION

$$P(\text{choosing a white ball}) = \frac{\text{number of white balls}}{\text{total number of balls}}$$

$$= \frac{3}{14} \quad \longleftarrow \quad 3 + 5 + 6 = 14$$

iStockphoto/BrianAJackson

Alamy/Foodcollection.com

EXAMPLE 5

Find the probability of choosing a red or a blue jelly bean from a bag containing 4 yellow, 6 white, 6 red and 2 blue jelly beans.

SOLUTION

$$P(\text{red or blue jelly bean}) = \frac{8}{18} \quad \longleftarrow \quad \text{6 red + 2 blue}$$

$$= \frac{4}{9} \quad \longleftarrow \quad \text{Simplify the fraction if possible}$$

ISBN 9780170350969

EXERCISE 16–03

1 What is the probability of choosing a red ball from a bag containing 5 red and 7 blue balls? Select the correct answer A, B, C or D.

A $\frac{5}{7}$ B $\frac{7}{12}$ C $\frac{7}{5}$ D $\frac{5}{12}$

2 What is the probability of choosing a green counter from a bag containing 4 red, 8 white and 6 green counters? Select A, B, C or D.

A $\frac{6}{12}$ B $\frac{1}{3}$ C $\frac{6}{16}$ D $\frac{4}{18}$

3 Find the probability of each event.

a Throwing a head when you toss a coin.

b Rolling a 3 on a die.

c Choosing a girl in a class of 12 boys and 8 girls.

d Selecting a white marble from a bag containing 5 white and 6 black marbles.

e Choosing a hard-centred chocolate from a box containing 12 soft and 10 hard-centred chocolates.

f Choosing a number less than 3 from a set of numbers from 1 to 10.

g Rolling a number greater than 3 on a die.

h Choosing a red lolly from a bag containing 4 yellow, 3 green and 5 red lollies.

i Rolling an odd number on a die.

j Selecting a particular student from a class of 25 students.

4 A box of chocolates contained the following flavours: 3 strawberry, 5 orange, 4 nut, 2 vanilla, 6 caramel. What is the probability of selecting:

a a nut chocolate? b a caramel one?

c a vanilla or strawberry chocolate? d not a vanilla flavor?

5 A pack of cards has 4 suits: diamonds, hearts, spades and clubs. In each suit there are 13 cards: ace, 2, 3, 4, 5, 6, 7, 8, 9, 10, jack, queen and king.

If a pack of cards is shuffled, find the probability of selecting:

a a king b a heart c an ace

d a number less than 4 e the 5 of clubs f a diamond

g the queen of spades h a 7 or an 8 i a king or queen

6 Find the probability of each of these events.

a Rolling a 5 on a die

b Throwing a tail when a coin is tossed

c Choosing a white ball from a bag containing 3 white, 6 red and 7 black balls

d Winning the lottery if I buy 5 tickets and 10 000 tickets are sold

e Choosing a chocolate with a soft centre from a box of chocolates containing 10 chocolates with hard centres and 15 chocolates with soft centres

f Choosing a boy from a class of 24 boys and 16 girls

g Choosing a red jelly bean from a bag containing 6 white, 5 black, 8 purple, 4 green and 7 red jelly beans

16–04 The range of probability

- An **impossible** event cannot occur so the probability of an impossible event is **0**.
- A **certain** event will definitely occur so the probability of a certain event is **1**.

These are the 2 extremes for probability. So the value of P(E) is from 0 to 1 for all events.

An **even chance** has a probability of $\frac{1}{2}$.

Impossible Unlikely Even chance Likely Certain

0 $\frac{1}{2}$ 1

EXAMPLE 6

Order the probability of each event on a number line.

a A baby being born on the weekend.

b Rolling a sum of 15 on a pair of dice.

c The sun rising in the east tomorrow morning.

d Selecting a black card from a normal deck of playing cards.

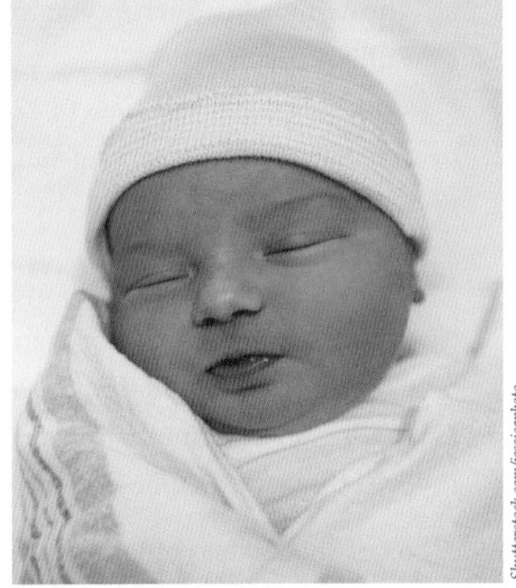

Shutterstock.com/jessicaphoto

SOLUTION

a There are only two weekend days (Saturday and Sunday) out of the 7 days of the week, so this is unlikely ($P(E)$ is less than $\frac{1}{2}$)

b This is impossible ($P(E) = 0$)

c This is certain ($P(E) = 1$)

d This has an even chance ($P(E) = \frac{1}{2}$)

b a d c

0 $\frac{1}{2}$ 1

Sum of probabilities

If you roll a die, the sample space is {1, 2, 3, 4, 5, 6}.

$P(1) = \frac{1}{6}, P(2) = \frac{1}{6}, P(3) = \frac{1}{6}, P(4) = \frac{1}{6}, P(5) = \frac{1}{6}, P(6) = \frac{1}{6}$

The sum of probabilities of every event in the sample space:

$\frac{1}{6} + \frac{1}{6} + \frac{1}{6} + \frac{1}{6} + \frac{1}{6} + \frac{1}{6} = \frac{6}{6} = 1$

The sum of all probabilities of all outcomes within a sample space is always 1.

ISBN 9780170350969

EXERCISE 16–04

1 What best describes the probability of rolling a number less than 7 on a die? Select the correct answer A, B, C or D.

A impossible B likely C even chance D certain

2 What is the probability from Question 1? Select A, B, C or D.

A 0 B $\frac{1}{2}$ C 1 D $\frac{2}{3}$

3 Draw a probability number line and place the following terms on it.

certain, even chance, impossible, unlikely, likely

4 Use a probability term to describe the probability of each of these events occurring.

a Snowing in Darwin tomorrow

b Throwing a head when a coin is tossed

c Rolling a 1 on a die

d Winning Lotto

e A traffic light showing red rather than green

f Wining a raffle of 100 with 2 tickets

g Rolling an odd number on a die

h Choosing a red shirt from a wardrobe containing all red shirts

i Picking a white ball from a bag with all red and blue balls

j Choosing a soft-centred chocolate from a box with mostly soft centres

5 Plot the events a–j in Question 4 on a number line.

6 Match the words below with these predicted events.

impossible, unlikely, even chance, likely, certain

a The sun will shine on a summer day.

b The sun will set in the west today.

c I will kick the next goal if I am the goal kicker.

d I will roll the number 5 on a die.

e I will choose a marble from a bag full of lollies.

f I will become a black-belt karate master without any training.

g It will be cold in Canberra in July.

h I will toss a head or a tail when I throw a coin.

i I will pass the next Maths test if I am good at Maths.

j I will watch TV tonight.

7 For the experiments below, list the possible outcomes and calculate their probabilities. Check that the sum of these probabilities is 1.

a Selecting a student from a class of 18 boys and 12 girls (Hint: Find P(boy) and P(girl) and then find their sum.)

b Selecting a ball from a bag containing 6 white balls and 5 red balls

c Selecting an item from a wardrobe containing 5 shirts, 8 trousers and 12 dresses

d Selecting a lolly from a box of lollies containing 15 mints, 20 sherbets and 12 toffees

16–05 Experimental probability

WORDBANK

experimental probability Probability based on the results of an experiment or past statistics.

frequency The number of times something happens.

trial One run or go of a repeated chance experiment, for example, one roll of a die.

In experimental probability:

$$P(E) = \frac{\text{number of times an event occurs}}{\text{total number of trials of experiment}}$$

EXAMPLE 7

A die was rolled 30 times, with the results shown in the frequency table.

Score	Frequency
1	2
2	4
3	5
4	7
5	4
6	8

Alamy/Lucie Lang

✱ Notice that for 30 rolls, 30 ÷ 6 = 5, so we would expect to roll 5 of each number from 1 to 6.
In experimental probability the number expected does not usually happen because we need to do a very large number of trials before we get close to the real probability.

What was the experimental probability of rolling:

a 4? b 5? c an odd number?

SOLUTION

a $\frac{7}{30}$ ← From the table, 4 was rolled 7 times out of 30.

b $\frac{4}{30}$ ← From the table, 5 was rolled 4 times out of 30.

c $\frac{2+5+4}{30} = \frac{11}{30}$ ← The odd numbers are 1, 3, 5 and they were rolled 2, 5, 4 times.

ISBN 9780170350969

EXERCISE 16–05

1 What is the experimental probability of throwing a head if I toss a coin 8 times and it lands on tails 3 times? Select the correct answer A, B, C or D.

A $\frac{3}{8}$ B $\frac{5}{3}$ C $\frac{3}{5}$ D $\frac{5}{8}$

2 In Question 1, what is the experimental probability of tossing a tail? Select A, B, C or D.

A $\frac{3}{8}$ B $\frac{5}{3}$ C $\frac{3}{5}$ D $\frac{5}{8}$

3 a If you tossed a coin 20 times, how many heads would you expect?

b Do you think this would happen?

4 a If you threw a coin 40 times, how many heads would you expect?

b How many tails would you expect?

c Now throw a coin 40 times and record the results in a table.

d Did you get the results you expected? Why, or why not?

Shutterstock.com/Maryna Pleshkun

5 a If you rolled a die 60 times how many of each number would you expect?

b Roll a die 60 times and record the results in a table.

c Did you get the results you expected?

d What was your experimental probability of rolling:

i 5? ii 3? iii an even number? iv 5 or 6?

6 Conduct the following experiment.

Place 10 red, 10 blue and 10 yellow counters in a box. Each student is to select a counter, record its colour and place the counter back in the box.

a How many would you expect of each colour if there are 30 students in the class?

b Did this happen?

d Write down the experimental probability of selecting each colour.

CODE PUZZLE

Use the table to decode the words and phrases from this chapter.

A	B	C	D	E	F	G	H	I	J	K	L	M
1	2	3	4	5	6	7	8	9	10	11	12	13

N	O	P	Q	R	S	T	U	V	W	X	Y	Z
14	15	16	17	18	19	20	21	22	23	24	25	26

1 3-8-1-14-3-5

2 3-5-18-20-1-9-14

3 9-13-16-15-19-19-9-2-12-5

4 21-14-12-9-11-5-12-25

5 5-22-5-14-20

6 15-21-20-3-15-13-5

7 16-18-15-2-1-2-9-12-9-20-25

8 20-8-5-15-18-5-20-9-3-1-12

9 5-17-21-1-12-12-25 12-9-11-5-12-25

10 19-1-13-16-12-5 19-16-1-3-5

ISBN 9780170350969

PRACTICE TEST 16

Part A General topics

Calculators are not allowed.

1 Evaluate 1984 ÷ 8.

2 Convert 0.03 to a fraction.

3 How many axes of symmetry does a scalene triangle have?

4 Copy this diagram and mark a pair of corresponding angles.

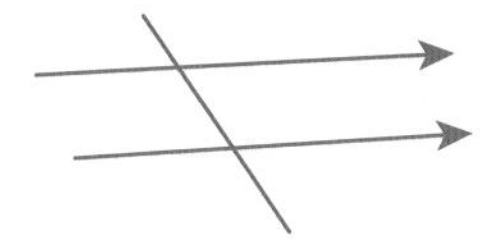

5 Convert $\frac{20}{6}$ to a simple mixed numeral.

6 Evaluate 1037.85 ÷ 100.

7 Write an algebraic expression for 'triple a number x, plus 6'.

8 On a stem-and-leaf plot, what part of the graph would the tens digit be?

9 Find the mode of 5, 8, 7, 6, 5, 4, 9.

10 What is the range of the scores in Question 9?

Part B Probability

Calculators are allowed.

16–01 The language of chance

11 Which probability term describes the position of the arrow on the probability scale below? Select the correct answer A, B, C or D.

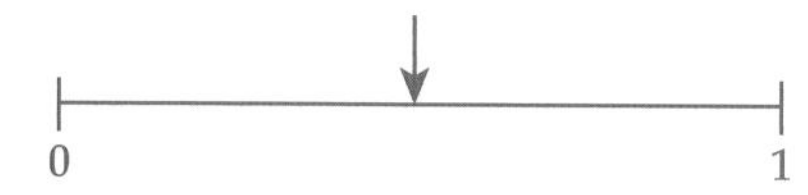

A impossible B even chance C likely D certain

12 What word(s) describes the probability of rolling a number greater than 2 on a die? Select A, B, C or D.

A certain B likely C unlikely D even chance

16–02 Sample spaces

13 What is the sample space for tossing two coins? Select A, B, C or D.

A {H, T} B {HH, HT, TH, TT} C {H, H, T, T} D {HH, TT, HT}

16–03 Probability

14 What is the probability of choosing a red marble form a bag containing 6 red and 12 blue marbles?

15 In a bag containing 4 red, 3 yellow and 5 white jelly beans, what is the probability of choosing a yellow jelly bean?

ISBN 9780170350969

16-04 The range of probability

16 What is the probability of a likely event? Select A, B, C or D.

A between 0 and $\frac{1}{2}$ B 1 C 0 D between $\frac{1}{2}$ and 1

17 What is:

a the smallest possible probability value?

b the sum of probabilities of all outcomes in a sample space?

16-05 Experimental probability

18 I toss a coin 26 times.

a How many tails would you expect?

b Will this definitely happen?

19 Tegan rolls a die 40 times and gets the results below:

Number	1	2	3	4	5	6
Frequency	5	6	8	7	4	10

What is the experimental probability of rolling:

a an odd number?

b a number greater than 4?

 ISBN 9780170350969

PERCENTAGES AND RATIOS

17

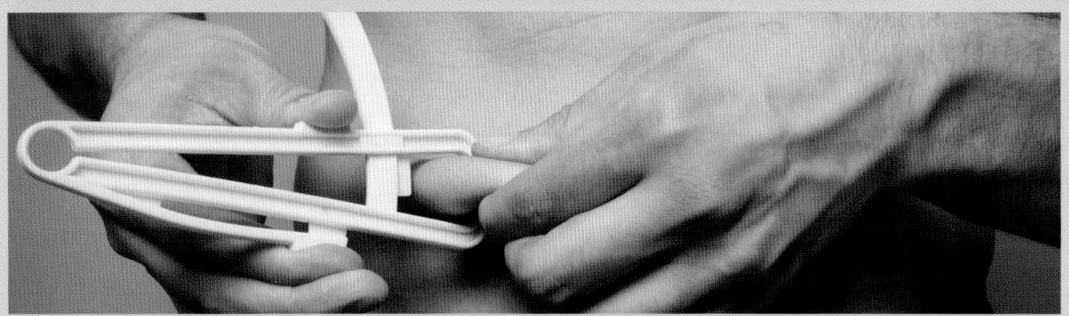

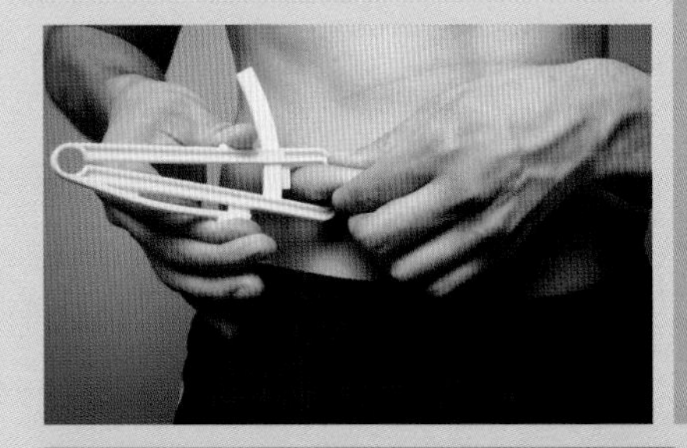

WHAT'S IN CHAPTER 17?

17-01 Percentages
17-02 Percentages and fractions
17-03 Percentages and decimals
17-04 Percentage of a quantity
17-05 Ratios
17-06 Equivalent ratios
17-07 Simplifying ratios

IN THIS CHAPTER YOU WILL:

- learn about percentages as being a special fraction 'out of 100'
- convert percentages to fractions
- convert percentages to decimals
- find a percentage of a number or metric amount
- learn about ratios that compare parts or shares
- find equivalent ratios
- simplify ratios

Shutterstock.com/LoloStock

17-01 Percentages

WORDBANK

Percentage A number that is written out of 100. The symbol is %. So 12% means $\frac{12}{100}$.

As $100\% = \frac{100}{100} = 1$ then $50\% = \frac{50}{100} = \frac{1}{2}$ and $25\% = \frac{25}{100} = \frac{1}{4}$. These are common percentages that you will use a lot.

To convert a percentage to a fraction:

- write the percentage over 100
- simplify the fraction if possible by dividing numerator and denominator by the HCF.

EXAMPLE 1

Convert to fractions and simplify each fraction where possible.

a 20% b 45% c 80% d $12\frac{1}{2}\%$

SOLUTION

a $20\% = \frac{20}{100}$

$= \frac{1}{5}$ ← ÷ numerator and denominator by 20

b $45\% = \frac{45}{100}$

$= \frac{9}{20}$ ← ÷ by 5

c $80\% = \frac{80}{100}$

$= \frac{4}{5}$ ← ÷ 20

d $12\frac{1}{2}\% = \frac{12\frac{1}{2}}{100}$

$= \frac{25}{200}$ ← × 2 to form whole numbers

$= \frac{1}{8}$ ← ÷ 25

ISBN 9780170350969

EXERCISE 17-01

1 What is 13% as a fraction? Select the correct answer A, B, C or D.

A $\frac{1}{13}$ B $\frac{13}{100}$ C $\frac{13}{1}$ D $\frac{100}{13}$

2 What is 60% as a simplified fraction? Select A, B, C or D.

A $\frac{60}{100}$ B $\frac{2}{5}$ C $\frac{6}{10}$ D $\frac{3}{5}$

3 Is each statement true or false?

a $21\% = \frac{1}{21}$ b $16\% = \frac{4}{25}$ c $33\% = \frac{33}{100}$ d $62\% = \frac{32}{50}$

4 Write each of the following as percentages.

a $\frac{15}{100}$ b $\frac{23}{100}$ c $\frac{54}{100}$ d $\frac{95}{100}$

5 Write each of these common percentages as a fraction.

a 50% b 75% c 25% d 100%

6 Convert each of the following percentages to a fraction.

a 30% b 42% c 70% d 85%

e 55% f 120% g 3% h 62%

i 5% j 75% k 40% l 18%

7 Copy and complete:

a $33\frac{1}{3}\% = \frac{33\frac{1}{3}}{100}$

$= \frac{33\frac{1}{3} \times \_\_}{100 \times 3}$

$= \frac{100}{300}$

$= \frac{\_\_}{3}$

b $37\frac{1}{2}\% = \frac{\_\_}{100}$

$= \frac{37\frac{1}{2} \times 2}{100 \times \_\_}$

$= \frac{\_\_}{200}$

$= \frac{\_\_}{8}$

8 Convert each of the following percentages to a fraction and simplify each fraction.

a $66\frac{2}{3}\%$ b $62\frac{1}{2}\%$ c $83\frac{1}{3}\%$ d $87\frac{1}{2}\%$

9 Write each percentage as a fraction or a whole number.

a 10% b 30% c 80% d 100%

e 150% f 200% g 175% h 40%

10 Copy and complete:

a If $10\% = \frac{1}{10}$ then $20\% = \frac{\_\_}{10}$ and $30\% = \frac{\_\_}{10}$.

b If $20\% = \frac{1}{5}$ then $40\% = \frac{\_\_}{5}$ and $60\% = \frac{\_\_}{5}$.

17-02 Percentages and fractions

To convert a fraction to a percentage, multiply the fraction by 100%.
This does not change the size of the fraction as 100% = 1.

EXAMPLE 2

Convert each fraction to a percentage.

a $\frac{13}{100}$ b $\frac{1}{2}$ c $\frac{3}{4}$ d $\frac{1}{3}$

SOLUTION

a $\frac{13}{100} = \frac{13}{100} \times 100\%$
$= 13\%$

b $\frac{1}{2} = \frac{1}{2} \times 100\%$
$= 50\%$

c $\frac{3}{4} = \frac{3}{4} \times 100\%$
$= 75\%$

d $\frac{1}{3} = \frac{1}{3} \times 100\%$
$= 33\frac{1}{3}\%$

Shutterstock.com/Paul Matthew Photography

ISBN 9780170350969

EXERCISE 17–02

1 What is $\frac{3}{4}$ as a percentage? Select the correct answer A, B, C or D.

A 34% B 30% C 40% D 75%

2 What is $\frac{2}{5}$ as a percentage? Select A, B, C or D.

A 20% B 40% C 50% D 25%

3 Write each fraction as a percentage.

a $\frac{24}{100}$ b $\frac{48}{100}$ c $\frac{65}{100}$ d $\frac{73}{100}$

e $\frac{54}{100}$ f $\frac{86}{100}$ g $\frac{90}{100}$ h $\frac{18}{100}$

4 Copy and complete:

a $\frac{1}{4} = \frac{1}{4} \times 100\%$
$= \_\_\_\_\%$

b $\frac{4}{5} = \frac{4}{5} \times \_\_\_\_\%$
$= 4 \times \_\_\_\_\%$
$= \_\_\_\_\%$

5 Convert each fraction to a percentage.

a $\frac{1}{4}$ b $\frac{1}{5}$ c $\frac{3}{10}$ d $\frac{2}{5}$

e $\frac{3}{8}$ f $\frac{4}{5}$ g $\frac{7}{10}$ h $\frac{3}{5}$

i $\frac{2}{3}$ j $\frac{7}{8}$ k $\frac{1}{6}$ l 1

6 a Convert each fraction to a percentage. Viktor wanted to give $\frac{1}{4}$ of his share of the business to his sister while Yulia wanted to give $\frac{3}{5}$ of her share to her son.

b Calculate how much the sister and son received if Viktor's share was \$24 000 and Yulia's share was \$30 000. (Use the fractions to calculate each amount.)

Fairfax/Orlando Chiodo

17-03 Percentages and decimals

To convert a percentage to a decimal:
- divide the percentage by 100
- move the decimal point 2 places left.

EXAMPLE 3

Convert to decimals:

a 35% b 80% c 6.5% d 180%

SOLUTION

a $35\% = 35 \div 100$
$= 0.35$

b $80\% = 80 \div 100$
$= 0.8$

c $6.5\% = 6.5 \div 100$
$= 0.065$

d $180\% = 180 \div 100$
$= 1.8$

Note that the decimal point moves 2 places left in each case.

To convert a decimal to a percentage:
- multiply the decimal by 100%. This does not change the size of the decimal as 100% = 1.
- move the decimal point 2 places right.

EXAMPLE 4

Convert each decimal to a percentage.

a 0.4 b 0.09 c 0.45 d 0.125

SOLUTION

a $0.4 = 0.4 \times 100\%$
$= 40\%$

b $0.09 = 0.09 \times 100\%$
$= 9\%$

c $0.45 = 0.45 \times 100\%$
$= 45\%$

d $0.125 = 0.125 \times 100\%$
$= 12.5\%$

Note that the decimal point moves 2 places right in each case.

ISBN 9780170350969

EXERCISE 17–03

1 What is 54% when written as a decimal? Select the correct answer A, B, C or D.

A 5.4 B 0.054 C 5.40 D 0.54

2 What is 0.8 written as a percentage? Select A, B, C or D.

A 80% B 8% C 0.8% D 0.08%

3 Is each statement true or false?

a 0.4 is equal to 4%.
b 0.05 is equal to 5%.
c 0.07 is equal to 70%.
d 0.48 is equal to 48%.
e 0.5 is equal to 50%.
f 0.06 is equal to 60%.

4 Convert each percentage to a decimal.

a 70% b 65% c 90% d 100%
e 85% f 24% g 92% h 58%
i 4.5% j 12.5% k 220% l 82.5%

5 Convert each decimal to a percentage.

a 0.3 b 0.04 c 0.26 d 0.5
e 0.47 f 0.64 g 0.6 h 0.95
i 0.625 j 0.55 k 1.4 l 2.5

6 Copy and complete this table.

Percentage		10%		20%	25%		50%	$66\frac{2}{3}\%$	
Fraction			$\frac{1}{8}$			$\frac{1}{3}$			
Decimal	0.05								0.75

7 A group of students was asked to vote for their favourite band. The results were graphed.

a What do the percentages add up to?
b Which band was the:
i most favoured? ii least favoured?
c What fraction of the group voted for U3?
d Convert this fraction to a decimal.
e If 400 students voted, how many voted for U3?

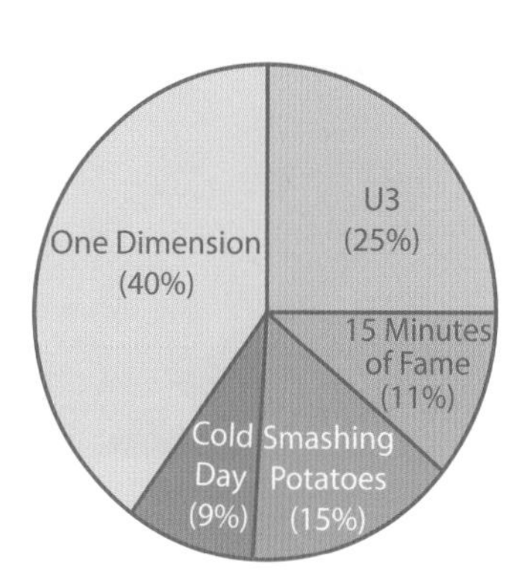

17-04 Percentage of a quantity

EXAMPLE 5

Find:

a 20% of \$4000

b 35% of 15 kg

SOLUTION

a $20\% \text{ of } \$4000 = \dfrac{20}{100} \times 4000$

$= \$800$

b $35\% \text{ of } 15 \text{ kg} = \dfrac{35}{100} \times 15$

$= 5.25 \text{ kg}$

EXAMPLE 6

Find $33\frac{1}{3}\%$ of 6 hours.

SOLUTION

$33\frac{1}{3}\% \text{ of } 6 \text{ hours} = \dfrac{1}{3} \times 6$ ← $33\frac{1}{3}\% = \dfrac{33\frac{1}{3}}{100} = \dfrac{1}{3}$

$= 2 \text{ hours}$

Fairfax/Tim Clayton

ISBN 9780170350969

EXERCISE 17–04

1 Find 25% of $840. Select the correct answer A, B, C or D.

A $420 B $220 C $210 D $440

2 Find $33\frac{1}{3}$% of 84 L. Select A, B, C or D.

A 42 L B 28 L C 18 L D 21 L

3 Is each statement true or false?

a 25% is equal to $\frac{1}{4}$

b 40% is equal to 0.45

c 65% is equal to $\frac{65}{100}$

d 90% is equal to 0.9

e 35% is equal to $\frac{7}{20}$

f 70% is equal to 0.7

4 Find:

a 25% of $200

b 40% of 80 km

c 45% of $5000

d 30% of 4 hours

e 50% of 1 day

f 80% of $600

g 75% of 24 m

h 90% of 50 km

i 150% of 6 hours

5 Simplify the percentages first and find:

a 12.5% of 8 m

b $33\frac{1}{3}$% of 12 cm

c 37.5% of $2000

d 8% of $700

e 5% of 400 L

f $66\frac{2}{3}$% of 54 kg

6 In Grania's moneybox were 200 coins. 10% of the coins were 50c pieces and the rest were $1 coins.

a How many coins were 50c pieces?

b What was the value of the 50c coins?

c What percentage of the coins were $1 pieces?

d How much money did Grania have in her moneybox altogether?

7 85% of the students participated at the school swimming carnival.

a If there were 1500 students at the carnival, how many students participated?

b What percentage of students did not participate?

c How many students did not participate?

17-05 Ratios

WORDBANK

ratio A comparison of quantities of the same kind in a particular order. The symbol for ratio is ':' , read as 'to'; for example the ratio of girls : boys is 3 : 2.

EXAMPLE 7

Write Dave's height to Ben's height as a ratio.

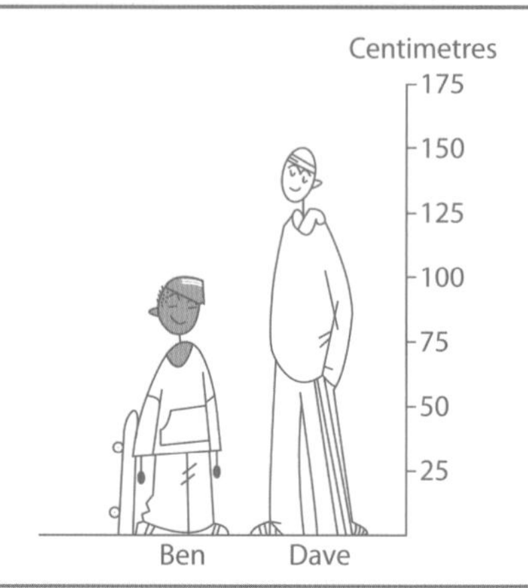

SOLUTION

Dave's height : Ben's height

= 150 cm : 100 cm

= 150 : 100

This could also be written as 150 to 100 or as a fraction $\frac{150}{100}$.

Because the units are the same, we can leave them out.

EXAMPLE 8

Shade in each diagram according to the ratio of colours given.

a pink : green = 3 : 5

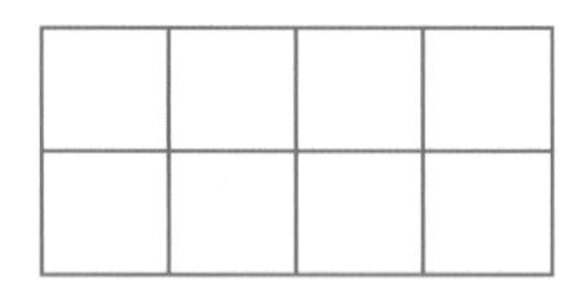

b purple : yellow : blue = 6 : 2 : 4

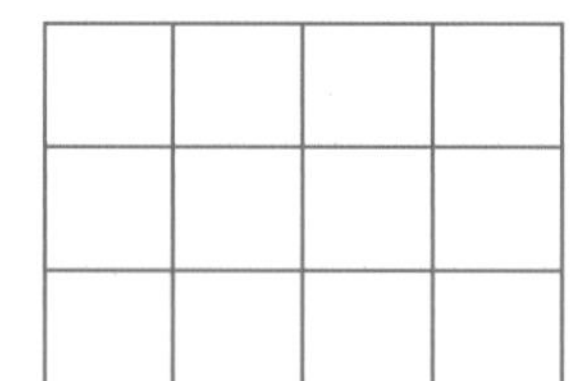

SOLUTION

a

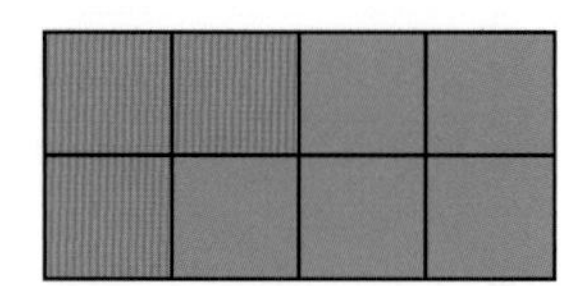

b

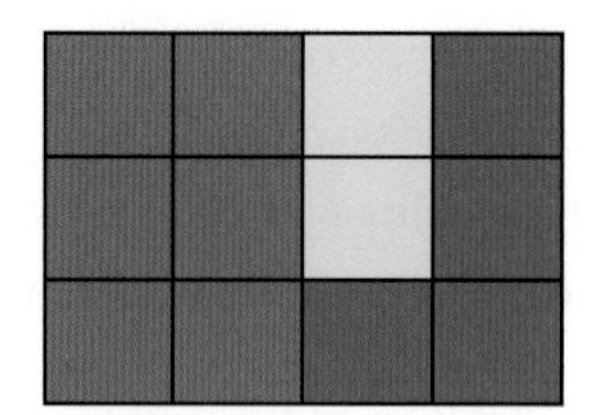

EXAMPLE 9

For the diagrams above, write down the ratios:

a green : pink

b pink : total shading

c yellow : blue : purple

SOLUTION

a 5 : 3

b 3 : 8

c 2 : 4 : 6

It is important to remember that the **order** of the ratio numbers is important. For example, pink : green is not the same as green : pink.

ISBN 9780170350969

EXERCISE 17–05

1 If there are 5 red and 2 blue marbles in a bag, what is the ratio of blue marbles : red marbles? Select the correct answer A, B, C or D.

A 5 : 2 B 2 : 7 C 5 : 7 D 2 : 5

2 For the bag of marbles in Question 1, what is the ratio of red marbles : total marbles? Select A, B, C or D.

A 5 : 2 B 2 : 7 C 5 : 7 D 2 : 5

3 Write each statement as a ratio.

a 4 cm to 7 cm b 5 m to 9 m c 23c to 41c
d $12 to $17 e 3 L to 5 L f 7 g to 9 g
g 12 min to 19 min h 5 hours to 11 hours i 19 mL to 20 mL

4 In a class of 30 students, 16 are boys. Write the ratio of:

a boys to girls b girls to boys
c boys to the total number in the class

5 For the scale below, write the ratio of:

a *AB* to *AC* b *AB* to *BC* c *CB* to *CA*

6 A bag of marbles has 5 red, 6 blue, 3 yellow and 4 black. What is the ratio of:

a red : blue b black : red c yellow : blue
d blue : black e black : red : blue f blue : black : yellow
g yellow : red : blue h yellow : not yellow i red : yellow : black : blue

7 For the shaded box below, write down the ratio:

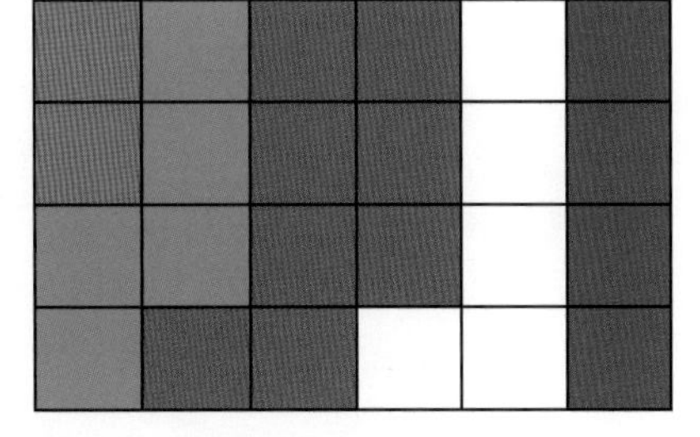

a pink : green b green : blue
c blue : purple d white : blue
e green : pink f blue : green
g white : purple h blue : pink
i purple : green j purple : blue

8 A box of chocolates has 12 soft centred, 16 hard centred and 8 nut chocolates. Write the ratio of:

a soft centres : hard centres b hard centres : nuts
c nuts : soft centres d hard centres : nuts : soft centres
e soft centres : not soft centres f nuts : not nuts

17-06 Equivalent ratios

WORDBANK

equivalent The same as something else. Equal to another amount.

Finding equivalent ratios is similar to finding equivalent fractions.

To find an equivalent ratio, multiply or divide each term in the ratio by the same number.

EXAMPLE 10

Copy and complete:

a $4 : 5 = 20 : \_\_\_\_$ b $120 : 80 = 12 : \_\_\_\_$ c $7 : 3 = \_\_\_\_ : 9$

SOLUTION

a $4 : 5 = 20 : \_\_\_\_$ ← $4 \times 5 = 20$
$4 : 5 = 20 : 25$ Do the same to 5 to complete the ratio: $5 \times 5 = 25$

b $120 : 80 = 12 : \_\_\_\_$ ← $120 \div 10 = 12$
$120 : 80 = 12 : 8$ Do the same to 80 to complete the ratio: $80 \div 10 = 8$

c $7 : 3 = \_\_\_\_ : 9$ ← $3 \times 3 = 9$
$7 : 3 = 21 : 9$ Do the same to 7 to complete the ratio: $7 \times 3 = 21$

Shutterstock.com/Andrii Opanasenko

ISBN 9780170350969

EXERCISE 17–06

1 Which ratio is equivalent to 5 : 6? Select the correct answer A, B, C or D.

A 6 : 5 B 10 : 30 C 20 : 24 D 6 : 12

2 Which ratio is equivalent to 12 : 100? Select A, B, C or D.

A 3 : 20 B 6 : 25 C 1 : 8 D 6 : 50

3 Copy and complete each pair of equivalent ratios.

a 1 : 4 = 5 : ____ b 5 : 8 = 50 : ____

c 3 : 4 = 21 : ____ d 3 : 2 = 9 : ____

e 3 : 1 = 15 : ___ f 3 : 5 = ___ : 30

g 2 : 9 = ___ : 45 h 8 : 6 = ___ : 18

i 5 : 2 = ____ : 14 j 6 : 5 = ____ : 30

4 Copy and complete:

a 30 : 25 = 6 : ____ b 60 : 80 = 6 : ____

c 28 : 36 = 7 : ____ d 45 : 20 = 9 : ____

e 150 : 50 = 15 : ____ f 24 : 48 = 6 : ____

g 20 : 65 = ____ : 13 h 84 : 12 = ____ : 1

i 30 : 39 = ____ : 13 j 27 : 18 = ____ : 2

5 Find simpler equivalent ratios for each of the following:

a 4 : 8 = 1 : ___ b 7 : 14 = 1 : ___ c 6 : 18 = 1 : ___

d 4 : 6 = ___ : 3 e 10 : 8 = ___ : 4 f 9 : 12 = 3 : ___

g 20 : 10 = ___ : 1 h 15 : 3 = ___ : 1 i 24 : 36 = ___ : 3

6 Write two equivalent ratios for each ratio.

a 4 : 9 b 2 : 6 c 4 : 5 : 6

Shutterstock.com/Dolores Giraldez Alonso

17-07 Simplifying ratios

Simplifying ratios is similar to simplifying fractions.

To simplify a ratio, divide each number in the ratio by the highest common factor (HCF).

EXAMPLE 11

Simplify the ratios:

a 18 : 24 b 4 : 6 : 12 c 28 : 24 : 16

SOLUTION

a $18 : 24 = \frac{18}{6} : \frac{24}{6} = 3 : 4$ ← The HCF is 6, so divide by 6

b $4 : 6 : 12 = \frac{4}{2} : \frac{6}{2} : \frac{12}{2} = 2 : 3 : 6$ ← The HCF is 2, so divide by 2

c $28 : 24 : 16 = \frac{28}{4} : \frac{24}{4} : \frac{16}{4} = 7 : 6 : 4$ ← The HCF is 4, so divide by 4

EXAMPLE 12

Simplify the ratios:

a 35 mm : 5 cm b 15 min : 1 hour c 4 kg : 800 g

SOLUTION

a 35 mm : 5 cm = 35 mm : 50 mm ← Make the units the same
= 35 : 50 ← ÷ 5
= 7 : 10

b 15 min : 1 hour = 15 min : 60 min ← Make the units the same
= 15 : 60
= 1 : 4 ← ÷ 15

c 4 kg : 800 g = 4000 g : 800 g
= 4000 : 800
= 5 : 1

 ISBN 9780170350969

EXERCISE 17–07

1 Simplify 18 : 36. Select the correct answer A, B, C or D.

A 2 : 4 B 3 : 6 C 9 : 18 D 1 : 2

2 Simplify 36 : 24. Select A, B, C or D.

A 24 : 16 B 3 : 2 C 12 : 8 D 6 : 4

3 Simplify each ratio.

a 6 : 8
b 4 : 12
c 5 : 25
d 8 : 28
e 4 : 6 : 10
f 6 : 9 : 15
g 12 : 24 : 60
h 18 : 20 : 32
i 60 : 45
j 250 : 25
k 48 : 64 : 16
l 72 : 36 : 27

4 Simplify each ratio by making the units the same.

a 15 mm : 2 cm
b 45 s : 2 min
c 40 cm : 1 m
d 120 min : 3 h
e 350 mg : 5 g
f 2 days : 72 h
g 1200 mm : 2 m
h 4500 g : 6 kg
i 8 m : 600 cm
j 2 h : 36 min
k 4500 mm : 500 cm : 10 m

5 For the shaded box below, find simplified ratios for:

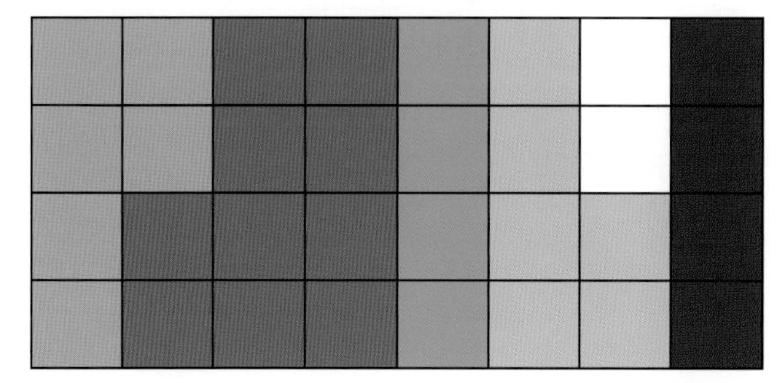

a pink : green
b purple : blue
c grey : pink
d green : blue
e pink : blue : green
f blue : grey : green
g grey : white : purple
h green : not green
i purple : grey : white
j blue : not blue
k pink : white : green : grey
l purple : not grey : pink : green

ISBN 9780170350969

FIND-A-WORD PUZZLE

Make a copy of this page, then find the words below in this grid of letters.

T	R	E	V	N	O	C	U	N	I	M	G	C	N	Q
S	N	G	P	E	G	N	J	O	D	Y	Z	M	F	O
D	Q	E	F	W	I	A	Y	I	P	N	Z	T	I	B
W	E	U	L	T	B	F	K	T	R	I	I	T	T	G
Y	A	C	A	A	I	D	X	C	O	V	A	F	L	R
T	Z	R	I	L	V	U	H	A	B	R	K	Z	G	L
J	Y	X	P	M	J	I	S	R	L	W	Q	S	N	L
D	E	M	R	M	A	A	U	F	E	H	W	B	A	G
H	I	R	I	C	S	L	Z	Q	M	D	A	T	U	C
S	Q	U	A	N	T	I	T	Y	E	O	Z	I	V	U
Q	L	S	V	P	U	I	F	D	W	H	R	U	P	A
S	C	A	L	E	M	S	N	K	X	T	G	G	G	L
N	C	D	W	P	D	O	B	Z	T	E	K	J	Y	L
A	A	X	X	P	H	F	C	C	E	M	O	P	K	Y
E	G	A	T	N	E	C	R	E	P	P	Y	E	L	W

COMPARE CONVERT DECIMAL EQUIVALENT
FRACTION METHOD PERCENTAGE PROBLEM
QUANTITY RATIO SCALE SIMPLIFY
UNITARY

ISBN 9780170350969

PRACTICE TEST 17

Part A General topics

Calculators are not allowed.

1 Evaluate 810.9 ÷ 100.

2 Complete this pattern: 15, 11, 7, 3, ____

3 How many letters of the word MATHEMATICS are vowels?

4 What is the size of this angle?

5 Copy and complete: 5.6 kg = ____ g

6 Evaluate $\sqrt{49}$.

7 List the first 5 multiples of 9.

8 Copy and complete: 18 – 2 × ____ = 4.

9 How many axes of symmetry does a parallelogram have?

10 Evaluate 54.2 × 1.1.

Part B: Percentages and ratios

Calculators are allowed.

17-01 Percentages

11 Write 45% as a simplified fraction. Select the correct answer A, B, C or D.

A $\frac{45}{100}$ B $\frac{9}{45}$ C $4\frac{1}{2}$ D $\frac{9}{20}$

17-02 Percentages and fractions

12 What is $\frac{7}{20}$ as a percentage? Select A, B, C or D.

A 7% B 20% C 140% D 35%

17-03 Percentages and decimals

13 What is 0.865 as a percentage? Select A, B, C or D.

A 8.65% B 865% C 86.5% D 0.865%

14 Write each number as a percentage.

a $\frac{3}{4}$ b 0.7 c $\frac{1}{8}$

15 Write each number as a decimal.

a 25% b 60% c 120%

17-04 Percentage of a quantity

16 Find 35% of $2500.

17 Find $33\frac{1}{3}$% of 12 km.

17–05 Ratios

18 There are 13 green and red lollies in a jar. Six of them are green. What is the ratio of red lollies : green lollies in the jar?

17–06 Equivalent ratios

19 Copy and complete: 4 : 9 = ___ : 36.

17–07 Simplifying ratios

20 Simplify 24 mm : 3 cm. Select A, B, C or D.

A 4 : 5 B 24 : 36 C 2 : 12 D 6 : 9

ISBN 9780170350969

CHAPTER 1

Exercise 1-01

1 D
2 C
3 a +4 b −60 c +20 d +18
e −28 f −10 g +100 h −24
4 a +8 b −7
c −7, −5, −3, −1, 0, 2, 4, 6, 8
5 a

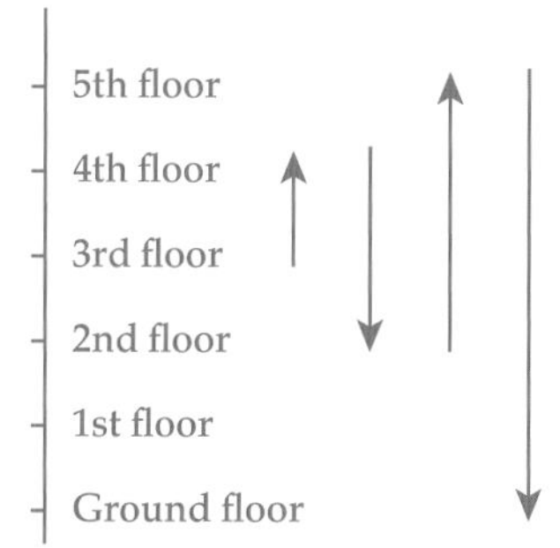

b 36 stairs c 60 stairs
6 6, 5, 4, 2, 0, −3, −9, −11

Exercise 1-02

1 B
2 A
3 a −3 −2 −1 0 1 2 3 4
b −8 −6 −4 −2 0 2 4
c −4 −2 0 2 4 6 8 10
d −12 −10 −8 −6 −4 −2 0 2 4
e −24 −20 −16 −12 −8 −4 0 4 8
4 a 26 b −16 c −16, −5, 0, 8, 11, 13, 26
5 a −8 −6 −4 −2 0 2 4 6 8
b −8 −7 −6 −5 −4 −3 −2 −1 0 1 2 3 4 5 6 7 8
c −8 −6 −4 −2 0 2 4 6 8
6 a INTEGERS b ARE
c INTERESTING
7 a > b < c > d >
e > f > g < h <
i < j < k < l >
8 a −2, −1, 0, 2, 4 b −5, −4, −3, 3, 6
c −12, −7, −3, −1, 0 d −8, −4, −1, 3, 8
e −25, −20, 0, 5, 35 f −23, −18, −12, 4, 29
9 a 7, 5, 3, −2, −3 b 4, 0, −1, −6, −8
c 8, 6, −1, −4, −11 d 12, 8, −6, −9, −15
e 7, −4, −9, −14, −28 f 32, 10, −48, −52, −90

Exercise 1-03

1 C
2 B
3 a 2 b −2 c −6 d 4
e −4 f −10 g 3 h −3
i −13 j −5 k 5 l −7
4 a 2 b −2 c −26 d −3
e 3 f −39 g 17 h −17
i −47 j 29 k −29 l −61
5 a 9 b −31 c 14 d 16
e 27 f −75 g −42 h 9
i −87 j 15 k −54 l −13
6 7 jumps
7 3 times
8 8 steps forward

Exercise 1-04

1 A
2 D
3 a 7 b −7 c −1 d 8
e −8 f −4 g 6 h −6
i −4 j 11 k −11 l −3
4 a 15 b −15 c −9 d 28
e −28 f −18 g 26 h −26
i −10 j 33 k −33 l 15
5 $111
6 $272
7 47 steps

Exercise 1-05

1 C
2 D
3 a A horizontal grid reference followed by a vertical grid reference used to locate a position.
b Horizontal
4 a B3 b D2 c E1
d C1 e E4
5 a Stuart b Tahine c Eve
d Jade e Sam
6 A street map, shopping centre map.
7 a B5 b D4 c C1
8 The treasure is at E3.

Exercise 1-06

1 A
2 B
3 a A number plane is a grid used to locate and plot points. Each point's location is described by two coordinates.
b We write coordinates inside brackets with a comma in between the numbers.
c Where the x and y axes meet.
d (0, 0)
4 x; right; y; up
5 $A(8, 10)$, $B(9, 9)$, $C(6, 8)$, $D(7, 3)$, $E(1, 9)$, $F(1, 4)$

6

$E(1, 8)$, $F(5, 9)$, $I(6, 8)$, $H(2, 7)$, $D(2, 6)$, $J(5, 6)$, $C(3, 5)$, $G(8, 4)$, $A(2, 3)$, $B(4, 2)$

7 a A triangle

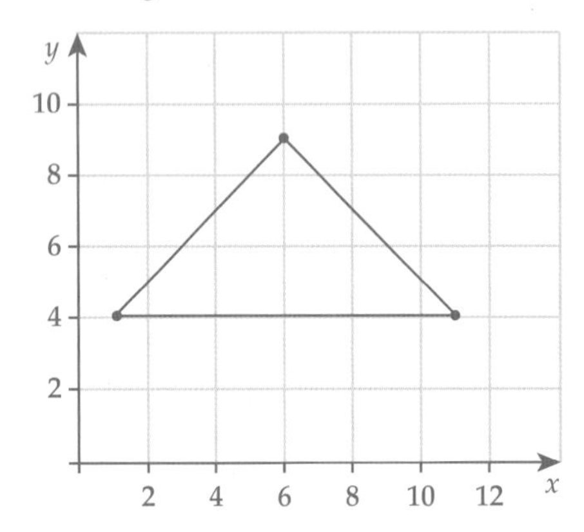

b A rectangle

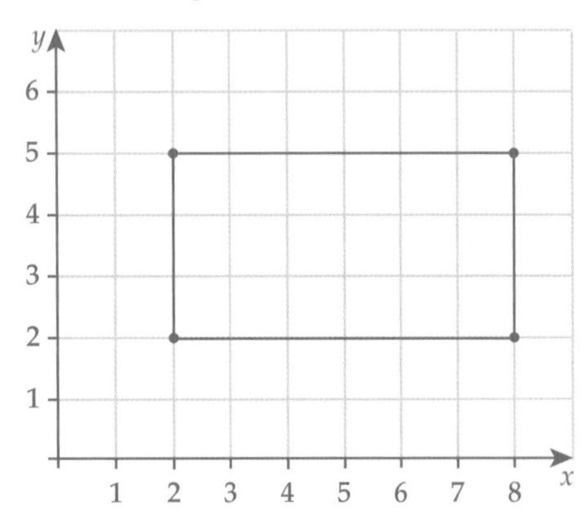

c A trapezium

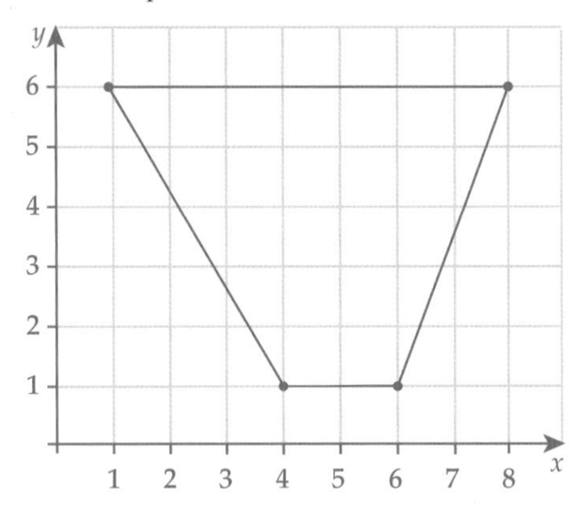

d A hexagon

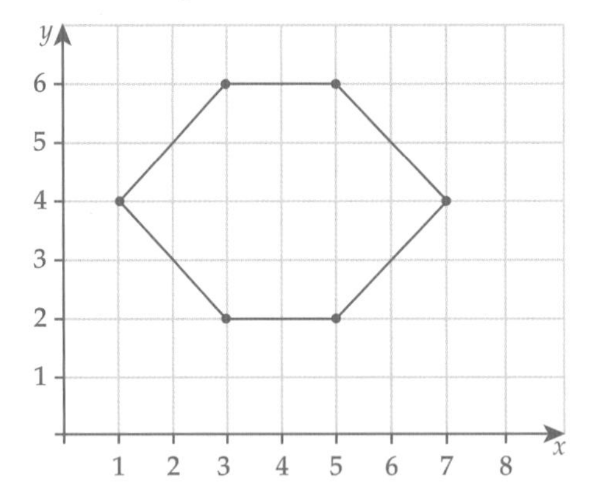

e A right-angled triangle

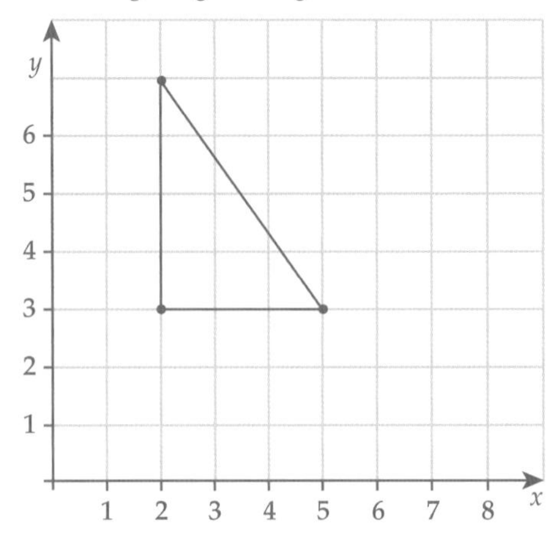

ISBN 9780170350969

f A parallelogram

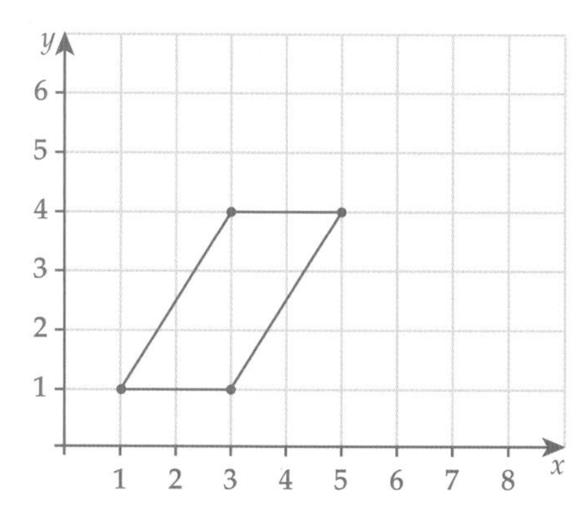

8 a $A(2, 5)$, $B(1, 5)$, $C(2, 6)$, $D(4, 7)$, $E(6, 8)$, $F(5, 7)$, $G(7, 8)$, $H(6, 7)$, $I(8, 6)$, $J(9, 5)$, $K(11, 3)$, $L(10, 5)$, $M(11, 7)$, $N(9, 5)$, $O(7, 4)$, $P(6, 3)$, $Q(7, 2)$, $R(5, 3)$, $S(6, 2)$, $T(2, 4)$, $U(1, 5)$

b A fish

Exercise 1-07

1 B
2 D
3 Anticlockwise
4 $A = (-5, 10)$; $B = (5, 6)$; $C = (-2, -8)$; $D = (5, -10)$; $E = (-7, 4)$; $F = (7, 2)$; $G = (-7, -4)$; $H = (7, -8)$
5 a 3rd b 4th c 1st d 2nd
6

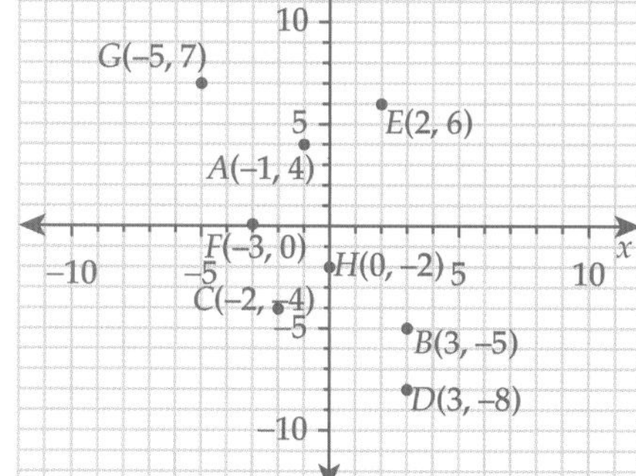

7 A and G: (–1, 4) and (–5, 7)
8 a

y
10
5
−10
−5
5
10
x
−5
−10

b 1st quadrant.

Language activity

1 PIN POINT
2 LEVEL LINE
3 NEXT NUMBER
4 SLACK SLOPE
5 ASKEW AXIS
6 PAPER PLANE

Practice test 1

1 45
2 168
3 620 mm
4 1 year 8 months
5 mL
6 25
7 $\frac{3}{5}$
8 East
9 4
10 31
11 C
12 D
13 B
14 a –2 b –16 c –6
15 C
16 a 6 b –15 c 3
17 a –5 + 8 – 11 b 8 steps left
18 a A2 b B1
19 vertical; horizontal; x; y
20 $A(-7, 2)$, $B(6, 6)$, $C(6, -5)$
21 A 2nd, B 1st, C 4th
22 A trapezium

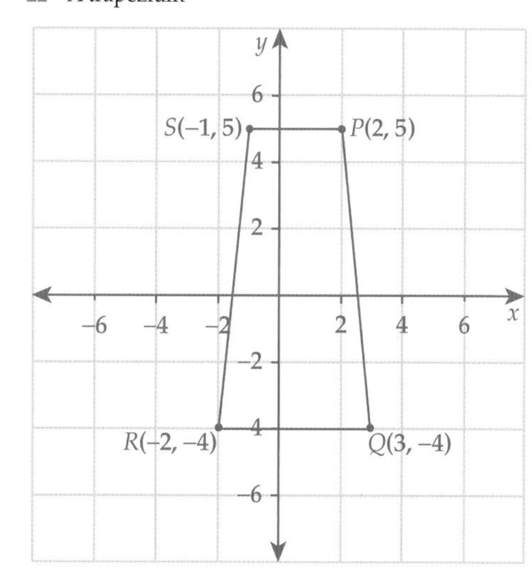

CHAPTER 2

Exercise 2-01

1 C
2 D
3 a 3 b Capital c Yes

ISBN 9780170350969

4 a $\angle PQR$ or $\angle RQP$ b $\angle XYZ$ or $\angle ZYX$
c $\angle TSR$ or $\angle RST$ d $\angle DEF$ or $\angle FED$
e $\angle MLN$ or $\angle NLM$ f $\angle CAB$ or $\angle BAC$
5 Teacher to check.
6 C

Exercise 2-02

1 B
2 A
3 a reflex b acute c straight
d acute e obtuse f revolution
g right h obtuse
4 Teacher to check.
5 a acute b obtuse
c reflex d reflex
e acute f straight
g obtuse h obtuse
i revolution j acute
k reflex l obtuse
6 acute, right, obtuse, straight, reflex

Exercise 2-03

1 B
2 A
3 a 48° b 90° c 47° d 50°
e 321° f 158°
4 Teacher to check.
5 Teacher to check.

Exercise 2-04

1 B
2 D
3 a False b True c True d True
e True f True
4 a 10° b 15° c 78° d 43°
5 a 130° b 55° c 102° d 17°
6 a

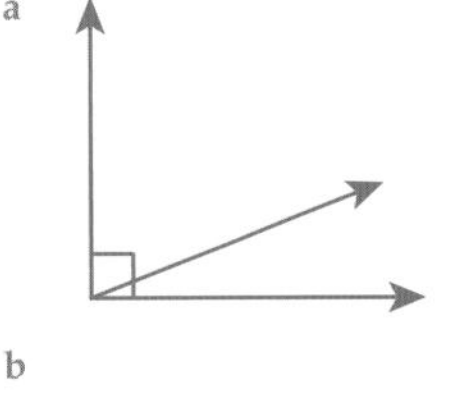

b

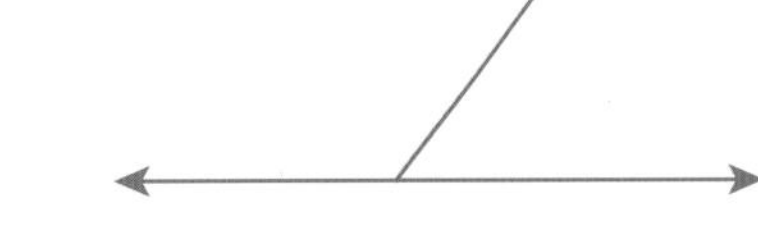

7 a 62; complementary angles
b 74; supplementary angles
c 44; supplementary angles
d 24; complementary angles
8 a $\angle COP$, $\angle POQ$ b $\angle ABD$, $\angle DBC$

Exercise 2-05

1 A
2 B
3 a Angles at a point add up to 360°.

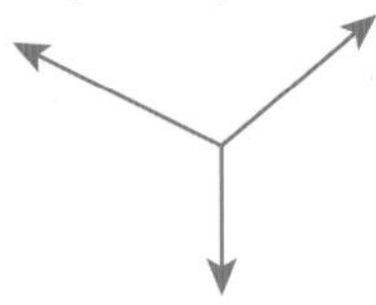

b Vertically opposite angles are equal.

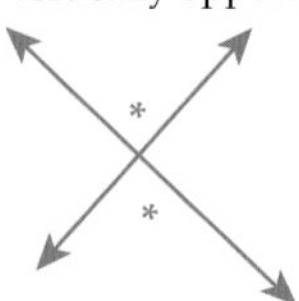

4 a False b False c True d True
5 a $x = 105$ b $y = 112$ c $n = 178$ d $z = 61$
e $d = 64$ f $y = 118$
6 a $\angle XKY$ b $\angle WKX$

Exercise 2-06

1 D
2 a f b c c c
3 a False b True c False d True
4 Teacher to check.
5 a $x = 72$ b $y = 98$ c $a = 114$
d $b = 82$; Reason for each: corresponding angles on parallel lines.
6 $m = 42$

Exercise 2-07

1 C
2 a g b d c d
3 a True b True c False d True
4 Teacher to check.
5 a $x = 69$ b $y = 105$ c $a = 83$
d $b = 99$; Reason for each: alternate angles on parallel lines.
6 $n = 82$

Exercise 2-08

1 B
2 a d b d c c
3 a False b False c True d False
4 Teacher to check.
5 a $x = 108$ b $y = 52$ c $a = 83$
d $b = 69$; Reason for each: co-interior angles on parallel lines.
6 $a = 126$: vertically opposite angles; $b = 126$: alternate angles on parallel lines; $c = 126$: vertically opposite angles; $d = 54$: straight angle

Practice test 2

1 38 m
2 $\frac{2}{3}$
3 $1400
4 rectangular prism
5 0.65

 ISBN 9780170350969

6 5 m
7 $1.15
8 6:30 p.m.
9 0.2, 0.25, 0.4, 0.46, 0.92
10

11 D
12 D
13 C
14 Teacher to check
15 a 62° b 36°
16 a 148° b 54°
17 $n = 34$
18 $c = 77$
19 $m = 162$
20 $a = 106$
21 $x = 75$
22 $a = 136, b = 44, c = 136$

CHAPTER 3

Exercise 3-01

1 D
2 B
3 a 50 b 80 c 160 d 2590
4 a 700 b 200 c 3900 d 5400
5 a 1000 b 7000 c 25 000 d 328 000
6 Exact answers
a 120 b 15 c 3093 d 3608
e 115 f 19.75 g 2896 h 33.89
7 $17 500
8 a 3940 km b 3900 km
c 4000 km
9 a $64 000 b $64 200 c $64 220 d $60 000
10 a 80 000 b $4000 c 8000
11 23 582 000

Exercise 3-02

1 C
2 A
3 a 807 b 543 c 103 d 140
e 4049 f 4810 g 202 h 779
i 3069 j 3145 k 1129 l 960
4 625
5 357
6 $279
7 a 1021 b $3334
8 $36
9 18 693
10 a $652 or similar
b $667.02

Exercise 3-03

1 D
2 A
3 a 28 b 21 c 25 d 24
e 2923 f 7339 g 54 h 187
i 4529 j 4554 k 3355 l 4764
4 2857
5 $167
6 a 400 km b 382 km
7 4952
8 30
9 a $22 b $28
10 a 700 b 658

Exercise 3-04

1 C
2 B
3 a 12 b 25 c 42 d 56
e 80 f 108 g 20 h 33
i 60 j 6 k 14 l 24
m 36 n 50 o 66 p 84
q 8 r 18 s 30 t 44
u 48 v 5 w 32 x 54
4 a 18 b 35 c 24 d 64
e 90 f 72 g 50 h 77
i 96 j 9 k 24 l 40
m 27 n 40 o 88 p 132
q 3 r 10 s 60 t 121
u 144 v 30 w 72 x 18
5

×	1	2	3	4	5	6	7	8	9	10
1	1	2	3	4	5	6	7	8	9	10
2	2	4	6	8	10	12	14	16	18	20
3	3	6	9	12	15	18	21	24	27	30
4	4	8	12	16	20	24	28	32	36	40
5	5	10	15	20	25	30	35	40	45	50
6	6	12	18	24	30	36	42	48	54	60
7	7	14	21	28	35	42	49	56	63	70
8	8	16	24	32	40	48	56	64	72	80
9	9	18	27	36	45	54	63	72	81	90
10	10	20	30	40	50	60	70	80	90	100

Exercise 3-05

1 B
2 C
3 a 4, 8, 12, 16, 20, 24 b 5, 10, 15, 20, 25, 30
c 20 d 20
4 a 6, 12, 18, 24, 30, 36, 42
b 9, 18, 27, 36, 45, 54, 63 c 18
5 a 10 b 12 c 45 d 12
e 100 f 60

ISBN 9780170350969

6 a 6 b 12 c 18 d 20
7 a 15 b 10 c 5 d 6th
8 a 4 b Only the 20th student
9 12 minutes

Exercise 3-06

1 A
2 C
3 a 340 b 476 c 472 d 1404
e 2920 f 3360 g 5060 h 31 302
4 a 34 b 86 c 110 d 56
5 a 244 b 60 c 172 d 76
6 a 9024 b 23 571 c 27 030 d 16 473
7 a 324 b 225 c 423 d 468
8 126
9 a 290 b 750 c 1300 d 4800
e 590 f 64 000 g 2000 h 47 000
10 a 160 b 70 c 130 d 460
11 828
12 a 60 b 1440 c 10 080

Language activity

BARRA CUDA GOT MORE SLEEP
Barra has bulging eyes!

Practice test 3

1 \$72
2 $\frac{1}{4}$
3 7 692 000
4 metre
5 324
6 9
7 3:27 p.m.
8 25 m
9 80
10 1, 5, 25
11 D
12 C
13 A
14 a 19 b 381 c 1437
15 55
16 a 48 b 28 c 81 d 60
17 12
18 a 3, 6, 9, 12, 15, 18 b 4, 8, 12, 16, 20, 24
c 12
19 48
20 15 minutes
21 a 162 b 17 296
22 6912

CHAPTER 4

Exercise 4-01

1 B
2 D
3 a

b $20 \div 4 = 5$

4

5 a 5 b 6 c 2 d 3
e 5 f 3 g 2 h 10
i 5 j 4 k 4 l 2
m 5 n 14 o 5 p 15
q 7 r 8 s 4 t 5
u 4 v 8 w 8 x 9
6 a 7 b 8 c 4 d 9
e 6 f 2 g 7 h 3
i 11 j 2 k 11 l 2
m 3 n 10 o 7 p 4
q 4 r 4 s 6 t 11
u 6 v 6 w 9 x 6
7 a 9 b 12 c 12 d 12
e 11 f 11
8 \$0.70

Exercise 4-02

1 D
2 B
3 a 22 b 94 c 121 d 121
e 117 f 751 g 3255 h 567
i 7214 j 2991 k 257 l 382
4 \$15
5 \$56
6 a 60 b 30 c 145 d 88
e 25 f 700 g 7 h 52
7 a 40 b 70 c 16 d 36
e 12 f 80 g 18 h 56
8 16

ISBN 9780170350969

9 27

10 40

Exercise 4-03

1 D

2 C

3 a $22\frac{1}{4}$ b $93\frac{5}{6}$ c $135\frac{2}{7}$ d $68\frac{1}{2}$

e $1163\frac{1}{5}$ f $673\frac{2}{9}$ g $3248\frac{1}{3}$ h $567\frac{4}{5}$

i 7219 j $2987\frac{2}{3}$ k $275\frac{3}{4}$ l $379\frac{7}{9}$

4 $33

5 $57

6 a 442 cm b 2 cm

7 $651.11

8 a 29 b 1

9 $5500

Exercise 4-04

1 C

2 B

3 Divisible by:

a 2, 3 and 6 b 2

c 2, 3 and 6 d 2, 3 and 6

e 3 f Not divisible.

4 Divisible by:

a 4 b 4 and 8

c 4 and 8 d 4

e 4 and 8 f 4

5 Divisible by:

a 5 and 10 b 5

c 5 and 10 d Neither

e 5 and 10 f 5 and 10

6 Divisible by 9:

a Yes b No c Yes

d No e Yes f No

7 a Yes b No c No

8 a Yes b Yes c Yes

9 a Yes b 64 c Yes d 48

10 12, 24, 36, 48

Exercise 4-05

1 A

2 C

3 a 1, 2, 5, 10 b 1, 2, 4, 8, 16

c 1, 3, 9 d 1, 2, 4, 5, 10, 20

e 1, 2, 4, 6, 8, 12, 24, 48

f 1, 2, 3, 4, 6, 9, 12, 18, 36

g 1, 2, 3, 4, 5, 6, 10, 12, 20, 30, 60

h 1, 2, 3, 4, 6, 8, 12, 24

i 1, 2, 4, 5, 8, 10, 20, 40

j 1, 2, 3, 6, 9, 18, 27, 54

4 C

5 B

6 a 1, 2, 3, 4, 6, 12 b 1, 3, 5, 15

c 1, 3 d 3

7 a 1, 2, 3, 5, 6, 10, 15, 30

b 1, 2, 3, 6, 7, 14, 21, 42

c 1, 2, 3, 6 d 6

8 a 2 b 4 c 15

d 10 e 30 f 4

9 12 and 18 or similar

10 9

Exercise 4-06

1 B

2 A

3 a 1, 2, 3, 4, 6, 12; composite

b 1, 23; prime c 1, 31; prime

d 1, 3, 5, 15; composite e 1, 17; prime

f 1, 2, 19, 38; composite

g 1, 2, 3, 6, 7, 14, 42; composite

h 1, 3, 19, 57; composite

i 1, 7, 13, 91; composite

4 62, 63, 64, 65, 66, 68, 69, 70, 72, 74, 75, 76, 77, 78

5 g 2, 3, 5, 7, 11, 13, 17, 19, 23, 29, 31, 37, 41, 43, 47, 53, 59, 61, 67, 71, 73, 79, 83, 89, 97, 101, 103, 107, 109, 113.

6 3 and 5, 5 and 7, 11 and 13, 17 and 19, 29 and 31, 41 and 43, 59 and 61, 71 and 73, 101 and 103, 107 and 109.

7 a 41 b 230

Exercise 4-07

1 B

2

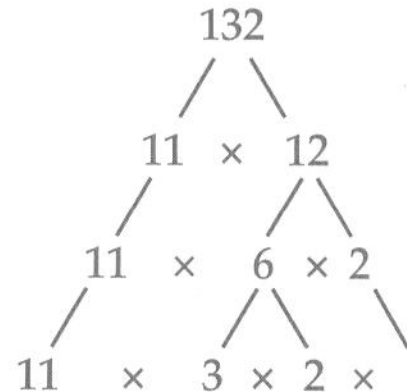

3 (other factor trees possible)

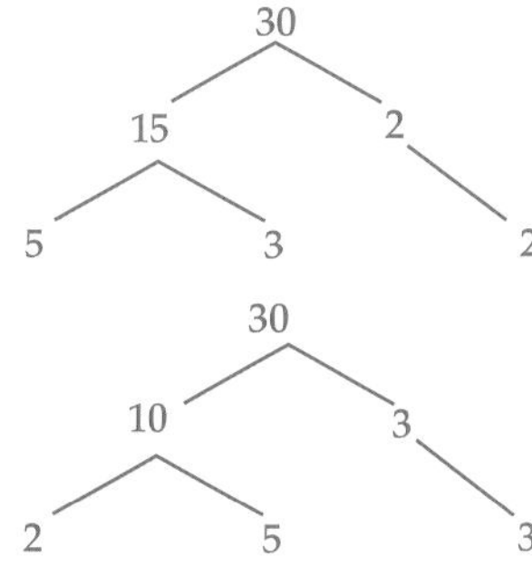

4 (Other factor trees possible)

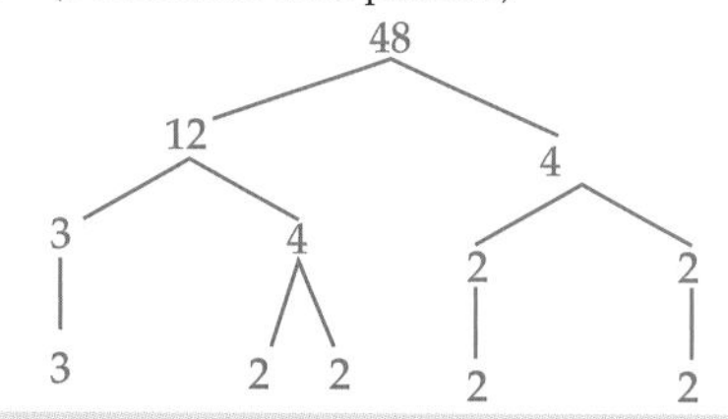

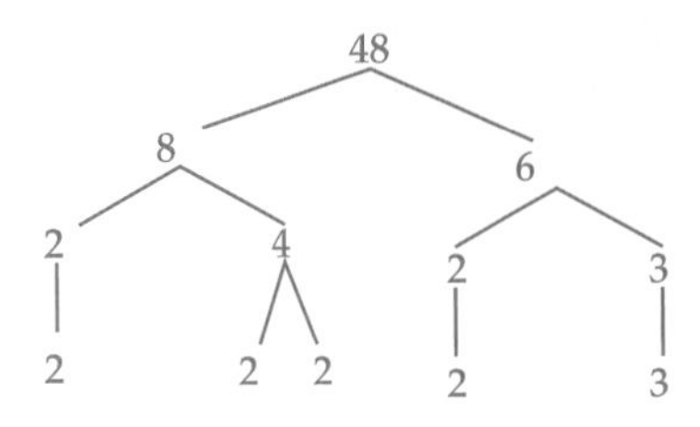

5 a $88 = 2 \times 2 \times 2 \times 11$ b $63 = 3 \times 3 \times 7$
c $45 = 3 \times 3 \times 5$ d $51 = 3 \times 17$
e $132 = 2 \times 2 \times 3 \times 11$
f $270 = 2 \times 3 \times 3 \times 3 \times 5$
g $396 = 2 \times 2 \times 3 \times 3 \times 11$
h $218 = 2 \times 109$

6 $1260 = 2 \times 3 \times 7 \times 5 \times 3 \times 2$

7 a $2 \times 2 \times 2 \times 2 \times 3$ b $2 \times 2 \times 2 \times 5 \times 5$
c $2 \times 2 \times 5 \times 23$ d $2 \times 2 \times 2 \times 89$
e $2 \times 7 \times 7$ f $2 \times 2 \times 2 \times 2 \times 3 \times 3$
g $5 \times 5 \times 13$ h $3 \times 3 \times 3 \times 5$

Language activity

1 DIVISION
2 QUOTIENT
3 SHORT
4 TREE
5 REMAINDER
6 DIVISIBILITY TEST
7 FACTOR
8 HIGHEST COMMON FACTOR
9 COMPOSITE
10 PRIME

Practice test 4

1 31
2 12 m
3 octagon
4 90
5 \$12
6 $\frac{3}{8}$
7 300
8 8, 4, 1, 0, −1, −3
9 6
10 96
11 C
12 B
13 C
14 a $20\frac{5}{8}$ b $57\frac{4}{5}$ c $352\frac{6}{7}$
15 a Divisible by 2, 3 and 6
b Divisible by 3
c Divisible by 2, 3 and 6
16 a Divisible by 4 b Divisible by 4 and 8
c Divisible by 4 and 8
17 D
18 a 1, 5, 11, 55
b 1, 2, 4, 5, 8, 10, 16, 20, 40, 80
19 4
20 2
21 a 23, 29, 31, 37
b 21, 22, 24, 25, 26, 27, 28, 30, 32, 33, 34, 35, 36, 38, 39
22 a

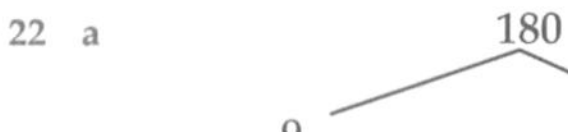
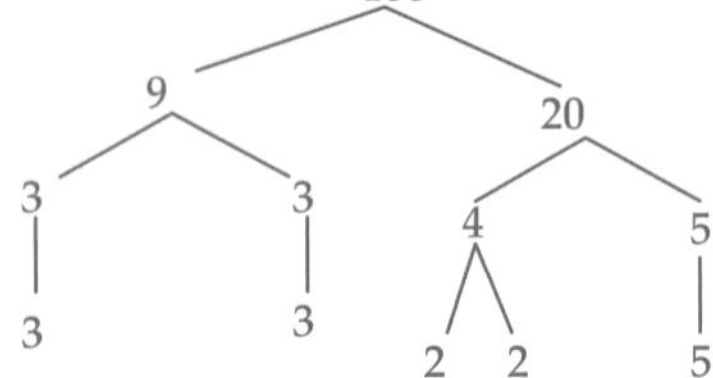

b $180 = 2 \times 2 \times 3 \times 3 \times 5$

CHAPTER 5

Exercise 5-01

1 C
2 B
3 a 3^4 b 2^5 c 5^3 d 4^6
e 7^3 f 6^4 g 9^5 h 12^4
i 23^5
4 a 81 b 32 c 125 d 4096
e 343 f 1296 g 59 049 h 20 736
i 6 436 343
5 a i 6 ii 4 b i 4 ii 5
c i 3 ii 6 d i 2 ii 7
e i 5 ii 4 f i 8 ii 3
g i 9 ii 4 h i 10 ii 5
i i 15 ii 4
6 a $6 \times 6 \times 6 \times 6$
b $4 \times 4 \times 4 \times 4 \times 4$
c $3 \times 3 \times 3 \times 3 \times 3 \times 3$
d $2 \times 2 \times 2 \times 2 \times 2 \times 2 \times 2$
e $5 \times 5 \times 5 \times 5$
f $8 \times 8 \times 8$
g $9 \times 9 \times 9 \times 9$
h $10 \times 10 \times 10 \times 10 \times 10$
i $15 \times 15 \times 15 \times 15$
7 a 1296 b 1024 c 729 d 128
e 625 f 512 g 6561 h 100 000
i 50 625
8 a 72 b 3888 c 8000 d 64 827
e 216 000 f 839 808
9 a 1000 b 100 000 c 100
d 1 000 000 e 100 000 000
f 10 000 000 g 1 000 000 000
h 10 000 i 10
10 They all begin with 1 followed by a number of 0s; the power shows the number of 0s.

 ISBN 9780170350969

Exercise 5-02

1 D
2 B
3 a 9 b 64 c 121 d 49
e 25 f 81 g 100 h 144
i 16
4 a 289 b 324 c 529 d 729
e 1024 f 2025 g 3844 h 5625
i 8100
5 a 3 b 6 c 8 d 10
e 9 f 1 g 12 h 2
i 11
6 a 19 b 15 c 17 d 24
e 28 f 32
7 a 400 b 40 000 c 4 000 000
8 The number of zeros doubled.
9 a 30 b 300 c 3000
10 The number of zeros halved.
11 8000

Exercise 5-03

1 B
2 A
3 a 27 b 216 c 1331 d 343
e –125 f 729 g 2197 h –1728
i –64 j 4096 k 9261 l –2744
4 power; three; cube root
5 a 2 b 6 c –3 d 7
e –4 f 8 g –10 h 11
i –1 j 15 k –6 l 21
6 a 8000 b 8 000 000
7 It was multiplied by three.
8 a 20 b 200
9 It was divided by three.
10 400

Exercise 5-04

1 B
2 A
3 a ×
b Whichever appears first from left to right
c ÷
d Whichever appears first from left to right
4 a × b ÷ c + d ×
e × f ÷ g × h ÷
i ÷
5 a 27 b 9 c 7 d 3
e 8 f 48 g 23 h 8
i 60
6 a 13 b 5 c 15 d 27
e 13 f 24 g 60 h 3
i 15
7 a 32 b 22 c 71 d 3
e 18 f 11 g 175 h 120
i 3 j 23 k 63 l 44
8 $23.50
9 $245
10 a 67 b $837.50

Exercise 5-05

1 B
2 C
3 a 1 b 2 c 3 d 4
4 a $\frac{3}{10}$ b $\frac{3}{100}$ c $\frac{3}{1000}$ d $\frac{3}{10\,000}$
5 a $\frac{6}{10}$ b $\frac{8}{100}$ c $\frac{9}{10\,000}$ d $\frac{5}{1000}$
e $\frac{4}{100}$ f $\frac{56}{100}$ g $\frac{78}{1000}$ h $\frac{342}{1000}$
i $\frac{56}{10\,000}$ j $1\frac{2}{10}$ k $2\frac{34}{100}$ l $4\frac{84}{1000}$
m $3\frac{25}{100}$ n $1\frac{8}{100}$ o $5\frac{8}{1000}$ p $2\frac{762}{10\,000}$
6 a $\frac{3}{5}$ b $\frac{2}{25}$ c $\frac{9}{10\,000}$ d $\frac{1}{200}$
e $\frac{1}{25}$ f $\frac{14}{25}$ g $\frac{39}{500}$ h $\frac{171}{500}$
i $\frac{7}{1250}$ j $1\frac{1}{5}$ k $2\frac{17}{50}$ l $4\frac{21}{250}$
m $3\frac{1}{4}$ n $1\frac{2}{25}$ o $5\frac{1}{125}$ p $2\frac{381}{5000}$
7 a 0.3 b 0.07 c 0.004 d 0.9
e 0.23 f 0.035 g 0.451 h 0.0083
i 1.3 j 2.05 k 4.027 l 4.52
m 9.4 n 0.456 o 0.078 p 3.0023

Exercise 5-06

1 C
2 A
3 a True b False c True d False
4 a < b > c < d >
e < f < g > h >
i < j < k < l >
5 a 0.06, 0.6, 0.632, 0.64, 0.699
b 0.05, 0.505, 0.54, 0.55, 0.589
c 0.9, 0.96, 0.989, 0.99, 0.9999
d 2.011, 2.0111, 2.1, 2.11, 2.111
6 a 0.899, 0.888, 0.83, 0.82, 0.8
b 0.33, 0.311, 0.305, 0.3, 0.003
c 1.696, 1.666, 1.65, 1.6, 1.006
d 3.2292, 3.219, 3.209, 3.2, 3.002
7
0 1
8 a Ebony 43.18, Mitch 43.8, Bobbie 45.4, Ben 48.6, Jiva 48.9, Sam 49.68, Gemma 52.15, Mala 52.65
b Ebony

Exercise 5-07

1 A
2 D
3 a 1.45 b 0.483 c 1.28 d 2.232
e 2.765 f 2.229 g 21.367 h 23.903
i 24.6542

4 a 24.8 b 18.87 c 11.63 d 132.72
e 52.57 f 29.25 g 53.34 h 840.54
i 27.825
5 a 1860.335 b 228.133 c 27.348 d 71.863
6 a $25.18 b $24.82
7 a 9.95 m b 5.05 m
8 600

Practice test 5

1 $\frac{7}{10}$

2

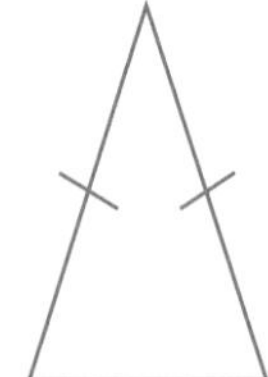

3 4.12
4 15 m^2
5 A shape with 4 straight sides
6 8, 16, 24, 32, 40
7 109
8 4:15 a.m.
9 32
10

11 C
12 B
13 D
14 a 125 b 343 c −64
15 a 3 b −6 c 9
16 a 0 b 100
17 a $\frac{7}{100}$ b $\frac{54}{1000} = \frac{27}{500}$
c $\frac{8}{10\,000} = \frac{1}{1250}$
18 a 0.09 b 0.017 c 0.345
19 0.435, 0.045, 0.04, 0.0042, 0.004
20 148.465
21 7.85

CHAPTER 6

Exercise 6-01

1 D
2 A
3 a 640 b 52 480 c 9.6 d 238 000
e 870 f 567.8 g 8 750 000
h 5 900 000 i 7100
4 a 0.056 b 0.0089 c 0.007 895
d 0.085 e 249.76 f 0.233
g 5.68 h 0.000 76
i 0.846 25
5 point; left; 10 000 000; decimal; right
6 a 1830 b 183 000
7 a 100 000 b 10 000
8 12 500 000

Exercise 6-02

1 B
2 C
3 a 2 b 1 c 2 d 2
e 3
4 a 25.5 b 19 c 87.22 d 0.24
e 21.384 f 72.36 g 43.36 h 1276.8
i 302.852
5 a 13.8 b 135.75 c 18.4 d 221.58
e 44.896
6 a 561.6 b 689.31 c 446.4 d 1607.2
e 1845.2 f 4546.8 g 16 501.8 h 660.44
i 1769.31
7 a $264.75 b $114.75
8 $24.97

Exercise 6-03

1 C
2 A
3 a 2 b 3 c 4 d 4
4 a 0.08 b 0.015 c 0.0048 d 0.0063
5 a 0.84 b 0.228 c 1.96 d 1.826
e 2.288 f 6.0588 g 4.696 h 98.208
i 0.512
6 a 2 b 2 c 3 d 2
e 4 f 2 g 3 h 3
i 2
7 a 7.68 b 42.84 c 12.054 d 491.91
e 5.222 f 191.85 g 138.912 h 10.752
i 460.23
8 a 5.6 m b $72.52
9 a 49.6 m b $301.06 c $481.06

Exercise 6-04

1 C
2 A
3 a 2.4 b 1.2 c 3.2 d 3.1
e 3.02 f 9.71 g 90.71 h 104.01
i 71.03
4 Answers should be: a 9.65 b 80.9
c 57.5 d 651.02 e 84.02 f 80.6
5 a 11.45 b 157.88 c 35.515 d 226.1
e 123.045 f 59.1 g 7.357 h 10.716
i 157.12
6 $4214.26
7 $402.06
8 $81.14

ISBN 9780170350969

Exercise 6-05

1 B
2 C
3 a 1 b 2 c 1 d 2
e 2 f 1
4 a 81 b 11 365 c 610.3 d 3214.1
e 9460.5 f 760.21
5 a 486.8 ÷ 4 = 121.7
b 3755.84 ÷ 5 = 751.168
c 3465.3 ÷ 4 = 1155.1
d 5696.4 ÷ 4 = 1424.1
6 a 60.59 b 1228.1 c 10 403.1 d 910.3
e 7030.2 f 7030.4 g 3160.4 h 432 179
i 1753.18
7 a 1723 pieces b Yes
c $49 105.50
8 a 15 b 0.7 c $28.40

Exercise 6-06

1 A
2 C
3 right; 5; down; up
4 a 4.6 b 8.2 c 12.5 d 18.9
e 7.4 f 24.0 g 124.5 h 7.8
5 a 6.28 b 8.36 c 12.65 d 124.57
e 5.87 f 16.66 g 23.50 h 283.24
6 a 56.78 b 56.784 c 56.7839
d 56.783 93
7 a $207.076 886 2 b $207.08
8 $5687.236

Exercise 6-07

1 A
2 C
3 Medium
4 a 6.5 kg for $10.50 b 1 kg for $17.50
c 4 L for $7.10 d 2.5 kg for $20.65
e 6 for $2.60 f 3 L for $5.25
5 a Bryony b Jayden
c Karina d Philippa
e Caitlin
6 Toptus

Exercise 6-08

1 B
2 D
3 $3498.73
4 630.616
5 267.68
6 29.8
7 a $53.30 b $46.70
8 11 760.8 km
9 a $231.20 b $68.80
10 a 5.6 m b $134.12
11 a 20 m b $122

Language activity

BABE 1 IS ON THE FARM AND BABE 2 IS IN THE CITY.
The film Babe 2 was set in the city.

Practice test 6

1 50
2 1835
3 128 cm
4 24
5 mL
6 A triangle with all sides different.

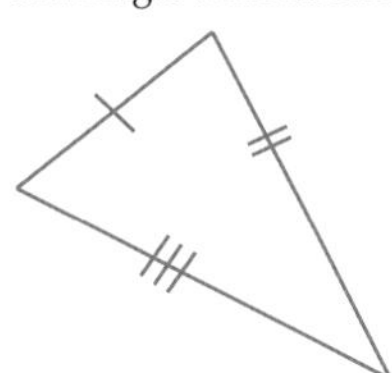

7 6
8 West
9 7:50 p.m.
10 52
11 D
12 D
13 A
14 a 0.063 b 8.72 c 399.33
15 B
16 $14.30
17 a 73.7 b 318.5
18 a 280 b No
19 a 5.7 b 12.4 c 156.3
20 $3.20 per dozen
21 a $8.85 b $41.15

CHAPTER 7

Exercise 7-01

1 B
2 D
3 a $\frac{1}{4}$ b $\frac{1}{2}$ c $\frac{3}{5}$ d $\frac{2}{6}$ or $\frac{1}{3}$
e $\frac{4}{4}$ f $\frac{3}{4}$ g $\frac{1}{2}$ h $\frac{1}{3}$
4 a $\frac{3}{4}$ b $\frac{1}{2}$ c $\frac{2}{5}$ d $\frac{4}{6}$ or $\frac{2}{3}$
e $\frac{0}{4}$ f $\frac{1}{4}$ g $\frac{1}{2}$ h $\frac{2}{3}$
5

6 $\frac{1}{3}$

7 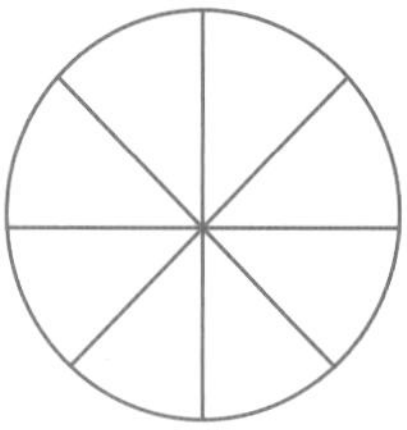

8 $\frac{12}{31}$

Exercise 7-02

1 B
2 C
3 a True b True c False d False
e True f False
4 a 0.5 b 0.25 c 0.625 d 0.6
e 1.4 f 0.875 g 0.8 h 0.3
i 0.125 j 1.5 k 0.1875 l 0.45
m 1.75 n 0.65 o 1.6 p 1.375
5 a $\frac{3}{10}$ b $\frac{1}{4}$ c $\frac{9}{100}$ d $\frac{3}{5}$
e $\frac{7}{25}$ f $\frac{19}{20}$ g $\frac{3}{20}$ h $\frac{7}{10}$

Exercise 7-03

1 D
2 C
3 numerator; denominator; numerator; denominator
4 a $\frac{1}{4}=\frac{3}{12}$ b $\frac{5}{6}=\frac{25}{30}$ c $\frac{2}{3}=\frac{8}{12}$ d $\frac{7}{10}=\frac{56}{80}$
e $\frac{7}{8}=\frac{63}{72}$ f $\frac{3}{4}=\frac{75}{100}$ g $\frac{20}{24}=\frac{5}{6}$ h $\frac{15}{40}=\frac{3}{8}$
i $\frac{18}{32}=\frac{9}{16}$
5 a No
b Jared ate 24, Prem ate 24 and Grace ate 32 chocolates.
6 a False b True c True d False
e True f True
7 a Yes b 80 beads were left.

Exercise 7-04

1 B
2 D
3 divide; denominator; simplest
4 a $\frac{7}{9}$ b $\frac{7}{8}$ c $\frac{5}{12}$ d $\frac{3}{4}$
e $\frac{2}{3}$ f $\frac{4}{7}$ g $\frac{4}{5}$ h $\frac{5}{6}$
5 a True b False c True d True
e False f True
6 $\frac{4}{5}$
7 a $\frac{37}{50}$ b $\frac{7}{10}$ c $\frac{5}{6}$
8 a $\frac{1}{3}$ b $\frac{9}{20}$ c $\frac{5}{12}$ d $\frac{1}{4}$
e $\frac{4}{5}$ f $\frac{1}{13}$

Exercise 7-05

1 C
2 A
3 a M b P c M d M
e P f I g P h P
i I j P k M l I
4 a An improper fraction has the numerator larger than or equal to the denominator.
b A mixed numeral is made up of a whole number and a fraction.
5 a $\frac{5}{3}$ b $\frac{5}{2}$ c $\frac{5}{4}$
6 a $1\frac{2}{3}$ b $2\frac{1}{2}$ c $1\frac{1}{4}$
7 a $1\frac{1}{5}$ b $2\frac{2}{3}$ c $1\frac{1}{4}$ d $2\frac{1}{4}$
e $1\frac{2}{3}$ f $1\frac{1}{5}$ g $1\frac{1}{9}$ h $3\frac{3}{4}$
8 a $\frac{10}{3}$ b $\frac{11}{4}$ c $\frac{19}{4}$ d $\frac{15}{8}$
e $\frac{20}{3}$ f $\frac{29}{6}$ g $\frac{27}{5}$ h $\frac{59}{8}$

Exercise 7-06

1 B
2 A
3 a $\frac{7}{10}$ b $\frac{5}{3}$ c $\frac{7}{8}$ d $\frac{2}{3}$
e $2\frac{1}{5}$ f $3\frac{4}{7}$
4 a $\frac{3}{5}$ b $\frac{9}{8}$ c $\frac{1}{4}$ d $\frac{1}{6}$
e $4\frac{2}{3}$ f $1\frac{1}{5}$
5 C
6 $\frac{1}{4}, \frac{2}{3}, \frac{5}{6}, \frac{11}{12}$
7 $\frac{5}{6}, \frac{3}{4}, \frac{2}{3}, \frac{1}{4}, \frac{1}{12}$
8 a 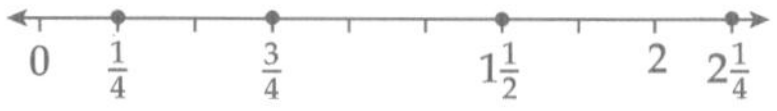

b 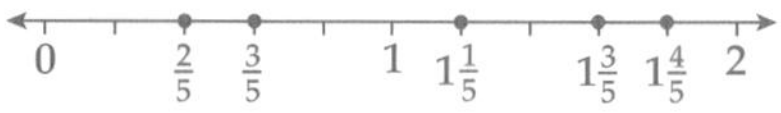

c 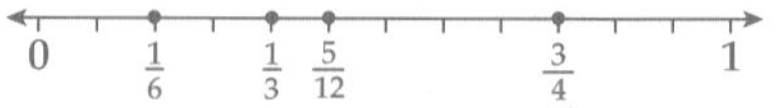

9 a 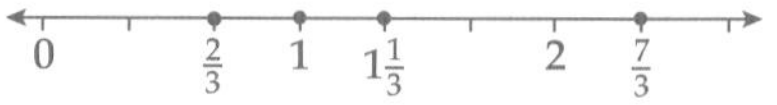

b 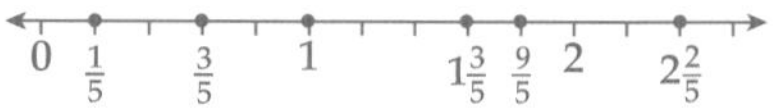

c 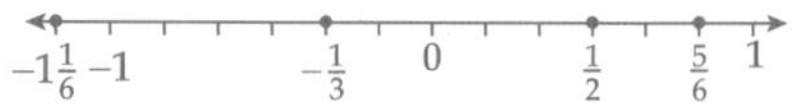

d 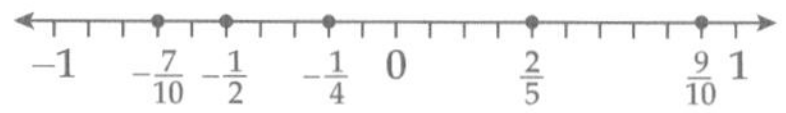

Exercise 7-07

1 C

2 B

3 a $\frac{2}{3}$ b $\frac{9}{10}$ c 1 d $\frac{4}{3}$ or $1\frac{1}{3}$

e $\frac{2}{5}$ f $\frac{1}{2}$ g $\frac{2}{5}$ h $\frac{1}{4}$

4 a $\frac{7}{10}$ b $\frac{11}{15}$ c $\frac{13}{20}$ d $\frac{14}{15}$

5 a $1\frac{7}{20}$ b $1\frac{7}{12}$ c $1\frac{7}{24}$ d $1\frac{22}{45}$

e $\frac{2}{9}$ f $\frac{1}{4}$ g $\frac{8}{15}$ h $\frac{3}{8}$

i $\frac{31}{40}$ j $\frac{19}{36}$ k $1\frac{11}{60}$ l $\frac{3}{28}$

6 a Quan 4 pieces, Reece 3 pieces, Emad 2 pieces.

b 3 pieces c $\frac{1}{4}$

Exercise 7-08

1 C

2 D

3 a $2\frac{1}{2}$ b $3\frac{3}{5}$ c $5\frac{3}{7}$ d $5\frac{3}{5}$

e $1\frac{1}{7}$ f $2\frac{1}{6}$ g $2\frac{1}{11}$ h $2\frac{3}{10}$

i $7\frac{4}{5}$ j $6\frac{9}{10}$

4 a $2\frac{5}{6}$ b $3\frac{1}{6}$ c $2\frac{3}{4}$ d $3\frac{7}{10}$

e $4\frac{1}{4}$ f $3\frac{7}{12}$ g $3\frac{11}{12}$ h $4\frac{5}{12}$

5 a $1\frac{5}{12}$ b $2\frac{1}{4}$ c $1\frac{3}{8}$ d $2\frac{1}{6}$

e $1\frac{1}{6}$ f $4\frac{2}{15}$ g $4\frac{1}{14}$ h $4\frac{1}{4}$

6 $4\frac{1}{4}$ chickens

Language activity

FRACTION; MIXED; NUMERAL; NUMERATOR; IMPROPER; COMMON; DENOMINATOR; PROPER

Practice test 7

1 Obtuse

2 0.018

3 –5, –2, 0, 1, 7

4 28

5 9410

6 212

7 22

8 4.63

9

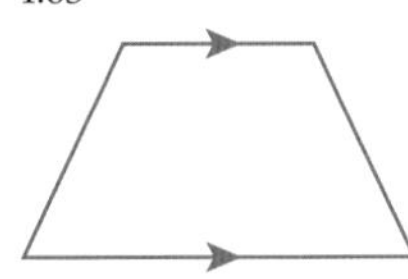

10 2, 3, 5, 7.

11 D

12 A

13 C

14 a True b False c False

15 a $\frac{2}{3}$ b $\frac{3}{4}$ c $\frac{11}{20}$

16 a $\frac{7}{12}$ b $\frac{3}{25}$

17 a $1\frac{5}{8}$ b $2\frac{2}{5}$

18 a $\frac{5}{3}$ b $\frac{15}{4}$

19 $\frac{3}{4}, \frac{2}{3}, \frac{5}{12}, \frac{2}{5}$

20 a $1\frac{7}{40}$ b $\frac{7}{12}$

21 a $5\frac{4}{15}$ b $2\frac{7}{8}$

CHAPTER 8

Exercise 8-01

1 C

2 A

3 a 7 b 7 c 20 d 8

e 5 f 22 g 3 h 30

i 15 j 50 k 75 l 12

4 a 40 min b \$15 c 40 km d \$60

e 24 m f 200 s g 12 books

h \$75 i 22 hours

j 56 mL k 800 g l \$36

5 a 2000 m b 1600 m c 576 m

6 a 100 pages b 64 pages

7 14

Exercise 8-02

1 B

2 A

3 a True b False c True d True

4 a $\frac{1}{2}$ b $\frac{2}{3}$ c $\frac{25}{48}$ d $\frac{1}{3}$

e $\frac{3}{14}$ f $\frac{3}{4}$ g $\frac{3}{5}$ h $\frac{2}{3}$

i 1 j $\frac{4}{15}$ k $\frac{1}{2}$ l 2

5 B

6 a $\frac{1}{4}+\frac{1}{4}+\frac{1}{4}=3\times\frac{1}{4}$
$=\frac{3}{4}$

b $\frac{3}{5}+\frac{3}{5}=2\times\frac{3}{5}$
$=\frac{6}{5}$
$=1\frac{1}{5}$

c $\frac{3}{8}+\frac{3}{8}+\frac{3}{8}+\frac{3}{8}+\frac{3}{8}=5\times\frac{3}{8}$
$=\frac{15}{8}$
$=1\frac{7}{8}$

d $5\times\frac{1}{7}=\frac{5}{7}$

e $7\times\frac{3}{20}=\frac{21}{20}$
$=1\frac{1}{10}$

f $8\times\frac{3}{5}=\frac{24}{5}$
$=4\frac{4}{5}$

Exercise 8-03

1 C
2 C
3 upside down; improper; upside down
4 a $1\frac{1}{2}$ b 5 c $1\frac{3}{5}$ d 10
e $\frac{1}{7}$ f $1\frac{1}{4}$ g $1\frac{1}{5}$ h $\frac{1}{9}$
5 a $\frac{3}{4}$ b $\frac{4}{9}$ c $\frac{3}{11}$ d $\frac{4}{7}$
e $\frac{3}{13}$ f $\frac{8}{29}$ g $\frac{5}{27}$ h $\frac{4}{25}$
i $\frac{5}{14}$

Exercise 8-04

1 B
2 C
3 multiply; reciprocal
4 a False b True c True d False
5 a $\frac{3}{8}\times\frac{4}{1}$ b $\frac{4}{7}\times\frac{14}{2}$ c $\frac{5}{8}\times\frac{16}{3}$ d $\frac{4}{5}\times\frac{3}{12}$
e $\frac{7}{8}\times\frac{24}{3}$ f $\frac{2}{3}\times\frac{9}{5}$ g $\frac{5}{7}\times\frac{21}{25}$ h $\frac{5}{8}\times\frac{4}{20}$
i $\frac{5}{7}\times\frac{8}{15}$ j $\frac{8}{3}\times\frac{9}{4}$ k $\frac{6}{5}\times\frac{10}{12}$ l $\frac{6}{9}\times\frac{12}{18}$
6 a $1\frac{1}{2}$ b 4 c $3\frac{1}{3}$ d $\frac{1}{5}$
e 7 f $1\frac{1}{5}$ g $\frac{3}{5}$ h $\frac{1}{8}$
i $\frac{8}{21}$ j 6 k 1 l $\frac{4}{9}$
7 a 12 b 10 c 21 d 15
e $\frac{1}{7}$ f $\frac{5}{49}$ g $\frac{1}{12}$ h $\frac{5}{12}$
8 B
9 Aisha received $9600 and the other daughters received $4200 each.
10 $\frac{11}{10}=1\frac{1}{10}$

Exercise 8-05

1 D
2 A
3 improper; numerators; denominators
4 a $1\frac{4}{5}$ b 4 c $1\frac{2}{3}$ d $\frac{1}{3}$
e 1 f $1\frac{3}{5}$ g $\frac{8}{21}$ h 3
i $2\frac{2}{7}$ j 12
5 a 2 b $1\frac{2}{3}$ c $1\frac{1}{2}$ d $1\frac{1}{3}$
e $6\frac{1}{4}$ f 2 g 3 h 9
i 6 j $4\frac{1}{2}$
6 17 outfits.

Exercise 8-06

1 B
2 A
3 improper; reciprocal
4 a 1 b $\frac{1}{2}$ c $\frac{1}{2}$ d $\frac{1}{3}$
e $\frac{12}{25}$ f $\frac{63}{80}$ g $4\frac{3}{8}$ h $\frac{22}{35}$
i $\frac{20}{81}$ j $\frac{9}{25}$
5 a 2 b $1\frac{1}{2}$ c 2 d $\frac{4}{5}$
e 2 f $1\frac{7}{20}$ g $1\frac{13}{36}$ h $2\frac{1}{4}$
i 12 j $\frac{18}{25}$
6 11

Practice test 8

1 A triangle with 3 equal sides
2 6
3 1, 2, 4, 5, 10, 20
4 0.032
5 $10
6 Yes
7 Yes
8 360°
9 60
10

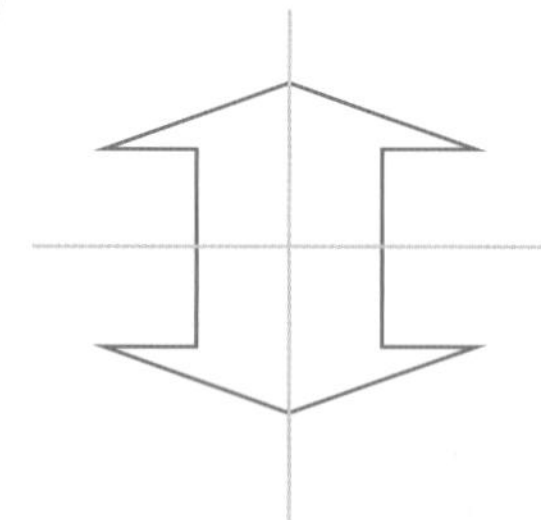

11 D
12 C
13 C
14 a $1\frac{1}{7}$ b 5 c $\frac{4}{9}$
15 a False b True
16 a $\frac{3}{20}$ b $1\frac{2}{3}$
17 A
18 a $4\frac{3}{8}$ b $14\frac{2}{3}$
19 a True b False
20 a $\frac{27}{64}$ b $1\frac{1}{3}$

CHAPTER 9

Exercise 9-01

1 C
2 C
3 a 8 b 4 c a d 6
e v f k
4 a 43 b 84 c 72 d 30
e 66 f 42 g 90 h 121

 ISBN 9780170350969

5 a True b False c True d True
e False f False
6 a 5; 10; 90 b 25; 100; 1200
7 a 70 b 100 c 600 d 700
e 270 f 800 g 60 h 60
8 a 10 b 7 c 4 d 10
9 a 176 b 108 c 120 d 240
e 153 f 104 g 198 h 288

Exercise 9-02

1 C
2 D
3 a, c, m, E, s, g
4 a $2r$ b e c $5n$ d $3m$
e $2st$ f c^2 g $4b$ h $6v$
5 a $m + 0 = m$ b $m \div 1 = m$
c $m - m = 0$ d $m + m + m = 3m$
6 a $3e$ b $10v$ c $2a + 2b$ d $12rs$
e s f $6ab$ g $2d$ h w
i $18a$ j $12ac$ k $4e - b$ l $3t + 4s$
m $4w$ n $3fg$ o $a + b$ p $42c^2$
7 a False b False c True d True
8 a $6 \times b$ b $a + 2 \times b$
c $m \times m$ d $m + 3 \times n$
e $5 \times m \times n$ f $r \times r \times r$
g $4 \times d$ h $6 \times a \times b \times c$
9 a $4 + 2n$ b $7ab - a$ c $12 - 3s$ d $a^2 - b^2$
e $2a - 3b$ f $20 + 3mn$
g $8 + 7c$ h $4 + d - d^2$

Exercise 9-03

1 C
2 B
3 a $-$ b $+$ c $\times$ d $+$
e $\div$ f $\times$ g $\times$ h $-$
i $\times$
4 a $4n$ b $n - 3$ c $2n$ d $n + 7$
e $3n - 9$ f $\frac{n}{4}$ g $n - 6$ h $\frac{n}{5}$
i $8 - n$ j n^2
5 a $2n + 3$ b $3j - 8$ c $a + 5$
d $7b$ e $2(s + t)$ f $10 - 2c$
g $3(a - 6)$ h $d + (a + 4)$
i $a - (b + c)$ j $b^2 - ac$
6 a the sum of a, b and c
b the product of 3 and b
c decrease 6 by g
d the quotient of r and 4
e twice a less b
f the quotient of triple b and c
g 5 times the sum of a and b
h the sum of 4 times b and 6
i 7 times v less 4
7 a $4(a + b + c)$ b $abc - 8$

Exercise 9-04

1 A
2 D
3 a 2 b 48 c –12 d 32
e –48 f 32 g 28 h –16
i 26 j –16 k 20 l 9
4 a –48 b 18 c 42 d –2
e –60 f 27 g 7 h 0
i 6 j 1 k 22 l –3
5 a 800 b 0 c 78 d 100
e 475 f –12 g 76 h 25
i 53
6 Should be -4×-4 not $-4 - 4$. Correct answer = –128.

Exercise 9-05

1 C
2 A
3 a $x = 7$ b $y = 13$ c $m = 7$ d $x = 24$
e $a = 21$ f $w = 9$ g $n = 13$ h $b = 35$
i $t = 24$
4 a $x = 5$ b $a = 3$ c $m = 9$ d $x = 10$
e $w = 5$ f $a = 7$ g $x = -3$ h $a = 4$
i $w = -15$ j $x = 24$ k $a = -3$ l $w = 4$
5 a $2n - 8 = 34$; $n = 21$ b $3n + 12 = 39$; $n = 9$
c $\frac{n}{2} + 15 = 24$; $n = 18$

Exercise 9-06

1 D
2 C
3 inverse; same; balanced; underneath
4 a subtraction b division
c multiplication d addition
5 a 3; $x = 5$ b 9; $m = 15$
c 3; $v = 12$ d 4; 4; $x = 20$
6 a $w = 3$ b $m = 16$ c $b = 6$ d $v = 50$
e $n = 9$ f $g = 12$ g $w = 42$ h $n = 45$
i $x = 88$ j $r = 63$ k $n = 11$ l $y = -16$
7 a $n - 13 = 54$; $n = 67$ b $\frac{n}{8} = 17$; $n = 136$

Exercise 9-07

1 C
2 D
3 a –2 b +1 c –3 d +2
e –3 f +6 g –5 h +3
i –6
4 a $a = 2$ b $g = 3$ c $m = 2$ d $v = 6$
e $x = 3$ f $c = 5$ g $a = 4$ h $e = 3$
i $v = 9$
5 a –1 b +2 c –3 d +2
e –3 f –6 g +4 h +8
i –4

6 a $x = 8$ b $m = 15$ c $n = 12$ d $s = 56$
e $w = 24$ f $x = 10$ g $x = 84$ h $s = 30$
i $m = 18$

7 a $3n + 4 = 49; n = 15$ b $\frac{n}{2} - 5 = 16; n = 42$

Language activity

Across: EXPRESSION
Down (from left to right): PATTERN; ALGEBRA; CONSTANT; INDEX; PRONUMERAL

Practice test 9

1 14
2 rectangle
3 13
4 $25
5 10
6 180
7 58
8 80
9 48
10 0.07, 0.7, 0.702, 0.75
11 D
12 D
13 a $3ab$ b y^2
14 $3(w + 5)$
15 a 14 b −58
16 5
17 a $x = 5$ b $a = 4$
18 a $x = 17$ b $x = 96$
19 a $w = 8$ b $x = 12$
20 Line 1

CHAPTER 10

Exercise 10-01

1 B
2 C
3 a quadrilateral b triangle
c quadrilateral (or trapezium)
d hexagon
e pentagon

4 a

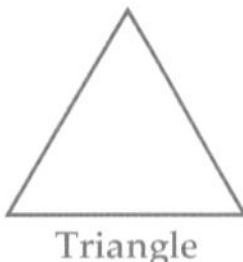

b

c

Pentagon

d

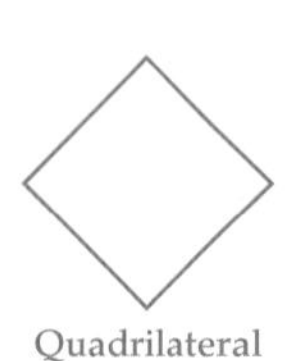

5 a octagon b decagon
6 pentagon
7 quadrilaterals (or rectangles, parallelograms), triangles
8 a triangle: roundabout
b quadrilateral: no entry
c octagon: stop sign
9 a regular pentagon
b irregular quadrilateral (or kite)
c irregular octagon
d irregular hexagon
e regular hexagon

Exercise 10-02

1 B
2 A
3

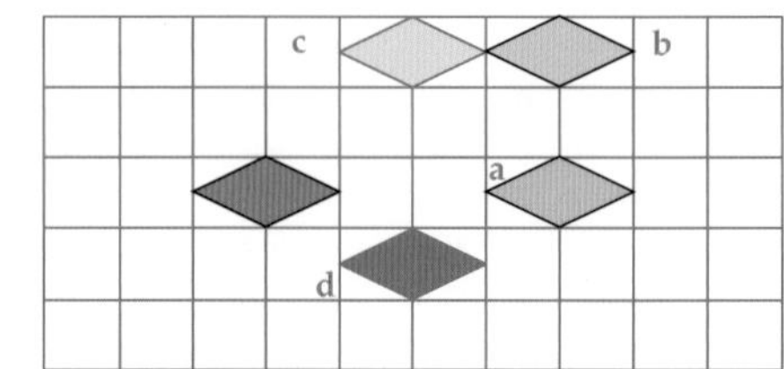

4

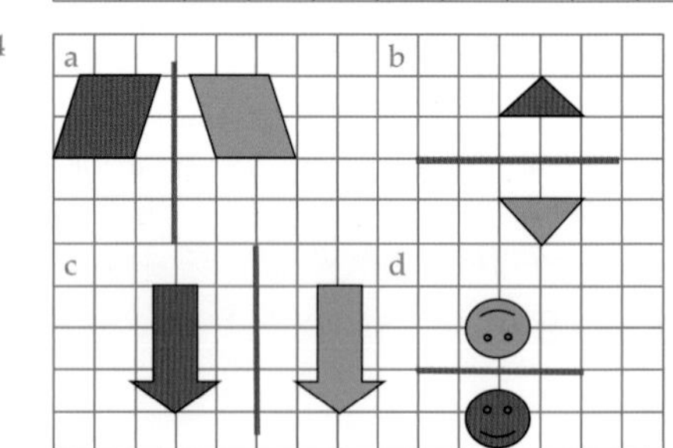

5 a

A
B
O

b

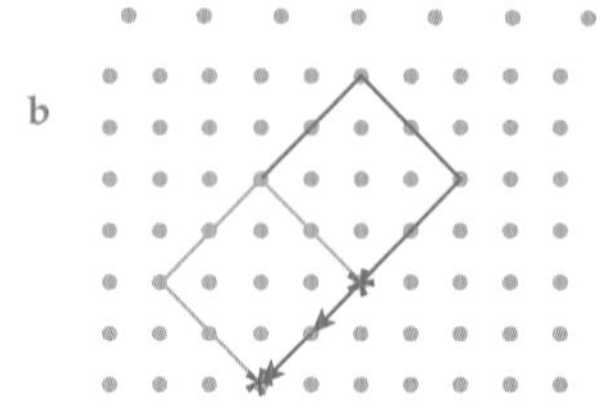

c

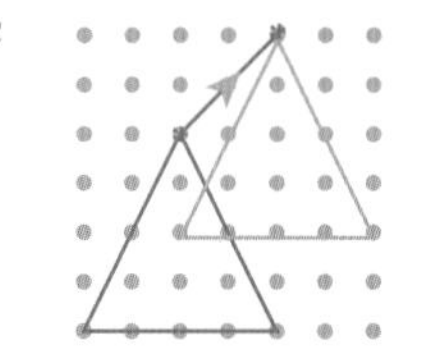

ISBN 9780170350969

d
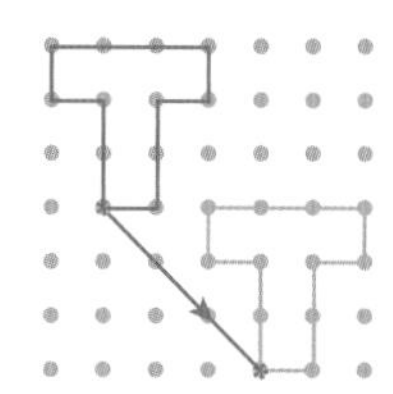

6 a
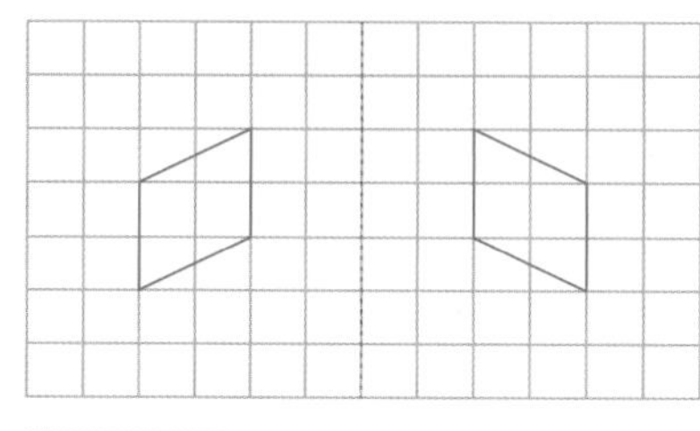

b
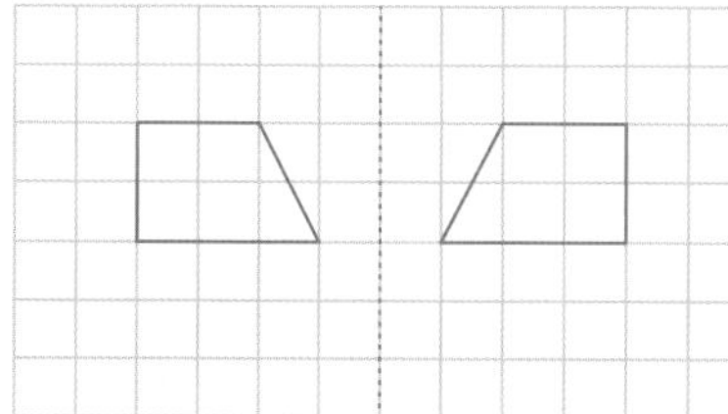

c
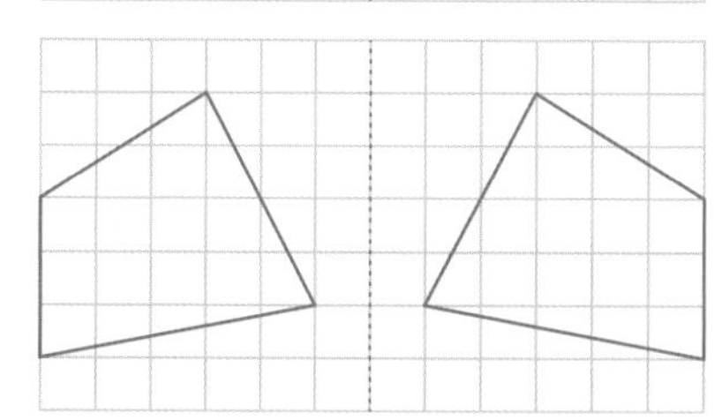

7 a reflection b translation

Exercise 10-03

1 D
2 A
3
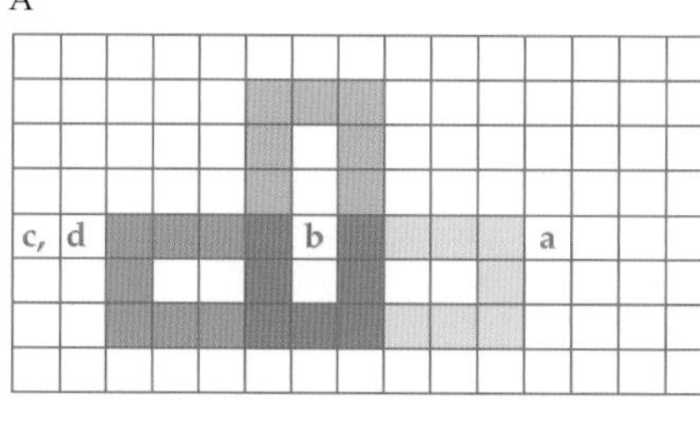

4 a
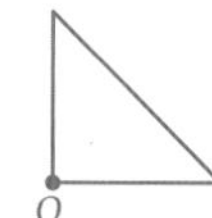

b
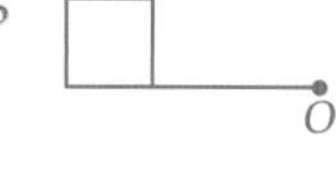

c
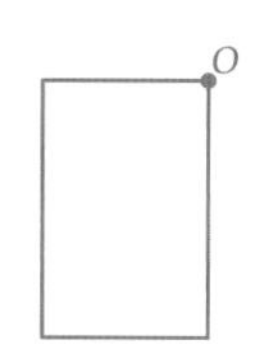

d
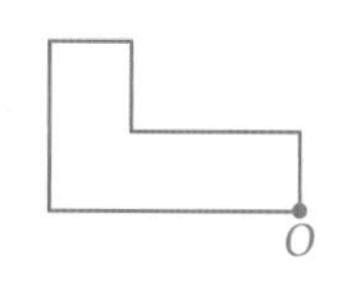

e
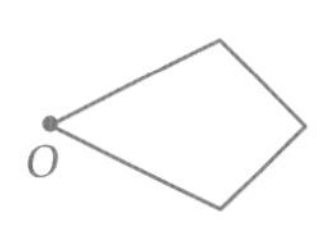

f
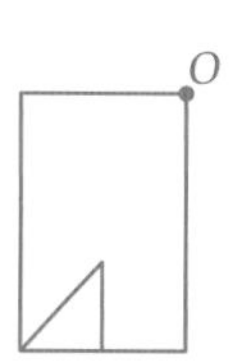

Exercise 10-04

1 A
2 D
3
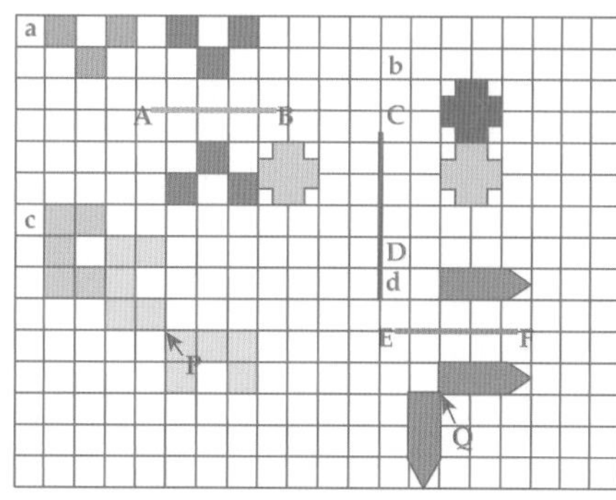

4 a i Reflection in line *AB*
ii Translation 3 units right and 1 down.
b i Rotation about *O* through 90° clockwise
ii Reflection in line *CD*
iii Translation 4 units right and 1 unit up
c i Translation 4 units right and 1 unit up
ii Rotation 90° clockwise
iii Reflection in line *EF*

5
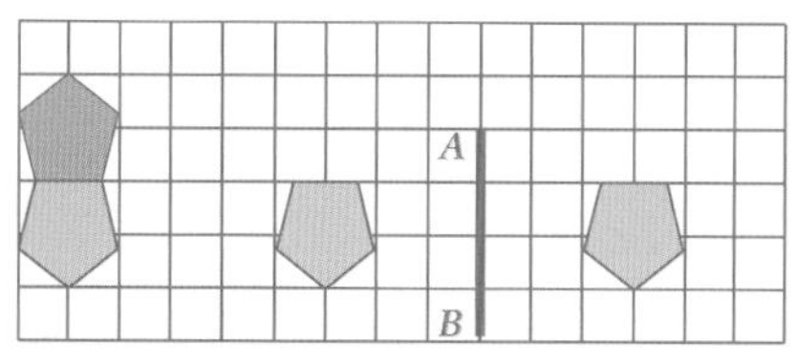

Exercise 10-05

1 D
2 C
3 a
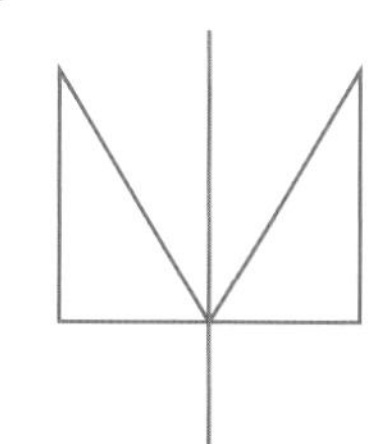

b

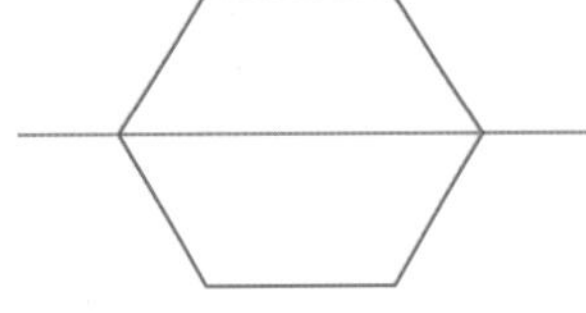

c

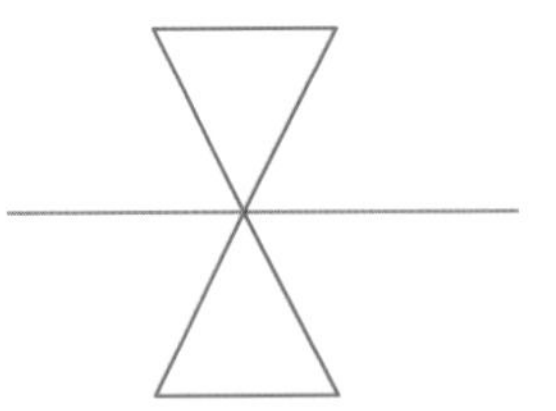

d

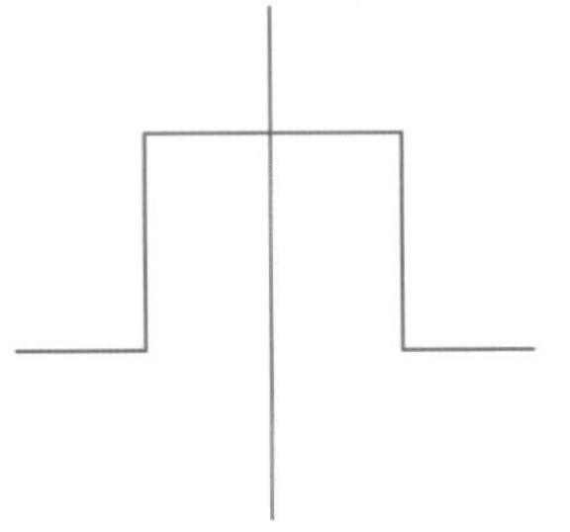

e

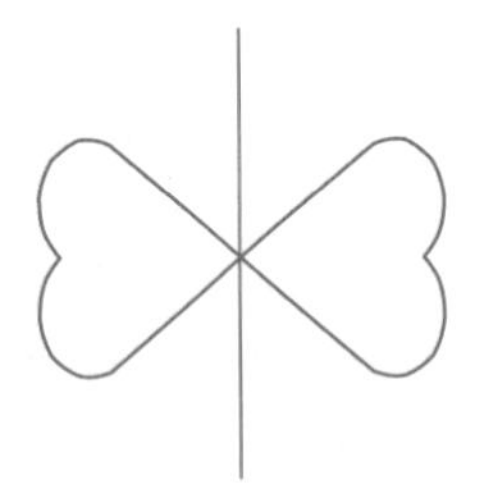

f

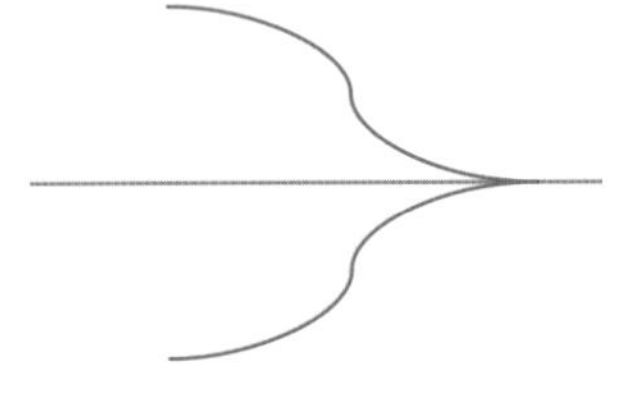

4 a

b

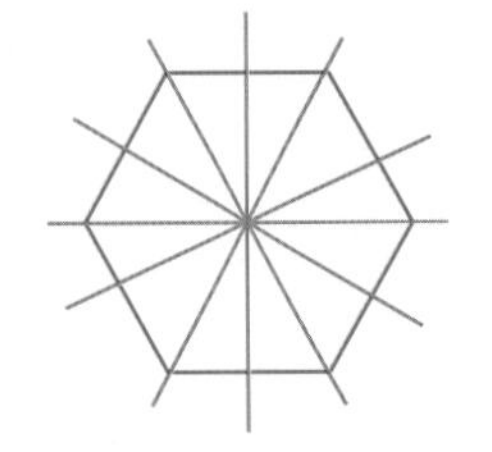

c

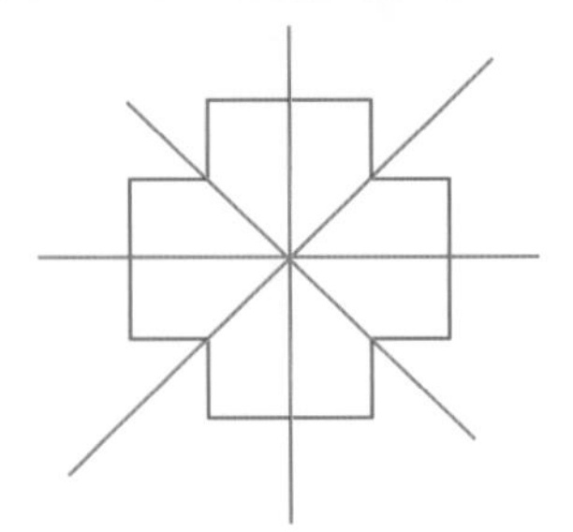

d

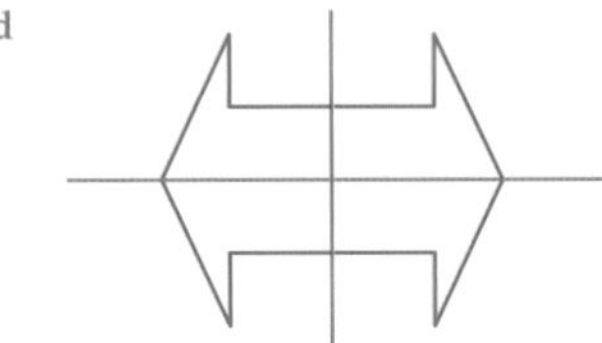

5 A B C D E F
G H I J K L
M N O infinite P Q R
S T U V W X
Y Z

Exercise 10-06

1 D
2 C
3 a Yes b Yes c Yes d No
e Yes f No g Yes h Yes
i Yes
4 a i 120° ii 3 b i 180° ii 2
c i 90° ii 4 d No
e i 90° ii 4 f No
g i 72° ii 5
h i any size rotation ii infinite
i i 60° ii 6
5 a 90° b 180° c 180° d 360°
e 60° f 90°
6 a 4 b 2 c 2 d 1
e 6 f 4
7 0, 1, 8

Exercise 10-07

1 B
2 D
3 a prism b pyramid
c neither d prism
e neither f pyramid
g prism h prism i neither
4 a rectangular prism b rectangular pyramid
d triangular prism f triangular pyramid
g hexagonal prism h rectangular prism

ISBN 9780170350969

5 a rectangle b hexagon
  c pentagon

6 a rectangular prism b hexagonal prism
  c pentagonal prism

7 A prism has identical cross-sections while a pyramid has edges that start at its base and meet at a point.

Language activity

1 One extra side; triangle, quadrilateral, pentagon
2 One extra side on the base; triangular prism, rectangular prism, pentagonal prism
3 One less side on the base; hexagonal prism, pentagonal prism, rectangular prism
4 One extra side on the base; rectangular pyramid, pentagonal pyramid, hexagonal pyramid

Practice test **10**

1 34
2 head or tail
3 $\frac{1}{2}$
4 108
5 0
6 90
7 10:20 p.m.
8 6.85
9 $\frac{4}{5}$
10 Teacher to check, 2×3^3
11 A
12 B
13 A
14 An image formed by flipping a shape over a line.
15 B
16

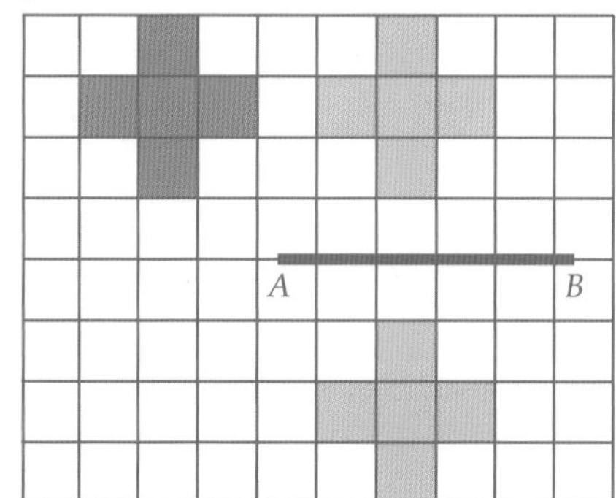

17 a

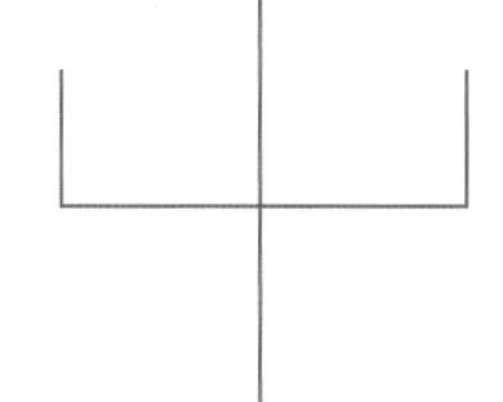

b

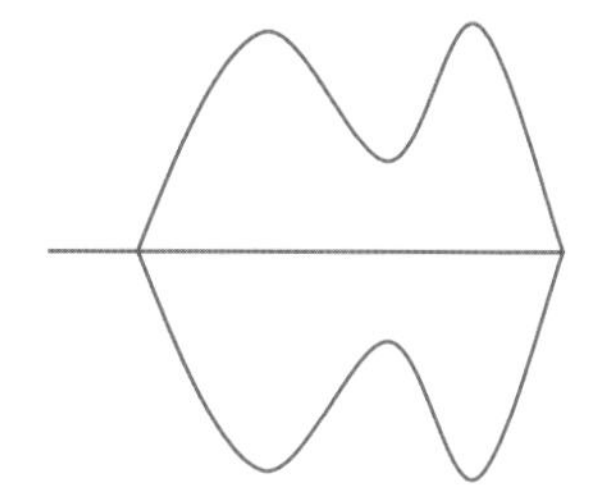

18 a

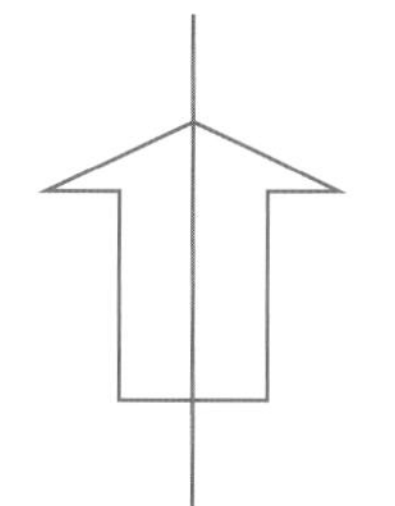

b

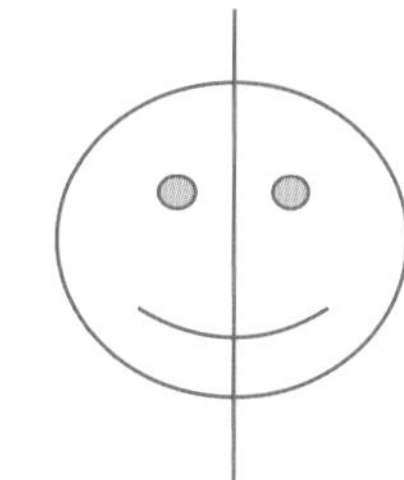

19 a Yes, 2 b Yes, 4

20 a rectangular prism
  b rectangular pyramid
  c hexagonal pyramid

CHAPTER 11

Exercise **11-01**

1 D
2 B
3 a isosceles b scalene
  c scalene
4 a

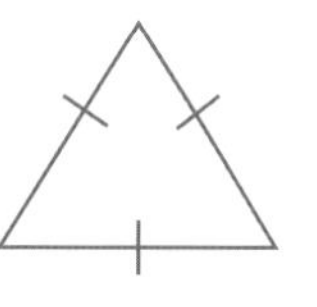

b

c

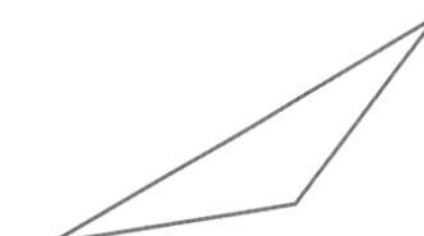

5 a ΔPQR b ΔXYZ c ΔMNP

6 a scalene, right-angled
  b isosceles, acute-angled
  c scalene, obtuse-angled

7 a

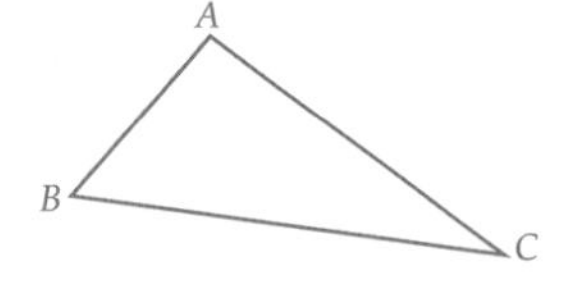

b

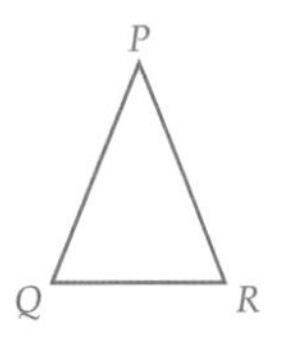

c

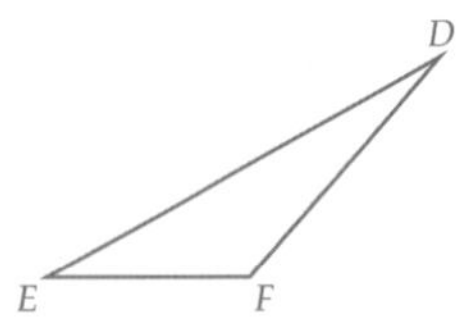

d

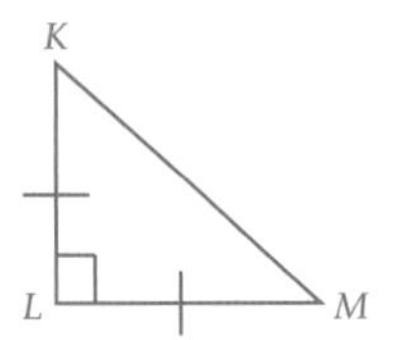

e

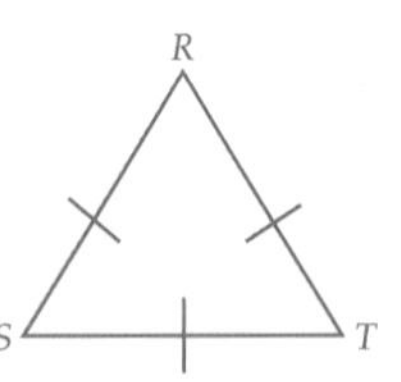

8 a False b False c True
d True e False f False

9 40: scalene, isosceles, equilateral, acute-angled

Exercise 11-02

1 C

2 D

3 a $n = 86$ b $b = 42$ c $a = 32$
d $x = 79$ e $y = 34$ f $c = 28$

4 a 60° b

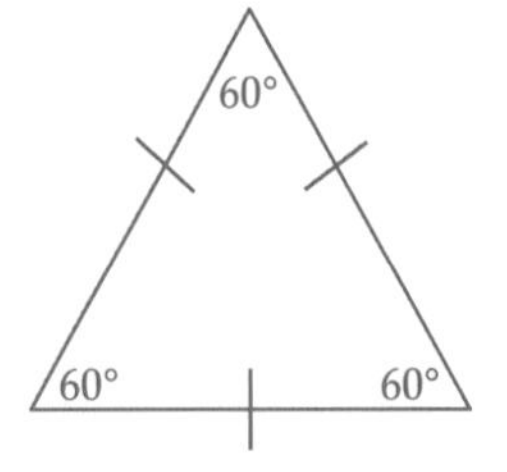

5 a $a = 73$ b $b = 67$ c $c = 63$

6

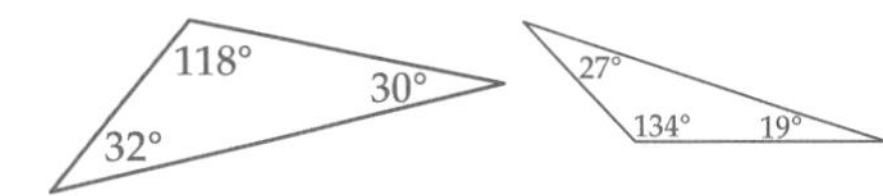

7

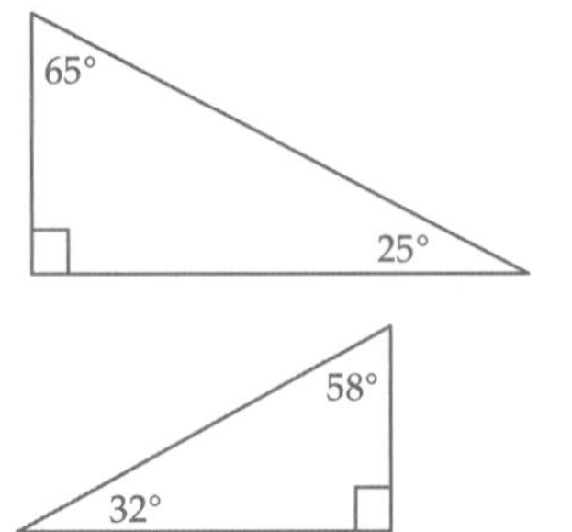

8 a

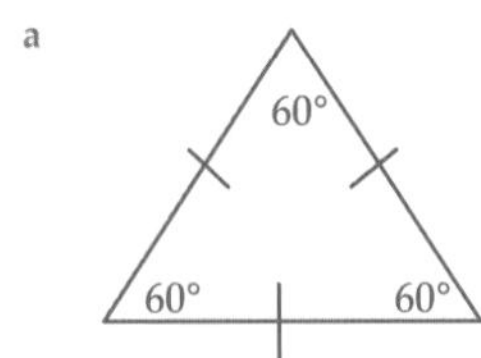

b

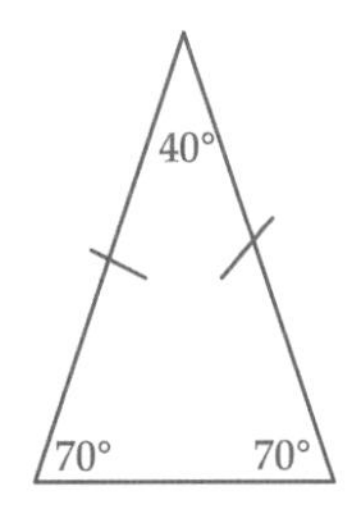

Exercise 11-03

1 D

2 A

3 a $\angle BAC$, $\angle ABC$, $\angle ACB$ b $\angle BCD$
c $\angle BCD = \angle BAC + \angle ABC$

4 a $x = 80$ b $y = 99$ c $q = 148$ d $n = 114$

5

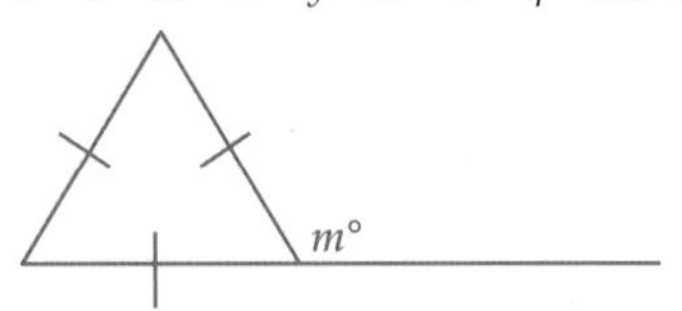

$m = 60 + 60$
$= 120$

Yes it will always be the same, as each angle in an equilateral triangle is 60°.

6 a $v = 70$ b $w = 125$

Exercise 11-04

1 C

2 A

ISBN 9780170350969

3 a

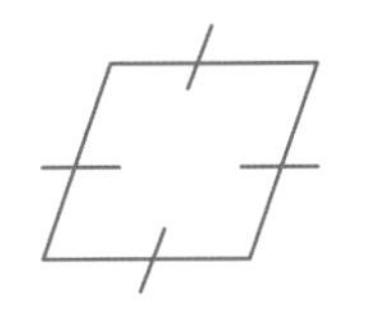

b

c

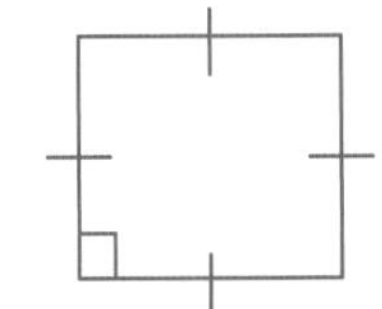

4 a and c

5 a rhombus b rectangle

c trapezium

d irregular quadrilateral

e parallelogram f square

6 a a quadrilateral with one pair of opposite sides parallel

b a quadrilateral with 4 sides equal and opposite sides parallel

c a quadrilateral with opposite sides both equal and parallel and all angles 90°

7 a True b False c True

d False e True f False

8

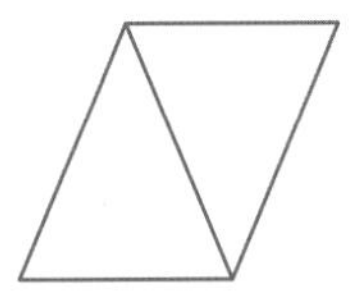

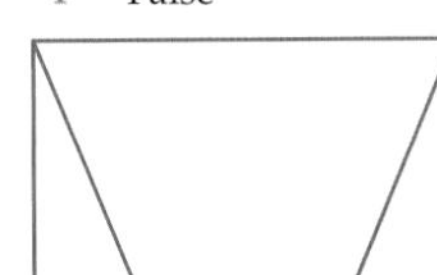

Exercise 11-05

1 B

2 A

3 a $n = 98$ b $b = 46$ c $a = 39$ d $x = 121$

e $y = 49$ f $c = 68.5$

4 a $m = 90$ b $n = 98$ c $a = 90$

5 82°

6

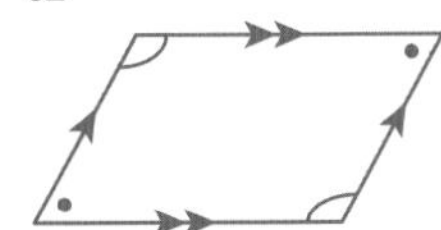

7 a both 105°

b 128° opposite the given angle, and the other two are both 52°

Exercise 11-06

1 C

2 B

3 a True b False c True d False

e False

4 a A square has all angles 90° and a rhombus does not.

b A parallelogram has opposite sides parallel and a quadrilateral does not.

c A rhombus has all 4 sides equal and a parallelogram does not.

d A parallelogram has both pairs of opposite sides parallel and a trapezium does not.

5

Quadrilateral	Angles 90°	Equal sides	Equal diagonals
parallelogram	No	No	No
rhombus	No	Yes	No
rectangle	Yes	No	Yes
kite	No	No	No
square	Yes	Yes	Yes
trapezium	No	No	No

6 a kite, rhombus, square

b

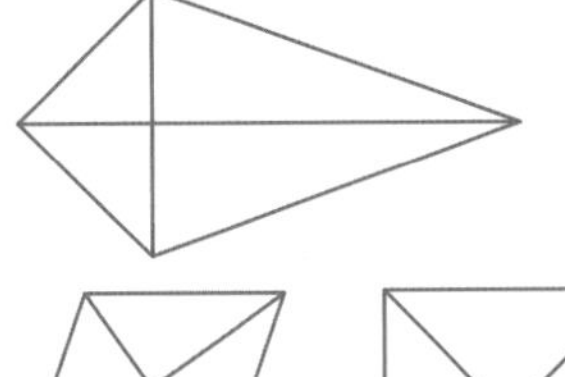

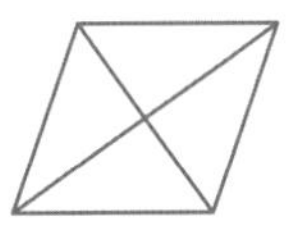

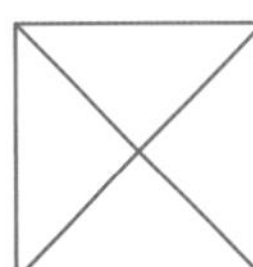

7

Quadrilateral	Sides	Angles	Diagonals
Parallelogram	Opposite sides equal and parallel	Opposite angles are equal	Diagonals bisect each other
Trapezium	One pair of opposite sides are parallel	All angles different	Diagonals are not equal
Rhombus	All sides equal	Opposite angles are equal	Diagonals bisect each other at right angles and bisect the angles of the rhombus
Rectangle	Opposite sides are equal	All angles are 90°	Diagonals are equal
Square	All sides equal in length	All angles are 90°	Diagonals bisect each other at right angles and bisect the angles of the square Diagonals are equal

Language activity

WHEN FINDING ANGLES IN QUADRILATERALS AND TRIANGLES, REMEMBER ANGLE SUM AND THE ANSWER WILL COME.

ISBN 9780170350969

Practice test 11

1 −14
2 0.2
3 6 cm
4 9
5 4th
6 $18ay$
7 125°
8 29.1
9 11:15 a.m.
10 $\frac{2}{21}$
11 A
12 C
13 C
14 C
15 $n = 114$
16 a rectangle b trapezium
 c square d rhombus
17 rectangle and square
18 $c = 118$
19 a True
 b False
 c True

CHAPTER 12

Exercise 12-01

1 C
2 B
3 a centimetres b millimetres
 c metres d litres
 e kilometres f millilitres
 g tonnes h milligrams
 i metres j kilograms
4

Millimetres	Centimetres	Metres	Kilometres
3000	300	3	0.003
55 000	5500	55	0.055
250 000	25 000	250	0.25
80 000 000	8 000 000	80 000	80
76 800	7680	76.8	0.0768
8200	820	8.2	0.0082
750 000	75 000	750	0.75

5 a 7.56 b 12.5 c 0.0268 d 850
 e 59 400 f 420 g 5.678 h 760 000
6 a 6 b 0.15 c 2.8 d 4500
 e 0.0035 f 7250 g 8000 h 4750
 i 9600 j 0.065 k 84 000 l 2800
7 4882.68 m
8 a True b False c False d True
 e False f False

Exercise 12-02

1 D
2 A
3 a mm b cm c km d cm
 e cm f km g cm h m
 i mm j km
4 a 32 mm b 16 mm c 42 mm d 29 mm
 e 20 mm f 53 mm
5 a 3 cm b 2 cm c 4 cm d 3 cm
 e 2 cm f 5 cm
6 a 50 km b 55 km c 6 km/h
 d Mt View to Koolgarlie
 e 14.7 km/h

Exercise 12-03

1 B
2 C
3 a 20 cm b 36 cm c 15 cm
 d 28 cm e 26 cm f 31 cm
4 a 28 cm b 25 m c 23 cm
5 a 24 m b 12.6 cm c 17 mm
 d 18.2 m e 16.2 cm f 13.6 mm

Exercise 12-04

1 B
2 D
3 a 36 mm b 30 cm c 30 m
 d 28 cm e 24 cm f 60 mm
4 a 195 mm b 33.2 m c 214 mm
 d 29.2 m e 40 m f 39 cm

Exercise 12-05

1 D
2 B
3 a 180 b 300 c 1500
4 a 48 b 90 c 2 d 21
 e 4 f 14 g 4 h 2
 i 100 j 366
5 a 120 b 210 c 77 d 45
6 Sunday, Monday, Tuesday, Wednesday, Thursday, Friday, Saturday
7 January, February, March, April, May, June, July, August, September, October, November, December
8 a 30 b 5000 c 450 d 56
9 10
10 10

ISBN 9780170350969

Exercise 12-06

1 C
2 D
3 a 0200 b 1515 c 1730 d 0550
e 0720 f 2130 g 0755 h 2315
4 a 0520 b 1815 c 0330 d 1925
e 1115 f 2108 g 0122 h 1220
i 2236 j 0545 k 1756 l 0028
5 a 3:20 a.m. b 2:50 p.m.
c 10:15 p.m. d 7:04 a.m.
e 9:52 p.m. f 4:32 a.m.
g 7:50 p.m. h 11:26 a.m.
i 11:45 p.m. j 12:36 a.m.
k 6:52 a.m. l 8:08 p.m.
6 a morning b 1:50 p.m.
c 5:00 a.m. d 12:25 a.m., 0025
7 5:34 p.m.

Exercise 12-07

1 C
2 A
3 a 2 h b 7 h c 3 h d 6 h
e 11 min f 26 min g 8 min h 23 min
4

Time	Hours and minutes	Time	Minutes and seconds
129 minutes	2 h 9 min	86 seconds	1 min 26 s
183 minutes	3 h 3 min	290 seconds	4 min 50 s
958 minutes	15 h 58 min	421 seconds	7 min 1 s

5 a 16 h 25 min b 9 h 9 min
c 12 h 35 min d 14 h 34 min
e 8 h 21 min f 15 h 21 min
6 7 h 55 min
7 5:49 pm
8 17 h 37 min
9 13 h 47 min

Exercise 12-08

1 B
2 C
3 a 50 min
b No, Bus 2 takes 30 min and Bus 3 takes 50 min
c Bus 1 d Bus 2 as it is quicker e Bus 1
4 a Go to Casino and change trains
b Go to Kew and change trains
c At Roseville and Kew
d At Casino and Kew
5 Teacher to check

Language activity

ACROSS: MILLENNIUM CENTURY PM DAY FORTNIGHT TIMELINE DECADE
DOWN: MINUTE LEAPYEAR WEEK MONTH YEAR DIGITAL HOUR ANALOGUE TIMEZONES AM SECOND

Practice test 12

1 –12
2 600
3 A triangle with two equal sides.
4 $50
5 0.054
6 3
7 $27a^2b$
8 $w = 72$
9 $\frac{1}{20}$
10 0.04, 0.4, 0.402, 0.45
11 C
12 D
13 D
14 13.8 cm
15 34.4 m
16 a 19 m b 48.2 m
17 312
18 3 h 43 min
19 1748
20 7:53 p.m.
21 18 h 7 min
22 19 min

CHAPTER 13

Exercise 13-01

1 C
2 a False b True c False d True
3 a 5 cm^2 b 6 cm^2 c 7 cm^2 d 6 cm^2
e 5 cm^2 f 6 cm^2 g 2 cm^2 h 2 cm^2
i 5 cm^2
4 a 4 cm^2 b 4 cm^2 c 7 cm^2 d 4 cm^2
e 4 cm^2 f 4 cm^2

Exercise 13-02

1 B
2 D
3 a 45 m^2 b 18.2 cm^2
c 23.94 m^2 d 46.24 cm^2
e 18.62 cm^2 f 105.78 m^2
4 a 17.64 cm^2 b 27.4576 m^2
c 5.6048 m^2
5 a 5.175 m^2 b 53.29 mm^2
c 0.002 548 m^2 or 2548 mm^2
d 3733.52 cm^2
6 a 17.5 m^2 b 87

Exercise 13-03

1 C
2 B
3 a $17.5\ m^2$ b $15.6\ cm^2$
c $13.02\ m^2$ d $1564\ m^2$
e $14.95\ m^2$ f $52.89\ m^2$
4 a $6.48\ cm^2$ b $89\ 342\ cm^2$
c $10.4\ m^2$
5 a $1.3275\ m^2$ b $9.245\ mm^2$
c $1380\ mm^2$ d $2911\ cm^2$
6 a $17.2\ m^2$ b $404.20

Exercise 13-04

1 A
2 C
3 a $28\ m^2$ b $32\ cm^2$
c $22.32\ m^2$ d $15.12\ m^2$
e $32.94\ cm^2$ f $425.88\ m^2$
4 a 8 m b 7 cm c 12.4 m
5 a $195\ cm^2$ b 4.5 m
c 6.8 cm d $25.37\ mm^2$
e 18.2 m f 21.6 cm

Exercise 13-05

1 A
2 B
3 a $132\ m^2$ b $30\ cm^2$ c $30\ m^2$
d $43.86\ m^2$ e $16.32\ m^2$ f $44.64\ m^2$
4 Area $= 12.4 \times 4.8 - 2 \times \frac{1}{2} \times 3.1 \times 4.8 = 44.64\ m^2$
5 a $174\ m^2$ b $46\ cm^2$ c $20.7\ m^2$
6 Teacher to check.

Exercise 13-06

1 C
2 B
3 a $36\ cm^3$ b $24\ cm^3$
c $48\ cm^3$ d $60\ cm^3$
e $32\ cm^3$ f $60\ cm^3$
4 a $10\ units^3$ b $18\ units^3$
c $54\ units^3$

Exercise 13-07

1 C
2 D
3 a $220\ m^3$ b $72\ cm^3$ c $112\ m^3$ d $176\ m^3$
e $614.125\ cm^3$ f $74.088\ m^3$
4 a $1755\ cm^3$ b $2744\ m^3$
c $7.8\ cm^3$ d $54.872\ cm^3$
e $4356\ m^3$
5 9 cm
6 8 m

Exercise 13-08

1 D
2 D
3 a litres b kilolitres
c millilitres d millilitres
e kilolitres f litres
4 Answers will vary, depending on the size of objects used.
5 a 6 b 45 c 8 d 9400
e 7.654 f 82 000 g 0.45 h 4.56
6 13 days
7 a $252\ m^3$, 252 000 L b $120\ cm^3$, 0.12 L
c $288\ m^3$, 288 000 L

Language activity

A MATHEMALIEN

Practice test 13

1 2.5
2 3500
3 20 m
4 $a = 135$, $b = 135$
5 7
6 $c = 14$
7 27
8

9 60
10 4
11 B
12 C
13 D
14 A
15 $42\ m^2$
16 $240\ cm^2$
17 14 m
18 $14.61\ m^2$
19 Line 4
20 C
21 $662.4\ m^3$
22 690 kL

CHAPTER 14

Exercise 14-01

1 a 10 b 5 c April d 15
e March and June f 5

ISBN 9780170350969

2 (sample answer)

Favourite ice-cream flavour

Chocolate

Strawberry

Vanilla

Choc chip

Blueberry

Mocha

Key: ☺ = 2 students

3 a Tim b Maria c 84 hours
 d Chitra and Tim

4 (sample answer)

Method of travel to school

Walking

Bus

Train

Car

Bike

Two or more methods

Key: = 4 students

5 Advantages: Attractive, Easy to read; Disadvantages: Inaccurate, time-consuming to draw.

Exercise 14-02

1 A

2 a Reilly b 6 c 9
 d Jones, Reilly, Broe and Lee e 2

3 a Column
 b Jana, Sarah, Nick and Gavin c Yes

4 a

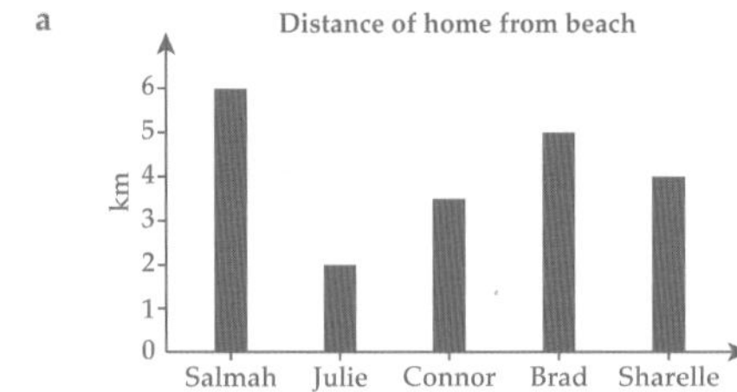

 b 2

5 a Maths b Art c 60
 d 53 e No
 f English, Geography, Maths, Science and Metalwork.

Exercise 14-03

1 a \$5 b \$10 c \$30

2 Teacher to check.

3 At 9 a.m. the patient's temperature was 39°, it then fell to 38.5° from 10 a.m. to 12 noon. It then fell more quickly to 37° by 1:30 p.m. where it remained until 3 p.m.

4 a A\$45 b A\$112 c NZ\$100

5 a 9 a.m. b 32 km c twice
 d 1:15 p.m. e 9 km/h

6 a 11°C, 12°C b noon and 3:20 p.m.
 c 11 a.m. to noon d 2°C/h
 e From 10 a.m. to 11 a.m. and from 1 p.m. to 2 p.m.

Exercise 14-04

1 D

2 C

3 a \$20 b \$40 c \$20
 d \$40 e \$60
 f yes, because he put $\frac{3}{8}$ of his pay into savings.

4

5 a 20c b 35c c False

6 Teacher to check.

Exercise 14-05

1 C

2 D

3 a 8 hours b $\frac{1}{4}$ c 6 hours
 d 2 hours e 6 hours

4 a food and clothing b \$160 c \$80
 d No

5 a Rent 108°, food 90°, entertainment 36°, clothes 54°, savings 72°

 b

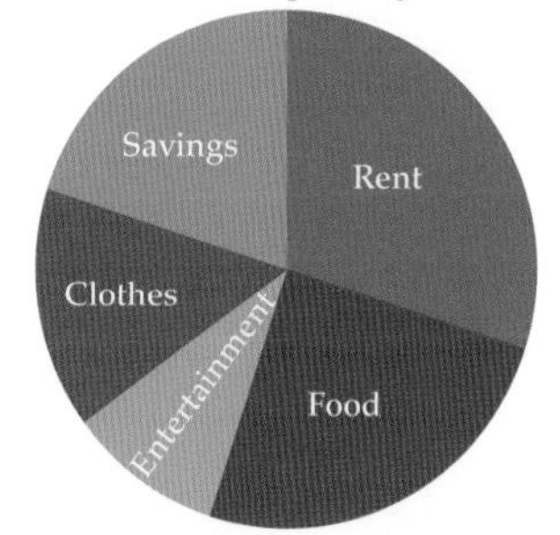

6 a house payments b $\frac{1}{3}$ c \$140
 d \$105

Exercise 14-06

1 D

2 The graph is misleading as the size of the picture is wrong. Both the radius and the height have been tripled which makes the picture much larger in volume than it should be. Only 1 dimension should be tripled.

3

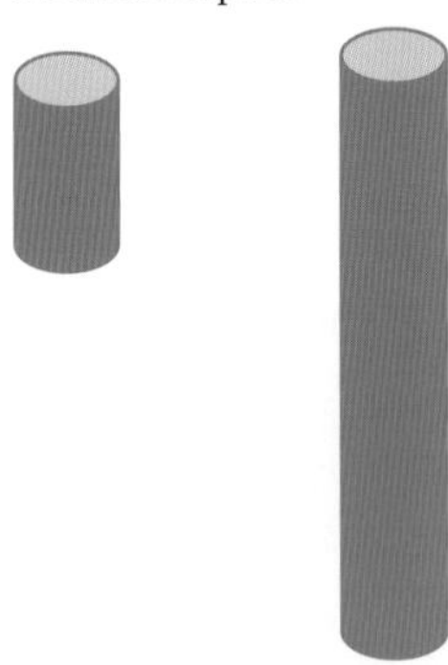

4 The graph is misleading because there is no title, the scale on the vertical axis is uneven and there is no scale on the horizontal axis.

5 a

27
24
21
18
15
12
9
6
3

b No, because we do not know the title or what is on the horizontal axis.

6 Yes, 1 TV picture is too small and one is too large.

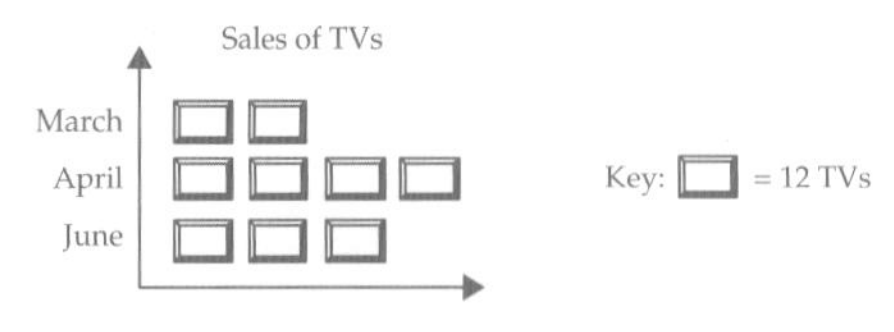

Practice test 14

1 $6b$

2 75

3 Teacher to check.

4 2028

5 0.03

6 385.1

7 No

8 360°

9 pentagon

10 23 cm

11 The Cowan family

12 The Boyd family

13 A

14 a 9 min b 24°C

15 Divide a rectangle into sections proportional to the data.

16 **Brian's working holiday**

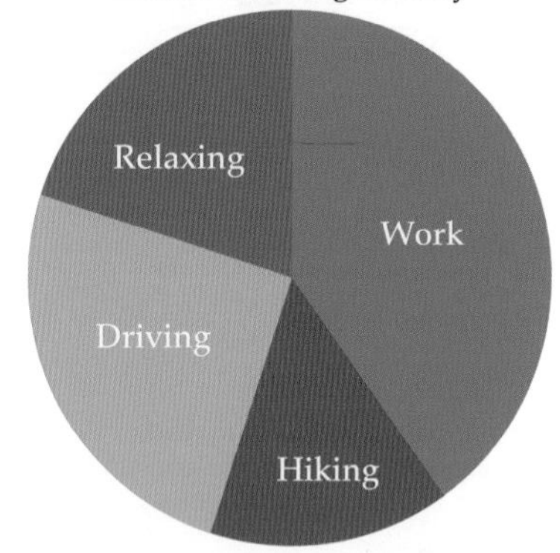

17 72°

18 No title, missing scales

CHAPTER 15

Exercise 15-01

1 C

2 B

3

Score	Tally	Frequency
47	||||	4
48	卌 ||	7
49	卌 |||	8
50	卌 卌 ||||	14
51	卌 ||	7
52	卌 ||	7
53	|||	3

ISBN 9780170350969

4

Score (000s)	Tally	Frequency
21	|||	3
22	||	2
23	𝍸 𝍸	10
24	||	2
25	||	2
26	𝍸	5
27	||	2
28	|||	3
29	|	1

5

Score	Tally	Frequency
2	|	1
3	|||	3
4	||	2
5	||	2
6	𝍸	5
7	𝍸 𝍸 ||	12
8	||	2
9	|||	3

6

Score	Tally	Frequency
0	𝍸	5
1	𝍸	5
2	𝍸 𝍸 𝍸 𝍸	20
3	𝍸 𝍸 |||	13
4	𝍸	5
5	||	2

7, 8 Teacher to check.

Exercise 15-02

1 B

2 D

3, 4 a

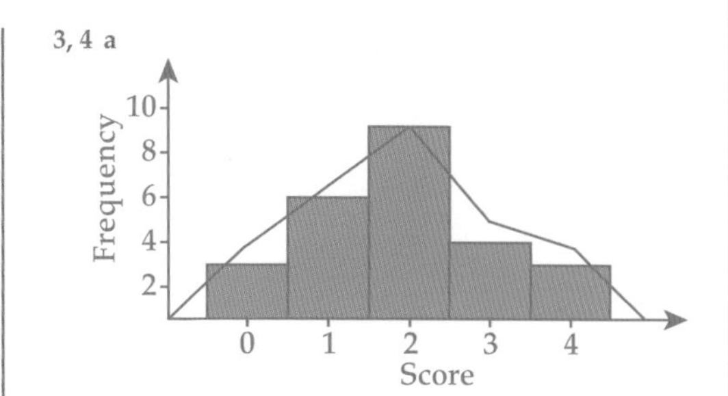

b

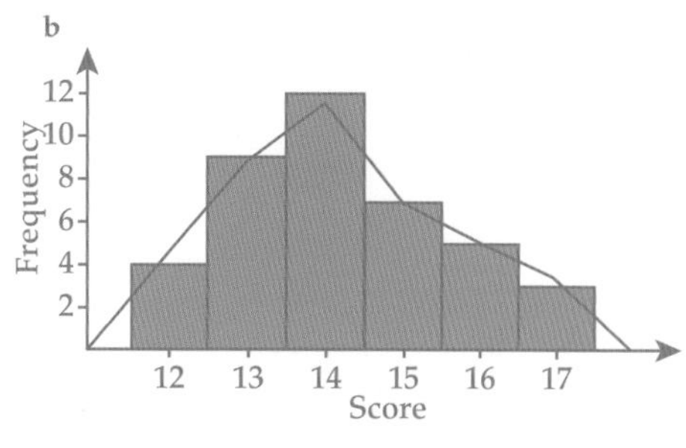

5 a

Score (x)	Tally	Frequency (f)
5	||	2
6	||||	4
7	𝍸 |	6
8	𝍸 𝍸	10
9	||||	4
10	|||	3
11	|	1
	Total	30

b, c

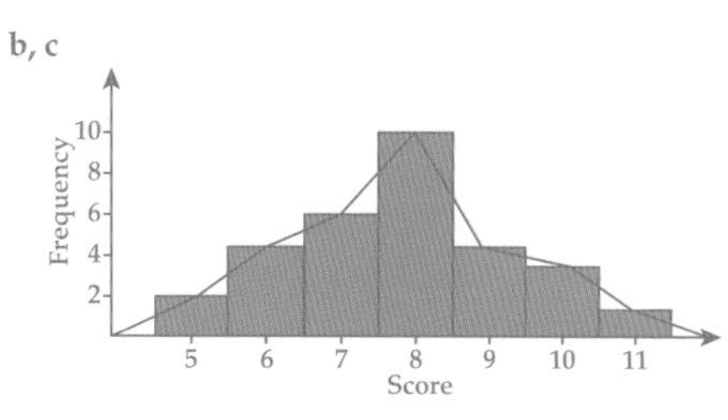

d 8 e 6

6 a

Score	Frequency
6	3
7	4
8	6
9	4
10	2
11	3
12	2

b

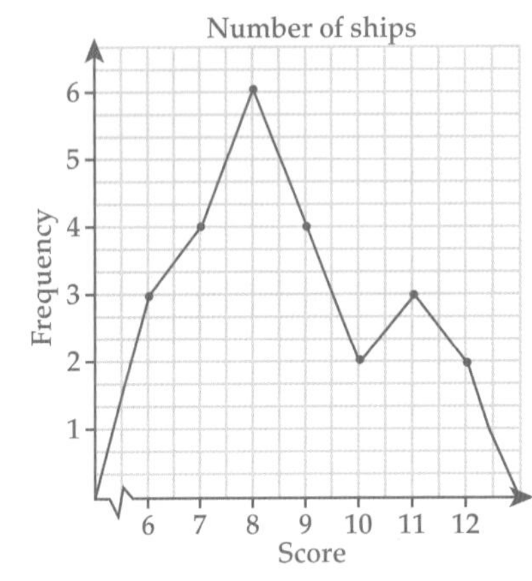

c 8 d 12

Exercise 15-03

1 A
2 D
3 a

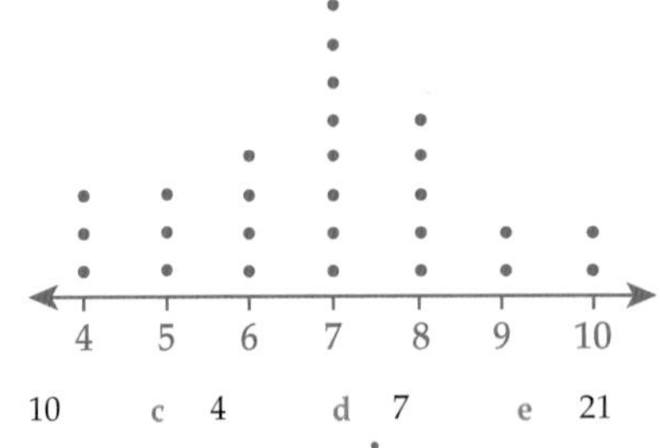

b 10 c 4 d 7 e 21

4 a

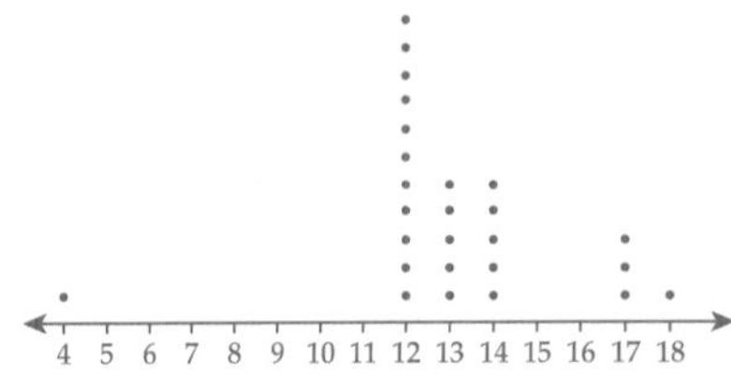

b 18 c 4 d 12 e Yes, 4
f Yes, from 12 to 14

5 a

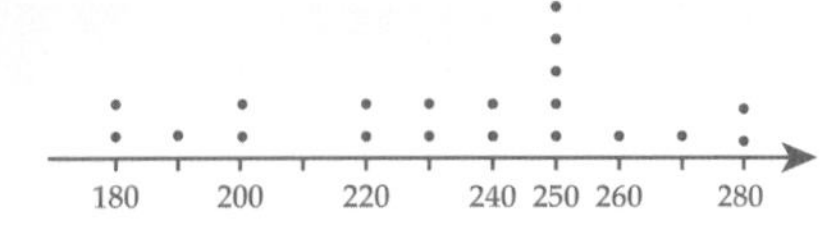

b $280 c $250 d $720

Exercise 15-04

1 C
2 a 12 b 47 c 9
d 2 e 27
3 a

Stem	Leaf
1	5 9
2	3 4 5 7 7 8 9
3	2 3 3 3 3 3 5 5 6 7 8 9
4	2 2 4 5 5 6 7 8 9

b Highest score = 49, lowest score = 15, most common score = 33

4 a

Stem	Leaf
3	9
4	277889
5	1122222222223455557889
6	112

b 52 toothpicks

5 a

McDavids

Stem	Leaf
28	0
29	5 9
30	7
31	4 7 8
32	6 7
33	2 5
34	

Burger Prince

Stem	Leaf
28	
29	8
30	6
31	
32	
33	
34	2 5 8
35	4 5 6 9
36	6 7

b Burger Prince c McDavids
d Burger Prince because their total sales were $3796 compared to McDavids total of $3470. This is obvious from the stem-and-leaf plot because Burger Prince's sales were higher.

Exercise 15-05

1 D
2 B
3 a Mean = 6.9, Mode = 7
b Mean = 5.2, Mode = 4
c Mean = 13.1, Mode = 13
d Mean = 26.1, Mode = 28
e Mean = 54.7, Mode = 54 and 55
4 It has 2 modes.
5 a 180.625 b Yes, 186
6 a Mean = 4.15, Mode = 2 b No

Exercise 15-06

1 D
2 A

ISBN 9780170350969

3 a Median = 6, Range = 7
b Median = 14, Range = 7
c Median = 24, Range = 8
d Median = 55, Range = 6
e Median = 110, Range = 19
f Median = 64, Range = 58

4 a Median = 75.5, Range = 75
b Mean = 66.03, Modes = 82, 86 and 98
c Tuesday, prices are cheaper on Tuesdays.
d On Friday, Saturday and Sunday, more people have time to go to the movies.

5 a 6 b 12 c 12.45 d 12
e The mean

6 Range =15.8, Median = 11.1

Exercise 15-07

1 B
2 A
3 a

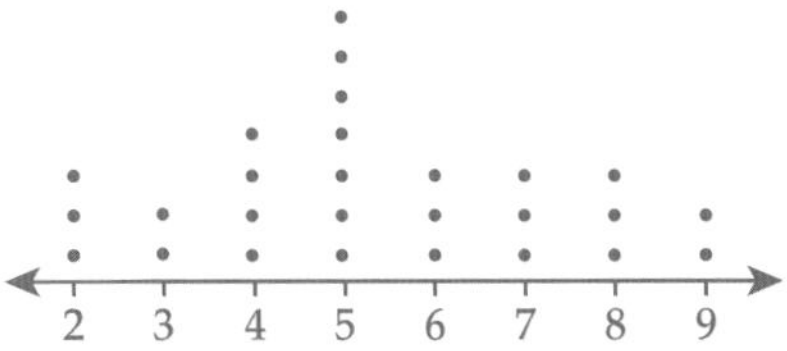

b 5.3 c Mode = 5, Range = 7, Median = 5
d The mean e The mode

4 a 60.14
b Mode = 63, Range = 35, Median = 62
c The mode

5 a

Stem	Leaf
12	1 5 8
13	2 5 5
14	2 6 8
15	2 6 8
16	3 8

b 143.5
c Mode = 135, Range = 47, Median = 144

6 a Anitha b Joe
c Anitha, as her scores are at the higher end.

Language activity

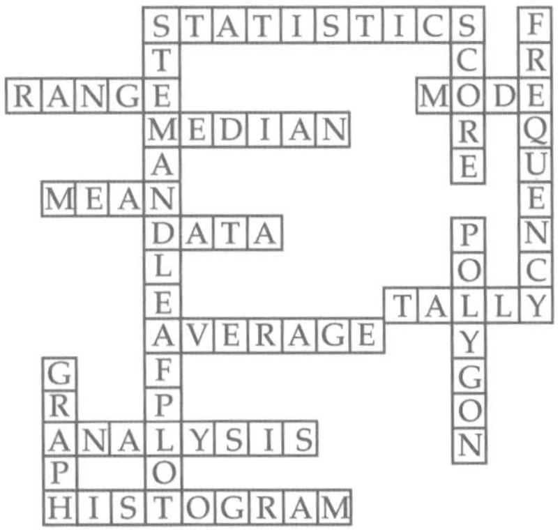

Practice test 15

1

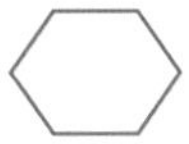

2 0.125
3 26
4 $y - 10$
5 trapezium
6 $90 = 2 \times 3 \times 3 \times 5$
7 $\frac{3}{10}$
8 $p = 69$
9 700
10 False
11 C
12 D
13 B
14 plot; mode
15 outlier
16 a 27 b 53 c 56.4
17 the mode
18 B
19 14
20 33
21 17
22 a 2.5 b 2

CHAPTER 16

Exercise 16-01

1 C
2 B
3 a Likely b Even chance
c Impossible d Even chance
e Certain f Unlikely
g Likely h Even chance
i Certain j Likely

4

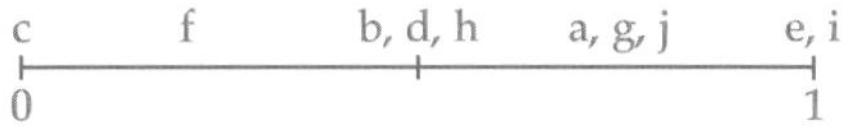

5 a True b True c False d True
e False f False

6 Teacher to check.

Exercise 16-02

1 A
2 a {Head, Tail}
b {a,b,c,d,e,f,g,h,i,j,k,l,m,n,o,p,q,r,s,t,u,v,w,x,y,z}
c {4, 5, 6, 7, 8, 9, 10}
d {2, 3, 4, 5, 6, 7, 8, 9, 10, 11, 12}
e {\$5, \$10, \$20, \$50, \$100}
f {a, e, i, o, u}
g {HH, HT, TH, TT}
h {8, 10, 12, 14, 16}
i {5c, 10c, 20c, 50c, \$1, \$2}
j {2, 3, 5, 7}

3 a 2 b 26 c 7 d 11
 e 5 f 5 g 4 h 5
 i 6 j 4
4 a {red, blue, green}; 3 outcomes b No
5 a {red, blue, green, yellow} b Yes
 c No d Yes
6 a {purple, green, orange} b No
 c No d Unlikely, or 1 in 4 e Yes
 f Even chance
7 Teacher to check.

Exercise 16-03

1 D
2 B
3 a $\frac{1}{2}$ b $\frac{1}{6}$ c $\frac{8}{20}=\frac{2}{5}$ d $\frac{5}{11}$
 e $\frac{10}{22}=\frac{5}{11}$ f $\frac{2}{10}=\frac{1}{5}$ g $\frac{3}{6}=\frac{1}{2}$ h $\frac{5}{12}$
 i $\frac{3}{6}=\frac{1}{2}$ j $\frac{1}{25}$
4 a $\frac{4}{20}=\frac{1}{5}$ b $\frac{6}{20}=\frac{3}{10}$ c $\frac{5}{20}=\frac{1}{4}$ d $\frac{18}{20}=\frac{9}{10}$
5 a $\frac{4}{52}=\frac{1}{13}$ b $\frac{13}{52}=\frac{1}{4}$ c $\frac{4}{52}=\frac{1}{13}$ d $\frac{12}{52}=\frac{3}{13}$
 e $\frac{1}{52}$ f $\frac{13}{52}=\frac{1}{4}$ g $\frac{1}{52}$ h $\frac{8}{52}=\frac{2}{13}$
 i $\frac{8}{52}=\frac{2}{13}$
6 a $\frac{1}{6}$ b $\frac{1}{2}$ c $\frac{3}{16}$
 d $\frac{5}{10\,000}=\frac{1}{2000}$ e $\frac{15}{25}=\frac{3}{5}$
 f $\frac{24}{40}=\frac{3}{5}$ g $\frac{7}{30}$

Exercise 16-04

1 D
2 C
3 impossible unlikely even chance likely certain
 0 $\frac{1}{2}$ 1
4 a impossible b even chance
 c unlikely d unlikely
 e even chance f unlikely
 g even chance h certain
 i impossible j likely
5 i a, c, d, f b, e, g j h
 0 $\frac{1}{2}$ 1
6 a likely b certain c likely
 d unlikely e impossible
 f impossible g likely
 h certain i likely j likely
7 a {boy, girl}; P(boy) = $\frac{18}{30}=\frac{3}{5}$,
 P(girl) = $\frac{12}{30}=\frac{2}{5}$; Yes sum is 1.
 b {white, red}; P(white) = $\frac{6}{11}$, P(red) = $\frac{5}{11}$;
 Yes sum is 1.
 c {shirts, trousers, dresses};
 P(shirt) = $\frac{5}{25}=\frac{1}{5}$, P(trousers) = $\frac{8}{25}$,
 P(dress) = $\frac{12}{25}$; Yes sum is 1.
 d {mints, sherbets, toffees}; P(mint) = $\frac{15}{47}$,
 P(sherbet) = $\frac{20}{47}$, P(toffee) = $\frac{12}{47}$; Yes sum is 1.

Exercise 16-05

1 D
2 A
3 a 10 b No
4 a 20 b 20
 c, d Teacher to check and discuss.
5 a 10 b Results will vary.
 c, d Teacher to check and discuss.
6 a 10 b No c Results will vary.

Language activity

1 chance
2 certain
3 impossible
4 unlikely
5 event
6 outcome
7 probability
8 theoretical
9 equally likely
10 sample space

Practice test 16

1 248
2 $\frac{3}{100}$
3 None
4 Teacher to check. For example,

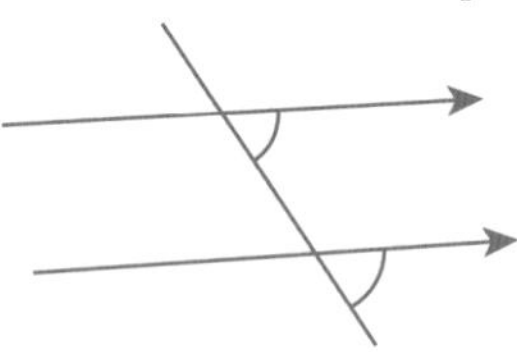

5 $3\frac{1}{3}$
6 1.037 85
7 $3x + 6$
8 The stem
9 5
10 5
11 B
12 B
13 B
14 $\frac{1}{3}$
15 $\frac{3}{12}=\frac{1}{4}$

ISBN 9780170350969

16 D

17 a 0 b 1

18 a 13 b No

19 a $\frac{17}{40}$ b $\frac{14}{40} = \frac{7}{20}$

CHAPTER 17

Exercise 17-01

1 B

2 D

3 a False b True c True d False

4 a 15% b 23% c 54% d 95%

5 a $\frac{1}{2}$ b $\frac{3}{4}$ c $\frac{1}{4}$ d 1

6 a $\frac{3}{10}$ b $\frac{21}{50}$ c $\frac{7}{10}$

d $\frac{17}{20}$ e $\frac{11}{20}$ f $1\frac{1}{5}$

g $\frac{3}{100}$

h $\frac{31}{50}$ i $\frac{1}{20}$ j $\frac{3}{4}$

k $\frac{2}{5}$ l $\frac{9}{50}$

7 a 3; 1 b $37\frac{1}{2}$; 75; 3

8 a $\frac{2}{3}$ b $\frac{5}{8}$ c $\frac{5}{6}$ d $\frac{7}{8}$

9 a $\frac{1}{10}$ b $\frac{3}{10}$ c $\frac{4}{5}$ d 1

e $1\frac{1}{2}$ f 2 g $1\frac{3}{4}$ h $\frac{2}{5}$

10 a 2; 3 b 2; 3

Exercise 17-02

1 D

2 B

3 a 24% b 48% c 65% d 73%

e 54% f 86% g 90% h 18%

4 a 25 b 100; 20; 80

5 a 25% b 20% c 30% d 40%

e 37.5% f 80% g 70% h 60%

i $66\frac{2}{3}$% j 87.5% k $16\frac{2}{3}$% l 100%

6 a 25%; 60%

b Sister: $6000, son: $18 000

Exercise 17-03

1 D

2 A

3 a False b True c False d True

e True f False

4 a 0.7 b 0.65 c 0.9 d 1.0

e 0.85 f 0.24 g 0.92 h 0.58

i 0.045 j 0.125 k 2.2 l 0.825

5 a 30% b 4% c 26% d 50%

e 47% f 64% g 60% h 95%

i 62.5% j 55% k 140% l 250%

6

Percentage	Fraction	Decimal
5%	$\frac{1}{20}$	0.05
10%	$\frac{1}{10}$	0.1
$12\frac{1}{2}$%	$\frac{1}{8}$	0.125
20%	$\frac{1}{5}$	0.2
25%	$\frac{1}{4}$	0.25
$33\frac{1}{3}$%	$\frac{1}{3}$	0.3
50%	$\frac{1}{2}$	0.5
$66\frac{2}{3}$%	$\frac{2}{3}$	0.6
75%	$\frac{3}{4}$	0.75

7 a 100%

b i One Dimension ii Cold Day

c $\frac{1}{4}$ d 0.25 e 100

Exercise 17-04

1 C

2 B

3 a True b False c True d True

e True f True

4 a $50 b 32 km c $2250 d 72 min

e 12 h f $480 g 18 m h 45 km

i 9 h

5 a 1 m b 4 cm c $750 d $56

e 20 L f 36 kg

6 a 20 b $10 c 90% d $190

7 a 1275 b 15% c 225

Exercise 17-05

1 D

2 C

3 a 4 : 7 b 5 : 9 c 23 : 41 d 12 : 17

e 3 : 5 f 7 : 9 g 12 : 19 h 5 : 11

i 19 : 20

4 a 16 : 14 b 14 : 16 c 16 : 30

5 a 3 : 8 b 3 : 5 c 5 : 8

6 a 5 : 6 b 4 : 5 c 3 : 6 d 6 : 4
e 4 : 5 : 6 f 6 : 4 : 3 g 3 : 5 : 6 h 3 : 15
i 5 : 3 : 4 : 6

7 a 2 : 5 b 5 : 4 c 4 : 8 d 5 : 4
e 5 : 2 f 4 : 5 g 5 : 8 h 4 : 2
i 8 : 5 j 8 : 4

8 a 12 : 16 b 16 : 8 c 8 : 12
d 16 : 8 : 12 e 12 : 24 f 8 : 28

Exercise 17-06

1 C
2 D
3 a 20 b 80 c 28 d 6
e 5 f 18 g 10 h 24
i 35 j 36
4 a 5 b 8 c 9 d 4
e 5 f 12 g 4 h 7
i 10 j 3
5 a 2 b 2 c 3 d 2
e 5 f 4 g 2 h 5
i 2
6 Teacher to check.

Exercise 17-07

1 D
2 B
3 a 3 : 4 b 1 : 3 c 1 : 5 d 2 : 7
e 2 : 3 : 5 f 2 : 3 : 5 g 1 : 2 : 5
h 9 : 10 : 16 i 4 : 3
j 10 : 1 k 3 : 4 : 1 l 8 : 4 : 3
4 a 3 : 4 b 3 : 8 c 2 : 5 d 2 : 3
e 7 : 100 f 1 : 3 g 3 : 5 h 3 : 4
i 4 : 3 j 10 : 3 k 9 : 10 : 20
5 a 3 : 2 b 5 : 3 c 2 : 3 d 2 : 3
e 3 : 3 : 2 f 3 : 2 : 2 g 2 : 1 : 5 h 1 : 7
i 5 : 2 : 1 j 3 : 13 k 3 : 1 : 2 : 2
l 5 : 14 : 3 : 2

Practice test 17

1 8.109
2 –1
3 4
4 90°
5 5600
6 7
7 9, 18, 27, 36, 45
8 7
9 none
10 59.62
11 D
12 D
13 C
14 a 75% b 70% c $12\frac{1}{2}\%$
15 a 0.25 b 0.6 c 1.2
16 $875
17 4 km
18 7 : 6
19 16
20 A

ISBN 9780170350969

INDEX

acute angle 22
acute-angled triangle 190
adding
  decimals 86–7
  fractions 126–7
  integers 6–7
  mixed numerals 128–9
  numbers 42–3
adjacent angles 26
algebraic expressions 154–5
  substitution into 156–7
alternate angles on parallel lines 32–3
analysing data 265–79
angle sum
  of a quadrilateral 198–9
  of a triangle 192–3
angles
  adjacent 26
  alternate 32–3
  at a point 28–9
  co-interior 34–5
  complementary 26–7
  corresponding 30
  measuring and drawing 24–5
  naming 20–1
  in a right angle 26
  on a straight line 26
  supplementary 26–7
  types of 22–3
  vertically opposite 28–9
area 228–9
  composite shapes 236–7
  parallelogram 234–5
  rectangle 230–1
  triangle 232–3
associative law 150–1
axis of symmetry 179

base 74
best buys 104–5

capacity
  metric units 208, 242
  and volume 242–3
certain event 284, 290
cluster 270, 271
co-interior angles on parallel
  lines 34–5
column graphs 250–1
commutative law 150
complementary angles 26–7
composite numbers 66–7
composite shapes
  area 236–7
  perimeter 214–15
composite transformations 176–8
conversion graphs 252, 254
convex quadrilateral 196
coordinates on maps 10–11
corresponding angles on parallel
  lines 30–1
cube (number) 78
cube root 78

decagon 168
decimal places 82
decimals
  adding 86–7
  dividing
    by another decimal 100–1
    by powers of 10 92–3
    by whole numbers 98–9
  and fractions 82–3, 114–15
  multiplying
    by another decimal 96–7
    by powers of 10 92–3
    by whole numbers 94–5
  ordering 84–5
  and percentages 302–3
  problems 106–7
  rounding 102–3
  subtracting 86–7
denominator 112, 124
distributive law 150
divided bar graphs 256–7
dividing
  decimals
    by another decimal 100–1
    by powers of 10 92–3
    by whole numbers 98–9
  fractions 140–1
  mixed numerals 144–5
  numbers 58–9
divisibility tests 62–3
division
  with remainders 60–1
  short 58–9
division facts 56–7
dot plots 270–1, 278–9

equal parts 112
equally likely event 286

equations 158–9
  one-step 160–1
  two-step 162–3
equilateral triangle 190
equivalent fractions 116–17
equivalent ratios 308–9
estimating 40–1
even chance 285, 290
event 284
experimental probability 292–3
exterior angle of a triangle 194–5

factor trees 68–9
factors 64–5
50-50 chance 284
fractions 112–13
  adding 126–9
  and decimals 82–3, 114–15
  dividing 140–1
  equivalent 116–17
  improper 120–3
  mixed numerals 120–3, 128–9, 142–5
  multiplying 136–7
  ordering 124–5
  and percentages 298–301
  proper 120
  of a quantity 134–5
  simplifying 118–19
  subtracting 126–9
frequency histograms 268–9
frequency polygons 268–9
frequency tables 266–7

graphs
  column 250–1
  divided bar 256–7
  line 252–3
  misleading 260–1
  picture 248–9
  sector 258–9
guess-and-check method 158–9

hexagon 168
highest common factor (HCF) 64–5, 310
histograms, frequency 268–9

impossible event 284, 290
improper fractions 120–3
index 64
index notation 64–5
integers 2–3
  adding 6–7
  ordering 4–5
  subtracting 8–9
inverse operations 160–3
irregular polygon 168
isosceles triangle 190

kite 196, 201

language of chance 284–5
laws of arithmetic 150–1
length
  measuring 210–11
  metric units 208
  see also perimeter
likely event 284
line graphs 252–5
line symmetry 179–80
lowest common multiple (LCM) 48–9, 124

mass, metric units 208
mean 272–4
measure of spread 276–7
measures of location 274–7
median 274, 276–7
metric system 208–9
misleading graphs 260–1
mixed numerals 120–3
  adding and subtracting 128–9
  dividing 144–5
  multiplying 142–3
mode 274–5
multiplication facts 46–7, 56
multiplying
  decimals
    by another decimal 96–7
    by powers of 10 92–3
    by whole numbers 94–5
  fractions 136–7
  mixed numerals 142–3
  numbers 50–1

negative integers 4, 6, 8
negative numbers 2, 14–15
non-convex quadrilateral 196
number line 4, 6
number plane 12–13
  with negative numbers 14–15

ISBN 9780170350969

numbers
  adding 42–3
  composite 66–7
  dividing 58–9
  multiplying 50–1
  prime 66–7
  subtracting 44–5
numerator 112

obtuse angle 22
obtuse-angled triangle 190
octagon 168
one-step equations 160–1
order of operations 80–1
ordering
  decimals 84–5
  fractions 124–5
  integers 4–5
origin 12
outcome 284
outliers 270, 271

parallel lines
  alternate angles 32–3
  co-interior angles 34
  corresponding angles 30–1
parallelogram 196, 200, 201
  area 234–5
pentagon 168
percentages
  and decimals 302–3
  and fractions 298–301
  of a quantity 204–5
perimeter 212–13
  composite shapes 214–15
picture graphs 248–9
place value 82
polygons 168–70
  frequency 268–9
positive integers 4, 6, 8
powers 74–5
prime numbers 66–7
prisms 183–4
  volume 238–41
probability 284, 288–9
  experimental 292–3
  range of 290–1
pronumerals 26, 152–3
proper fractions 120
protractor 24
pyramids 183–4

quadrants 14
quadrilaterals 168
  angle sum 198–9
  properties 200–2
  types of 196–7

range 276–7
ratios 206–7
  equivalent 308–9
  simplifying 310–11
reciprocals 138–9
rectangle 196, 200, 201
  area 230–1
  perimeter 212
rectangular prism 183
  volume 238–41
rectangular pyramid 183
reflection 171–3
reflex angle 22
regular polygon 168
revolution 22, 28
rhombus 196, 200, 201
right angle 22
right-angled triangle 190
rotation 174–5
rotational symmetry 181–2
rounding 40–1
  decimals 102–3
  time 220–1

sample spaces 286–7
scalene triangle 190
sector graphs 258–9
short division 58–9
simplifying
  fractions 118–19
  ratios 310–11
solving equations
  by guess-and-check method 158–9
  by inverse operations 160–3
square (geometry) 196, 200, 201
square (number) 76–7
square root 76–7
statistical graphs 247–61
stem-and-leaf plots 272–3, 278–9
step graphs 252–3
straight angle 22
substitution 156–7

ISBN 9780170350969

INDEX

subtracting
  decimals 86–7
  fractions 126–7
  integers 8–9
  mixed numerals 128–9
  numbers 44–5
sum of probabilities 290
supplementary angles 26–7
symmetry
  line 179–80
  rotational 181–2

tally marks 266
time 216–17
  24-hour time 218–19
  calculations 220–1
times tables 46–7, 56
timetables 222–3
transformations 171–5
  composite 176–8
translation 171–3
transversal 30
trapezium 196, 200
travel graphs 252
triangles 168, 190
  angle sum 192–3
  area 232–3
  exterior angle 194–5
  types of 190–1
24-hour time 218–19
two-step equations 162–3

unit cost 104
unlikely event 284
unordered stem-and-leaf plots 272

variables 152–3
vertex 20
vertically opposite angles 28–9
volume 238
  and capacity 242–3
  rectangular prism 238–41

x-axis 12
x-coordinate 12

y-axis 12
y-coordinate 12

ISBN 9780170350969